Lecture Notes in Computer Science

Edited by G. Goos, J. Hartmanis and J. van Leeuwen

Advisory Board: W. Brauer   D. Gries   J. Stoer

Springer
Berlin
Heidelberg
New York
Barcelona
Budapest
Hong Kong
London
Milan
Paris
Santa Clara
Singapore
Tokyo

Thomas Kropf (Ed.)

Formal
Hardware Verification

Methods and Systems in Comparison

 Springer

Series Editors

Gerhard Goos, Karlsruhe University, Germany

Juris Hartmanis, Cornell University, NY, USA

Jan van Leeuwen, Utrecht University, The Netherlands

Volume Editor

Thomas Kropf
Universität Karlsruhe, Institut für Rechnerentwurf und Fehlertoleranz
Postfach 6980, D-76128 Karlsruhe, Germany
E-mail: kropf@ira.uka.de

Cataloging-in-Publication data applied for

Die Deutsche Bibliothek - CIP-Einheitsaufnahme

Formal hardware verification : methods and systems in
comparison / Thomas Kropf (ed.). - Berlin ; Heidelberg ; New York ;
Barcelona ; Budapest ; Hong Kong ; London ; Milan ; Paris ; Santa
Clara ; Singapore ; Tokyo : Springer, 1997
  (Lecture notes in computer science ; 1287)
  ISBN 3-540-63475-4

CR Subject Classification (1991): B, F.3-4

ISSN 0302-9743
ISBN 3-540-63475-4 Springer-Verlag Berlin Heidelberg New York

Typesetting: Camera-ready by author
SPIN 10546448   06/3142 – 5 4 3 2 1 0   Printed on acid-free paper

Preface

Why you should read this book

This book presents an overview of many methods and systems for hardware verification. Its emphasis lies in the presentation of those approaches to hardware verification that have led to tools and systems usable to verify non-trivial circuits. Thus this book gives an excellent overview of what can be achieved today using formal methods for proving the correctness of circuits, especially as the contributions have been written by well-known research groups in this field.

What makes this book different

In contrast to conference papers, the sections in this book completely describe the respective approaches to verifying circuits. Thus not only are incremental advancements to the state of the art presented but a whole method is described. However, this is done not from a purely theoretical viewpoint but from one of applications. Thus the emphasis lies on the descriptions of which steps have to be taken in order to verify a certain circuit.

The book has a strong emphasis on comparability. It was written so as to facilitate evaluation of the strengths and weaknesses of each approach. This was achieved by two means. Firstly, each book section follows the same outline, leading to a uniform presentation of each approach. Secondly, all approaches have verified the same set of examples. For this, four common example circuits were taken from the IFIP WG10.5 set of hardware verification examples. These circuits, which are described in detail in the appendix, have been processed by all research groups. Therefore, it is easy to see how the different implementations and specifications have to be formalized in each approach. Moreover, it turns out that each approach has certain advantages that makes it best suited for certain classes of circuits. On the other hand, certain proof goals may not be verified with some approaches, which is also clearly stated in each section – as well as the description of additional circuits for which the respective approach turned out to be especially useful. This makes this book quite different from regular conference papers, where usually only well-suited examples are used in the experimental results section and weaknesses are often not stressed.

What this book is not about

This book is not an introductory textbook to hardware verification. Although all sections contain some theoretical background of the respective work, this may not be sufficient for newcomers to hardware verification. On the other hand, if people have some basic knowledge of automata theory, temporal logic, and predicate logic, this should be sufficient for understanding a good part of the book. Every section contains pointers to starting literature in the respective domain.

Additionally, the different approaches have not been compared and evaluated by the editor. It is in no way meant to lead to some kind of system ranking as there is no best verification system: different approaches are well suited for different kinds of verification problems. However, the aim has been to make these differences more transparent.

How the systems and tools have been chosen

The selection of the tools (and the respective groups of authors) were based on a personal (and probably unintentionally biased) evaluation of which hardware verification approaches have already matured into usable systems. This did not require that a tool be commercially available, however. On the contrary, many tools are (still) purely academic. Others are on the brink of being marketed or are already used successfully in industrial environments. To give an idea of the broad range of approaches, the most prominent representative of each research direction was identified to get a good coverage of what is being done at the moment. The respective research groups were then requested to add a section to this book. About a third of the contacted groups were not interested or unable to allocate enough spare time. All others are represented in this book, which thus still gives an excellent overview of the current state of the art in hardware verification.

Acknowledgements

First of all, I would like to thank all authors who contributed to this book. Writing whole book sections in parallel to (higher priority) daily work is an effort which cannot be overestimated – especially when taking into consideration that dedicated work had to be done to carry out the verification experiments for the common circuit examples. Mike Casey deserves credits for initiating the idea of the book at a European Design and Test Conference. I was also glad to get support of the members of the Hardware Verification Group at Karlsruhe for the extensive reviewing process. Klaus Schneider provided the TeX expertise required to successfully compile the various sources with all their different styles and packages.

July 1997 *Thomas Kropf*

Table of Contents

Verifying VHDL Designs with COSPAN
Kathi Fisler and Robert P. Kurshan 206

The C@S System
Klaus Schneider and Thomas Kropf 248

List of Contributors

Eduard Cerny
d'IRO
Université de Montréal
C.P. 6128
Succ. Centre-Ville
Montréal, H3C 3J7
Canada
cerny@iro.umontreal.ca

Francisco Corella
Hewlett-Packard Company
8000 Foothills Boulevard
Roseville, CA 95747-5649
USA
fcorella@rosemail.rose.hp.com

David Cyrluk
Computer Science Laboratory
SRI International
MS EL284,
333 Ravenswood Avenue
Menlo Park, CA, 94025-3493
USA
Cyrluk@csl.sri.com

Kathi Fisler
Department of Computer Science
Rice University
6100 S. Main, MS 132
Houston, TX 77005-1892
USA
kfisler@cs.rice.edu

Scott Hazelhurst
Department of Computer Science
University of the Witwatersrand
Johannesburg
Private Bag 3, 2050 Wits
South Africa
scott@cs.wits.ac.za

Thomas Kropf
Institute of Computer Design
and Fault Tolerance
University of Karlsruhe
Kaiserstr. 12
D-76128 Karlsruhe
Germany
kropf@ira.uka.de

Robert P. Kurshan
Bell Laboratories
700 Mountain Avenue
Room 2C-353
Murray Hill, NJ 07974-0636
USA
k@research.bell-labs.com

Michel Langevin
Nortel Technology
PO Box 3511 Station C
Ottawa Ontario K1Y 4H7
Canada
mlange@nortel.ca

Harald Rueß
Universität Ulm
Fakultät für Informatik
James-Franck-Ring
D-89069 Ulm, Germany
ruess@informatik.uni-ulm.de

Klaus Schneider
Institute of Computer Design
and Fault Tolerance
University of Karlsruhe
Kaiserstr. 12
D-76128 Karlsruhe
Germany
schneide@ira.uka.de

Carl-Johan H. Seger[1]
Department of Computer Science
University of British Columbia
2366 Main Mall, Vancouver, B.C.
Canada V6T 1Z1
cseger@ichips.intel.com

Xiaoyu Song
d'IRO
Université de Montréal
C.P. 6128
Succ. Centre-Ville
Montréal, H3C 3J7
Canada
song@iro.umontreal.ca

Mandayam Srivas
Computer Science Laboratory
SRI International
333 Ravenswood Avenue
Menlo Park California 94025
USA
Srivas@csl.sri.com

Jørgen Staunstrup
Dept. of Information Technology
B-344
Technical University of Denmark
DK-2800 Lyngby
Denmark
jst@it.dtu.dk
http://www.it.dtu.dk/~jst

Sofiène Tahar
Dept. of ECE
Concordia University
1455 de Maisonneuve Blvd. W.
Montréal, H3G 1M8
Canada
tahar@ece.concordia.ca

Zijian Zhou
Nortel Technology
PO Box 3511 Station C
Ottawa Ontario K1Y 4H7
Canada
zzhou@nortel.ca

---

[1] Carl Seger is currently with the
Intel Development Labs, JFT-102,
Intel Corporation, 2111 NE
25th Avenue, Hillsboro,
OR97124-5961, USA.

Symbolic Trajectory Evaluation

Scott Hazelhurst and Carl-Johan H. Seger

1. Introduction

The last decade has seen tremendous developments in the theory and practice of verification, particularly the verification of hardware. Verification remains a difficult and expensive task, and this is reflected in the diversity of approaches to verification, and the wide range of fruitful work that has been done.

Automatic model checking is very attractive because it is automatic — unfortunately there are fundamental computability and complexity results which limit the general applicability of the technique. Nevertheless, through considerable work on data structures (such as BDDs), the use of compositionality and abstraction, and the integration of theorem proving techniques, a number of tools using some or all of these ideas have been developed that can verify circuits of significant size.

We still see that different approaches are suitable for different problems, depending on the level of abstraction of the design, and the properties that must be checked.

Symbolic Trajectory Evaluation (STE) is a model checking approach designed to verify circuits with very large state spaces; it is more sensitive to the property being checked than to the size of the circuit. It has a natural approach to user-guided abstraction, and supports a simple yet effective compositional theory. Two important properties of STE are:

- It is suitable for verifying detailed designs of circuits at the gate or switch level (see Section 4. for details).
- STE provides accurate models of timing, which is reflected in the types of properties checked for.

This chapter presents the theory and practice of symbolic trajectory evaluation. The rest of this section presents a brief history of STE, followed by an outline of the rest of the chapter.

1.1 Overview of STE and Verification Method

In our verification methodology, the circuit implementation is modelled as a finite state machine. The specification is given as a set of assertions (which are each made up of temporal logic formulas), and verification consists of showing that each assertion is true of the model (that the model satisfies the formula).

Figure 1.1 gives an overview of the verification process from the point of view of our tool that performs the verification, the VossProver. On the left we see that the tool takes a circuit description (at either the gate or switch level) and automatically extracts a finite state machine model. On the right, we see how the user interacts with the system.

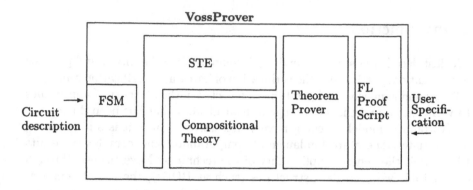

Fig. 1.1. Overview of verification

The proof of correctness is written in a script language, which is a general purpose programming language called FL. The theorem prover part of the VossProver provides a set of routines which can be used to infer verification results. As only these routines can be used for inferring results, the soundness of the end result is promoted. The theorem prover is a simple, specialised theorem prover providing the following:

- One inference rule which calls STE (the underlying model checking algorithm) to verify an assertion. STE manipulates the finite state machine directly.
- A number of inference rules which implement a compositional theory for combining simpler verification results into more complex ones.
- The ability to reason about data such as integers and vectors.

This verification methodology combines theorem proving and model checking. However, from the user's point of view there is only one notation for expressing correctness and a simple interface to the VossProver using its theorem prover and FL. Subsequent sections will describe both the theoretical foundations (STE and its compositional theory) and the practical aspects of the methodology, as well as present some experimental results.

1.2 History of Symbolic Trajectory Evaluation

Symbolic trajectory evaluation is a model checking technique that grew out of multi-level simulation on the one hand, and symbolic simulation on the other hand. It is the formal verification method closest to traditional simulation.

In [Brya91d, Brya91c], Bryant demonstrated the usefulness of ternary modelling for verifying a variety of circuits. The methodology is based on ternary simulation of VLSI circuits, where a third value U is added to the set {L, H} (we use H to represent high voltage or binary 1, and L to represent low voltage or binary 0) of possible signal values, indicating an unknown or indeterminate logic value. Assuming a monotonicity property of the simulation algorithm, one can ensure that any binary (i.e., L or H) values resulting when simulating patterns containing Us would also result when the U's are replaced by any combination of Ls and Hs. Thus, the number of patterns that must be simulated to verify a circuit can often be reduced dramatically by representing many different operating conditions by patterns containing Us.

Ternary modelling is a special case of a more general abstraction technique based on partially-ordered system models. That is, the actual state space of the circuit (in this case all possible combinations of binary values) is extended with values representing sets of circuit states, such that the resulting state set is partially ordered. With ternary simulation, a state with some nodes set to U covers those circuit states obtained by replacing the U values with all combinations of L and H. The state with all nodes set to U thus covers all actual circuit states. By extending the next-state function of the circuit to one over the expanded state set, we can verify circuit behaviour for a set of different operating conditions with a single simulation run. By suitable restrictions of the specification syntax and the extended next-state function, we can guarantee that any property verified on this more abstract form of simulation must also hold for the original circuit.

The concept of symbolic simulation was first proposed by researchers in the late 1970s as a method for evaluating register transfer language representations [CaJB79]. The early programs were very limited in their analytical power since their symbolic manipulation methods were weak. Consequently, symbolic simulation did not evolve much further until more efficient methods of manipulating symbols emerged. The development of Ordered Binary Decision Diagrams (OBDDs) for representing Boolean functions [Brya86] radically transformed symbolic simulation.

The first 'post-OBDD' symbolic simulators were simple extensions of traditional logic simulators [Brya85a]. In these symbolic simulators the input values could be arbitrary Boolean expressions over some Boolean variables rather than only Ls and Hs as in traditional logic simulators. Consequently, the results of the simulation were not single values but rather Boolean functions describing the behaviour of the circuit for the set of all possible data represented by the Boolean variables.

Since a symbolic simulator is based on a traditional logic simulator, it can use the same, quite accurate, electrical and timing models to compute the circuit behaviour. For example, a detailed switch-level model, capturing charge sharing and subtle strengths phenomena, and a timing model, capturing bounded delay assumptions, are well within reach. Also – and of great significance – the switch-level circuit used in the simulator can be extracted automatically from the physical layout of the circuit. Hence, the correctness results will link the physical layout with some higher level of specification.

Although ternary modelling, or its generalisation, allows us to cover many conditions with a single simulation sequence, it lacks the analytic power required for complete verification, except for restricted classes of circuits such as memories [Brya91c]. In [BeBS91], Beatty, Bryant and Seger showed that by combining ternary modelling with symbolic simulation, even more complex behaviours can be modelled with a single simulation run than when using ternary modelling alone. With ternary symbolic simulation, the simulation algorithm designed to operate on scalar values L, H, and U, is extended to operate on a set of symbolic values. Each symbolic value indicates the value of a signal for many different operating conditions, parameterised in terms of a set of symbolic Boolean variables. In essence, ternary symbolic simulation allows us to combine multiple ternary simulation sequences into a single symbolic sequence.

Symbolic trajectory evaluation [SeBr95] takes the notion of ternary symbolic simulation one step further by providing a concrete means of specifying and verifying the desired behaviour of the system operating over time. The specifications take the form of *symbolic trajectory formulas* mixing Boolean expressions and the temporal *next-time* operator. The Boolean expressions provide a convenient means of describing many different operating conditions in a compact form. By allowing only the most elementary of temporal operators, the class of properties we can express is relatively restricted, as compared to other temporal logics (e.g. [ClES86]). Nonetheless, our experience is that we can readily express many aspects of synchronous digital systems at various levels of abstraction. For example, it is quite adequate for expressing many of the subtleties of system operation, including clocking conventions and pipelining.

1.3 Outline of Chapter

Section 2. presents related work. Much of the related work is presented in detail in other chapters of this book, so only a few points of comparison are made to this work. Other related work is also discussed.

The fundamental theory of STE is presented in Section 3. First, STE's novel model representation is discussed, followed by descriptions of the temporal logic in which properties are expressed.

The discussion in Section 3. applies to any system being modelled. Section 4. gives a description of how hardware is modelled. The first part of the section describes the types of circuits that the VossProver can verify. The VossProver represents a circuit using a finite state machine (FSM). The FSM can be generated from a variety of hardware descriptions including transistor and gate netlists, behavioural and structural VHDL formats, and Voss's own gate-level format. The second part of the section shows how these circuit FSMs are represented, and how properties of circuits can be specified.

Section 5. describes how verification is accomplished. First, the theory is presented: symbolic trajectory evaluation and its associated compositional theory. Next, the section describes the front-end of the Voss verification system, and presents a verification methodology based on STE. This is followed by a discussion of tool-making, in particular lessons learned from combining model checking and theorem-proving approaches to verification. The section concludes by discussing the strengths and limitations of the methodology, what can and cannot be proved.

Section 6. presents some of the experimental evidence we have gained with our tools. We present some of our more recent work: the verification of an IEEE-compliant floating point multiplier, and a number of the IFIP WG10.5 benchmarks.

Section 7. concludes by summarising the main results and presents some ideas for future research.

2. Related Work

The components of our methodology are an underlying model checking algorithm, a theory for abstraction and compositionality, and the combining of theorem proving and model checking.

2.1 Model Checking Algorithms

For finite state systems, symbolic model checking methods are very popular and have had success in a number of applications. The use of BDDs has revolutionised model checking by providing a compact method for implicit state representation, thereby increasing by orders of magnitude the size of the state space that can be dealt with. Other approaches exist too [BoWo94, CoHe94]: however BDDs seem to be most effective for a large class of problems (see [Brya92] for a survey of different approaches).

The most well-known work based on symbolic model checking and BDDs has emerged at the end of the 1980s. A number of model checking algorithms for the modal μ-calculus and other logics have been developed [BCLM94]. The SMV verification system based on these ideas has successfully verified a range of systems [BCLM94, McMi92a].

The basic idea of these approaches is to represent the transition relation of the system under consideration with a BDD. A set of states is also represented with a BDD. Given a formula of a temporal logic (e.g. CTL), the model checking task is to compute the set of states that satisfy the formula. The operations defined on BDDs allow the computation of operations such as existential quantification, conjunction etc. Using these BDD operations, it is possible to compute the set of reachable states and the set of states satisfying a given formula.

Although these methods have had considerable success, the computational complexity and cost of model checking remains a significant stumbling block, as circuit verification by symbolic model checking remains intractable. A number of approaches have been suggested to improve the performance of the algorithm: compositional approaches; abstraction; and improving representational methods (for example, partitioning the next state relation [BCLM94]).

That BDDs revolutionised automatic model checking indicates the importance of good and appropriate data structures, and motivates the search for new ones. Considerable work is being done on extending BDD-style structures and developing new ones [BrCh94, ClFZ95a]).[1]

All these approaches to improve symbolic model checking need to be pursued. Circuits with wide data paths are not suitable for verification with SMV, which itself is unable to verify circuits with arithmetic data. However, by extending the method through the use of abstraction [ClGL94] or more sophisticated data structures [ClZh95] such circuits can be verified.

There are other symbolic model checking approaches, with different methods of representing state spaces and next state functions. For example, Jain and Gopalakrishnan have proposed a method based on symbolic simulation [JaGo94]. Rather than representing the state of the system with one boolean expression (i.e. a BDD), they convert such a formula to a parametric representation. This is then used by a symbolic simulator as input, and the output of the simulator is compared to the specification, which is given as a set of symbolic transitions. Although symbolic trajectory evaluation has different representations for state and specification, there is similarity in the approach to STE. Previous work showing the successful use of STE includes [Beat93, Darw94]. Other symbolic methods have been proposed in [BoFi89a, CoBM89b].

2.2 Abstraction

The main problem with model checking is the state explosion problem – the state space grows exponentially with system size. Two methods have some popularity in attacking this problem: compositional methods and abstraction. While they cannot solve the problem in general, they do offer significant improvements in performance.

---

[1] Some of these approaches are applicable to other model checking approaches too.

The direct method of verifying that a circuit has a property f is to show the model M satisfies f. The idea behind abstraction is that instead of verifying property f of model M, we verify property f_A of model M_A and the answer we get helps us answer the original problem. The system M_A is an abstraction of the system M.

One possibility is to build an abstraction M_A that is equivalent (e.g. bisimilar [Miln89a]) to M. This sometimes leads to performance advantages if the state space of M_A is smaller than M. This type of abstraction would more likely be used in model comparison (e.g. as in [HaTu93]).

Typically, the behaviour of an abstraction is not equivalent to the underlying model. The abstractions are *conservative* in that M_A satisfies f_A implies that M satisfies f (but not necessarily the converse). Some examples of abstraction methods are [GrLo93, Long93].

In hardware verification, abstraction is often particularly needed in dealing with the data path of circuits since large data paths increase the state space considerably. A drawback of abstraction is that it takes effort to both come up with the suitable abstraction (see [ClFJ93]) and to prove that the abstraction is conservative. For an example of this type of proof see [BuDi94].

Clarke *et al.* define abstractions and approximations [ClGL94]. They show how an approximation can be abstracted from the program text without having to construct the model of the system. They provide a number of possible abstractions: congruence modulo an integer (the use of the Chinese remainder theorem); representation by logarithm; single-bit and product abstraction; and symbolic abstraction. They show how this is used on a number of examples.

In our approach, the method of state space representation implicitly supports abstraction. Each assertion to be verified implicitly creates an abstraction of the model, which means that abstraction does not require any special skill. Using this method significant data paths can be dealt with, and we do not lose any timing information. An advantage of our logical framework is that we are able to distinguish between the case of the model not having the desired property, and the abstraction of the model being too weak.

2.3 Compositional Reasoning

Typically in compositional model checking, the model of the system is decomposed into a number of sub-models, properties are proved of these sub-models, and then using appropriate reasoning dependent on the technical framework, results are inferred of the overall system. Compositional reasoning has a number of advantages in verification [AnSW94]:

- Modularity: if a module of a system is replaced, only the module need be verified;
- In design or synthesis it is possible to have undefined parts of a system and still be able to reason about it;
- By decomposing the verification task, verification can be made simpler;

– Re-use of verification results is promoted.

The difficulty of compositional reasoning is that often it is the case that a particular component may not have a property that we desire of it when placed in a general environment. However, when placed in the context of the rest of the system, then the component does exhibit the property. See [AbLa95] for some discussion of the issues involved in this type of reasoning.

Another approach to compositional reasoning – modular verification – is based on defining a preorder relation, $\preceq$, between models [GrLo94, Long93]. This preorder is based on a simulation relationship between the models and has the property that if $M_1 \preceq M_2$ and M_2 satisfies ϕ then M_1 satisfies ϕ. Suppose we wish to show that a process M when placed in its environment, E, satisfies a property ϕ. While M may not in general satisfy ϕ, it may satisfy it whenever its environment satisfies another property ψ. Given the formula ψ, there exists a 'tableau' M_ψ which is the strongest element in the preorder which satisfies ψ. If $E \preceq M_\psi$ and $M \| M_\psi$ satisfies ϕ, then by the property of the preorder, $M \| E$ satisfies ϕ. The verification therefore includes proving the simulation relation and performing model checking. Both of these steps are automatic, using symbolic algorithms. This method is only applicable to finite state systems.

Aziz *et al.* propose a compositional method dependent on the formula being checked [ASSS94]. The model is represented as a composition of state machines. Given a formula to be checked, an equivalence relation is computed for each machine which preserves the truth of the formula. Using these equivalence relations, quotient machines are constructed and the composition of these machines computed. This composition will have a smaller state space than the original composition and can be used to determined the correctness of the formula.

In our approach, we do not partition the circuit or next state function – instead we rely on the abstraction discussed previously to support the compositional approach. Given an assertion to verify, the formulas involved determine the abstraction – essentially those parts of the circuit not contributing or contributing only trivially to the properties being checked are abstracted away. This means that the cost of verification is related to the formulas being checked, rather than the size of the circuit. Thus, our compositional approach is comparatively simple, since all the properties proved are valid of the overall circuit, and we need only focus on how to compose properties in different ways.

3. Basics

This section describes the theoretical basis of symbolic trajectory evaluation. In order to accomplish model checking we need to:

– develop a *model representation* for circuits under study;

- agree what it means to say a property is true or false of such models;
- design a language we shall use to describe properties (in this case a temporal logic);
- develop an algorithm for verifying whether a property is true of the model.

Section 3.1 describes the model structures, finite state machines, which can be used to represent circuits. (The discussion here is fairly abstract and applies to any system being modelled. Section 4. refines this discussion to circuit models.)

Sections 3.2 discusses what it means to say that a property is true or false of the model, and Section 3.3 presents the temporal logic in which specifications are given. Decision procedures for deciding the truth of temporal logic formulas are given in Section 5.

3.1 The Model Structure

The *model structure* $(\langle S, \sqsubseteq \rangle, \mathcal{R}, \mathbf{Y})$ represents the system under consideration.

- S, a complete lattice under the information ordering $\sqsubseteq$, represents the state space. Let X be the least element in S.
- $\mathcal{R} \subseteq S$, the set of *realisable states*
- $\mathbf{Y}: S \to S$ is a monotonic next state function: if $s \sqsubseteq t$ then $\mathbf{Y}(s) \sqsubseteq \mathbf{Y}(t)$.

The distinguishing feature of STE is its use of a lattice to represent the state space. The partial order of the lattice represents an information ordering or abstraction relation between states; a state is an abstraction of all the states above it in the information ordering. The higher up we go in the information ordering, the more information we have. The computational advantage of this is that, given the appropriate logical framework, if we prove a property holds of a state in the lattice, it holds of all states above it in the lattice. We first give abstract definitions and presentation, but it is important to bear in mind that circuits have natural representations as lattices, and the use of the information ordering allows us to easily abstract out the necessary information for a particular proof – Section 4.2 gives concrete details of this.

$\mathcal{R}$ represents those states which correspond to states the system could actually attain. $S - \mathcal{R}$ are the 'inconsistent' states, which arise as artifacts of the verification process. Which states are realisable and which are inconsistent is entirely up to the intuition of the modeller; the entire state space could be realisable, or only a part of it.

There is a technical requirement: $\mathcal{R}$ must be downward closed, so that if $x \in \mathcal{R}$, and $y \sqsubseteq x$ then $y \in \mathcal{R}$. This makes computation much easier and has a sensible intuitive basis. Intuitively, if a state is not realisable, it is because it is 'inconsistent'; any state above it in the information ordering must be even more 'inconsistent' and thus also not realisable. Conversely, if a state is 'consistent', then a state below it in the information ordering will also be 'consistent'.

Verification conditions will be of the form: do sequences of states that satisfy g also satisfy h? Distinguishing unrealisable behaviour from realisable behaviour allows the detection of cases where verification conditions are vacuously satisfied.

Although the next state function is inherently deterministic, the partial-order structure of the state space can model non-determinism to some extent. A useful analogy here is that a non-deterministic finite state machine can be modelled by a deterministic one — in the deterministic machine, a state represents a set of states of the non-deterministic machine. In the same way, in our partial-order setting, a state represents all the states above it in the partial order. By embedding a flat (a set without any structure) non-deterministic model in a lattice, the model becomes deterministic.

Branching time versus linear time One of the key issues of temporal logics is whether the logic is linear time or branching time [Stir92]. Since the next state function of the model is deterministic, and since in practice all temporal formulas used are finite, the question of whether the logic used is linear or branching time is rather a fine point. By using the model structure adopted here, non-deterministic paths that exist in a flat model structure are merged, losing information in the process. It is possible to ask the question whether in all runs of the system property g holds; however, the answer returned will be 'unknown.' So, it would not be accurate to characterise the logic proposed here as either linear time or branching time, since the distinction between the two is blurred. As the expressiveness of the logic and the type of non-determinism used in models is limited compared to many other verification approaches, this question of branching versus linear time semantics is not nearly as important as in other contexts.

3.2 The Quaternary Logic $\mathcal{Q}$

We now look at what it means to say that a property is true or false of a lattice model. The four values of $\mathcal{Q}$, the quaternary propositional logic, are used as the basis for this. The four values represent truth, falsity, undefined (or unknown) and overdefined (or inconsistent). Such a logic was proposed by Belnap [Beln77], and has since been elaborated upon and different application areas discussed in a number of other works [Fitt91, Viss84]. This section first gives some mathematical background, based on [Fitt89, RySa92], and then definitions are given and justified. There are two major motivations for using a four-valued logic rather than a two-valued logic. As it is easier to explain these motivations properly after the semantics of the temporal logic have been described, we ask readers to suspend judgement until Section 3.3.2 on page 19 on why we have chosen this approach, and to concentrate for the moment on what we have done.

A *bilattice* is a set together with two partial orders, $\preceq$ and $\leq$, such that the set is a complete lattice with respect to both partial orders. A bilattice is *distributive* if for both partial orders the meet distributes over the join and vice-versa. A bilattice is *interlaced* if the meets and joins of both partial orders are monotonic with respect to the other partial order [Fitt89].

In our application domain, we are interested in the interlaced bilattice

$$Q = \{\perp, \mathbf{f}, \mathbf{t}, \top\}$$

where the partial orders are shown in Figure 3.1. $\mathbf{f}$ and $\mathbf{t}$ represent the boolean values false and true, $\perp$ represents an unknown or neither-true-nor-false truth value, and $\top$ represents an inconsistent or both-true-and-false truth value. $\mathcal{B}$ denotes the set $\{\mathbf{f}, \mathbf{t}\}$ (so $\mathcal{B} \subset Q$).

The partial order $\preceq$ represents an information ordering on the truth domain, and the partial order $\leq$ represents a truth ordering. (Note, the ordering $\sqsubseteq$ is used for comparing *states* and the orderings $\preceq$ and $\leq$ are used to compare *truth values*). It is very important to emphasise at this point that *different* lattices are used to represent truth information and state information.

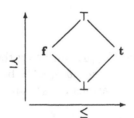

Fig. 3.1. The Bilattice Q

Informally, the information ordering indicates how much information the truth value contains: the minimal element $\perp$ contains no truth information; the mutually incommensurable elements $\mathbf{f}$ and $\mathbf{t}$ contain sufficient information to determine truth exactly; and the maximal element $\top$ contains inconsistent truth information. The truth ordering indicates how true a value is. The minimum element in the ordering is $\mathbf{f}$ (without question not true); and the maximum element is $\mathbf{t}$ (without question true). The two elements $\perp$ and $\top$ are intermediate in the ordering — in the first case, the lack of information places it between $\mathbf{f}$ and $\mathbf{t}$, and in the second case, inconsistent information does. Formally, the partial orders $\leq$ and $\preceq$ are relations on Q (i.e., subsets of $Q \times Q$).

For representing and operating on Q as a set of truth values, there are natural definitions for negation, conjunction and disjunction, namely the weak negation operation of the bilattice and the meet and join of the Q with respect to the truth ordering [Fitt89].

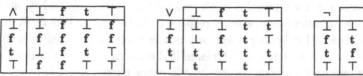

Fig. 3.2. Conjunction, Disjunction and Negation Operators for Q

The classic definitions of these operators are shown in Figure 3.2, and have the following pleasant properties, which make Q suitable for model-checking partially-ordered state spaces.

- The definitions are consistent with the definitions of conjunction, disjunction and negation on boolean values.
- These operations obey the natural distributive laws, and also obey De Morgan's laws (so, the definition of disjunction was redundant).
- Efficiency of implementation. The quaternary logic can be represented by a dual-rail encoding, i.e. a value in Q is represented by a pair of boolean values, where:
$$\bot = (F, F),$$
$$\mathbf{f} = (F, T),$$
$$\mathbf{t} = (T, F),$$
$$\top = (T, T).$$
If a is represented by the pair (a_1, a_2) and b by the pair (b_1, b_2) then $a \wedge b$ is represented by the pair $(a_1 \wedge b_1, a_2 \vee b_2)$, $a \vee b$ by the pair $(a_1 \vee b_1, a_2 \wedge b_2)$ and $\neg a = (a_2, a_1)$. These operations on Q can be implemented as one or two boolean operations.

Implication, $\Rightarrow$, is defined as a derived operator $a \Rightarrow b \overset{\text{def}}{=} \neg a \vee b$.

There is an intuitive explanation of the dual-rail encoding and the implementation of the operators. If q is encoded by the pair (a, b), a is evidence *for* the truth of q, and b is evidence *against* q. To compute $q_1 \wedge q_2$, we conjunct the evidence for q_1 and q_2 and take the disjunction of the evidence against. The computation of $q_1 \vee q_2$ is symmetric. And if a is the evidence for q and b the evidence against q, then b is the evidence for $\neg q$ and a is the evidence against $\neg q$. This view gives some support for the definitions that $\top \vee \bot = \mathbf{t}$ and $\top \wedge \bot = \mathbf{f}$.

3.3 The Temporal Logic

Now we know what it means to say that a property is true of a lattice model, we must develop the language for expressing properties, and we can use Q to guide this process. This section first presents the scalar version of TL, the fragment of TL not containing variables, and then presents the symbolic version of TL, which contains variables.

3.3.1 Scalar Version of TL. Given a model structure $(\langle S, \sqsubseteq \rangle, \mathcal{R}, \mathbf{Y})$, a Q-predicate over S is a function mapping from S to the bilattice Q. A Q-predicate, p is monotonic if $s \sqsubseteq t$ implies that $p(s) \preceq p(t)$ (monotonicity is defined with respect to the information ordering of Q). A Q-predicate is a generalised notion of predicate, and to simplify notation, the term 'predicate' is used in the rest of this discussion. Note that Q represents truth values and not state values; the only way truth and state values are related is by predicates.

Example 3.1. Take, as an example, the state space S given in Figure 3.3. The connecting lines indicate the partial order. Define $g, h : S \to Q$ by:

$$g(s) = \begin{cases} \bot & \text{when } s = s_0 \\ \mathbf{f} & \text{when } s \in \{s_1, s_2, s_4, s_5, s_6\} \\ \mathbf{t} & \text{when } s \in \{s_3, s_7, s_8\} \\ \top & \text{when } s = s_9 \end{cases}$$

and

$$h(s) = \begin{cases} \bot & \text{when } s \in \{s_0, s_2, s_6\} \\ \mathbf{f} & \text{when } s \in \{s_1, s_4, s_5\} \\ \mathbf{t} & \text{when } s \in \{s_3, s_7, s_8\} \\ \top & \text{when } s = s_9. \end{cases}$$

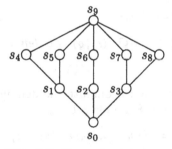

Fig. 3.3. Example statespace

Figure 3.4 depicts these definitions graphically. g and h are Q-predicates. Each state in the lattice on the left is annotated with the truth value that g attains on that state; similarly each state in the lattice on the right is annotated with the truth value that h attains on that state. The same state space and functions will be used in subsequent examples.

□

Note that in the example, s_3 is the weakest state (in other words, the state lowest in the ordering) for which $g(s) = \mathbf{t}$. In a sense, s_3 partially characterises

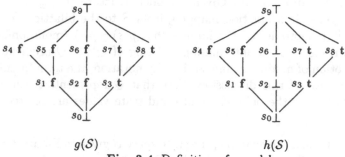

$$g(\mathcal{S}) \qquad\qquad h(\mathcal{S})$$

Fig. 3.4. Definition of g and h

g, and we use this idea as a building block for characterising predicates, motivating the next definition. Given a predicate p, we are interested in the pairs (s_q, q) where s_q is a weakest state for which $p(s) = q$; these pairs all called defining pairs.

Definition 3.1. $(s_q, q) \in \mathcal{S} \times \mathcal{Q}$ *is a* defining pair *for a predicate g if $g(s_q) = q$ and $\forall s \in \mathcal{S}, g(s) = q$ implies that $s_q \sqsubseteq s$.* □

In Example 3.1, $(s_3, \mathbf{t})$ is a defining pair for g. If $g(s) = \mathbf{t}$ then $s_3 \sqsubseteq s$. However, there is no defining pair $(s_{\mathbf{f}}, \mathbf{f})$ for g since there is no unique weakest element in $\mathcal{S}$ for which g takes on the value $\mathbf{f}$. On the other hand $(s_1, \mathbf{f})$ is a defining pair for h.

Definition 3.2. *If $g \colon \mathcal{S} \to \mathcal{Q}$ then*

$$D(g) = \{(s_q, q) \in \mathcal{S} \times \mathcal{Q} : (s_q, q) \text{ is a defining pair for } g\},$$

is the defining set *of g.* □

Using this definition it is easy to compute the defining sets of the functions g and h that were defined in Example 3.1.

$$\begin{aligned} D(g) &= \{(s_0, \bot), (s_3, \mathbf{t}), (s_9, \top)\} \\ D(h) &= \{(s_0, \bot), (s_1, \mathbf{f}), (s_3, \mathbf{t}), (s_9, \top)\} \end{aligned}$$

If a monotonic predicate has a defining pair for every element in its range, then its defining set uniquely characterises it. Such monotonic predicates are called simple predicates and form the basis of our temporal logic. The following notation is used in the next definition and elsewhere in the chapter: if $g : A \to B$ is a function then $g(A) = \{g(a) : a \in A\}$ is the *range* of g.

Definition 3.3. *A monotonic predicate $g \colon \mathcal{S} \to \mathcal{Q}$ is* simple *if for all $q \in g(\mathcal{S})$, there is a $(s_q, q) \in D(g)$.* □

In Example 3.1, h is simple since every element in the range of h has a defining pair. On the other hand, g is not simple since there is no defining pair $(s_f, \mathbf{f})$. Informally, g is not simple since we cannot use a single element of S to characterise the values for which $g(s) = \mathbf{f}$. We shall the set of simple predicates to construct the temporal logic.

Definition 3.4 (Syntax Scalar Extended Logic — TL). *Let G be the set of simple predicates. The set of scalar TL formulas is defined by the following abstract syntax*

$$\text{TL} ::= G \mid \text{TL} \wedge \text{TL} \mid \neg \text{TL} \mid \text{Next TL} \mid \text{TL Until TL}$$

□

Restricting the base of TL to be the simple predicates is only a syntactic restriction, since all monotonic predicates can be expressed as TL formulas [Haze96].

The semantics of a formula is given by the satisfaction relation Sat relating sequences of states to TL formulas. We consider sequences of states as ordered lists of states, and do not formally define them [Binm77]. Let S^ω be the set of all infinite sequences, thus $Sat : S^\omega \times \text{TL} \to \mathcal{Q}$. Given a sequence σ and a TL formula g, Sat returns the degree to which σ satisfies g.

Suppose g and h are TL formulas. Informally, if g is a simple predicate, a sequence satisfies it if g holds of the initial state of the sequence. Conjunction has a natural definition. A sequence satisfies $\neg g$ if it does not satisfy g. A sequence satisfies Next g if the sequence obtained by removing the first element of the sequence satisfies g. A sequence satisfies g Until h if there is a k such that the first $k - 1$ suffixes of the sequence satisfy g and the k-th suffix satisfies h.[2] Note that in the definitions below, $\wedge$ and $\neg$ (bold face symbols) are operations on TL formulas, whereas $\wedge$ and $\neg$ are operations on $\mathcal{Q}$. We use the following notation when using sequences.

1. Lower case Greek letters, $\sigma, \tau, \ldots$ are used to refer to sequences.
2. If $\sigma = s_0 s_1 s_2 \ldots$, then σ_i denotes s_i.
3. If $\sigma = s_0 s_1 s_2 \ldots$ is a sequence, $\sigma_{\geq i}$ refers to the sequence $s_i s_{i+1} \ldots$, which is a suffix of σ.
4. Superscripts are used to refer to different sequences, e.g. σ^1, σ^2. Although this conflicts with the usual use of superscript in mathematical text, there is little chance of confusion since 'squaring' states is not defined.
5. If s is a state which is a vector of elements, then $s[k]$ refers to the k-th component of s.

For example, $\sigma^3_{\geq i}$ refers to the suffix of the sequence σ^3 obtained by removing first i elements of σ^3. $(\sigma^3_{\geq i})_0[k] = \sigma^3_i[k]$ is the k-th component of the i-th element in the sequence σ^3.

---

[2] In the special case of g and h being simple predicates, this is equivalent to saying that g is true of the first $k - 1$ states in the sequence, and h is true of the k-th state.

Definition 3.5 (Semantics of TL). *Let* $\sigma = s_0 s_1 s_2 \ldots \in \mathcal{S}^\omega$:

1. *If* $g \in G$ *then* $\mathrm{Sat}(\sigma, g) \overset{\mathrm{def}}{=} g(s_0)$.
2. $\mathrm{Sat}(\sigma, g \wedge h) \overset{\mathrm{def}}{=} \mathrm{Sat}(\sigma, g) \wedge \mathrm{Sat}(\sigma, h)$
3. $\mathrm{Sat}(\sigma, \neg g) \overset{\mathrm{def}}{=} \neg \mathrm{Sat}(\sigma, g)$
4. $\mathrm{Sat}(\sigma, \mathtt{Next}\, g) \overset{\mathrm{def}}{=} \mathrm{Sat}(\sigma_{\geq 1}, g)$
5. $\mathrm{Sat}(\sigma, g \,\mathtt{Until}\, h) \overset{\mathrm{def}}{=} \bigvee_{i=0}^{\infty} \left(\left(\bigwedge_{j=0}^{i-1} \mathrm{Sat}(\sigma_{\geq j}, g) \right) \wedge \mathrm{Sat}(\sigma_{\geq i}, h) \right)$

□

The form of the definition of conjunction and until is a little different to the standard definition given for temporal logics. The standard definition of the until operator would look like: $\mathit{Sat}(\sigma, g\,\mathtt{Until}\,h)$ if there exists an i such that $\left(\bigwedge_{j=0}^{i-1} \mathit{Sat}(\sigma_{\geq j}, g) \right) \wedge \mathit{Sat}(\sigma_{\geq i}, h)$. The definition given here is essentially the same for a two-valued logic. In the four-valued case it is computationally very convenient (and we argue more elegant) to eliminate the unnecessary meta-logical level.

This is the strong version of the until operator: h must eventually hold (note that g need never hold). The until operator is defined as an infinite disjunction of conjunctions. That this is well defined comes from $\mathcal{Q}$ being a complete lattice with respect to the truth ordering. Recall that $\wedge$ is defined as the meet of the truth ordering, and $\vee$ is defined as the join. Moreover, in a complete lattice, *all* sets have a meet and join. Therefore each conjunction is well defined, and thus the disjunction of the conjunctions is too.

Using these operators we can define a number of other operators as shorthand.

Definition 3.6 (Other operators). *Some that we shall use are:*

- Disjunction: $g \vee h = \neg((\neg g) \wedge (\neg h))$.
- Implication: $g \Rightarrow h = (\neg g) \vee h$.
- Sometime: $\mathtt{Exists}\, g = \mathtt{t}\,\mathtt{Until}\, g$ *(some suffix of the sequence satisfies* g*)*.
- Always: $\mathtt{Global}\, g = \neg(\mathtt{Exists}\, \neg g)$ *(no suffix of the sequence does not satisfy* g*, hence all must satisfy* g*)*.
- Weak until: $g \,\mathtt{UntilW}\, h = (g \,\mathtt{Until}\, h) \vee (\mathtt{Global}\, g)$ *(this doesn't demand that* h *ever be satisfied)*.
- *The generalised* $\mathtt{Next}$ *operator is defined by:*

$$\mathtt{Next}^0 g = g$$
$$\mathtt{Next}^{k+1} g = \mathtt{Next}\,(\mathtt{Next}^k g)$$

- *The bounded always operator, defined by*

$$\mathtt{Global}\,[(a_0, b_0), \ldots, (a_n, b_n)]\, g = \bigwedge_{j=0}^{n} \left(\bigwedge_{k=a_j}^{b_j} \mathtt{Next}^k g \right),$$

asks whether g *holds between* a_j *and* b_j *for* $j = 0, \ldots, n$.

□

If $q = Sat(\sigma, g)$ then we say that σ satisfies g with truth value q, and if $q \preceq Sat(\sigma, g)$, then we say that σ satisfies g with at least truth value q.

One of the key properties of the satisfaction relation is that it is monotonic, i.e. if $\sigma^1 \sqsubseteq \sigma^2$, then $Sat(\sigma^1, g) \preceq Sat(\sigma^2, g)$.

3.3.2 Motivation for Four-Valued Logic. The first reason for using a four-valued logic is based on our view of how one should reason about partially-ordered state spaces. The information ordering on states is an abstraction relation, thus if s_1 and s_2 are above s_0 in the information ordering, s_0 represents an abstraction of s_1 and s_2. Suppose a simple predicate g is true of s_1 ($g(s_1) = \mathbf{t}$) and false of s_2 ($g(s_2) = \mathbf{f}$). What value would be reasonable to give to $g(s_0)$? If we view the abstraction relation as a *conservative* abstraction it would be possible to say $g(s_0) = \mathbf{f}$ (although as we shall shortly see this causes a significant technical problem). However, we argue this is unsatisfactory as such an assignment does not accurately capture the state of knowledge about a system. By conflating lack of information and knowledge of falseness of a proposition we lose information. Both are useful to know, and we should separate out the concerns. This means that we need a logical value describing an unknown truth value, denoted $\perp$, which gives at least a three-valued logic. But not only does a four-valued logic give a nicer logical structure, once one accepts the need for inconsistent states, a similar argument shows that a fourth value, $\top$, denoting an inconsistent truth value, is also needed.

The second argument is more technical in nature but clearly shows why a two-valued framework is not powerful enough. We require that our predicates be monotonic with respect to states (and more generally sequences of states). Making this restriction not only gives us an efficient model checking algorithm, but we argue is essential from a reasoning point of view as is now explained. Suppose we have a sequence of states $s_1, \ldots, s_n$ such that $s_i \sqsubseteq s_{i+1}$ for $i = 1, \ldots, n - 1$ (viz. a chain). It makes sense that the truth value of a predicate will change as we go up the sequence; but it would be nonsensical in terms of our understanding of the information ordering for *any* changes to permissible. For example, if the truth value of a predicate swung from $\mathbf{t}$ to $\mathbf{f}$ and back again as we progressed up the sequence, our intuition and motivation for the information ordering would be untenable. Thus it is reasonable to assume that only monotonic predicates are of interest (and indeed all TL formulas are monotonic).

So suppose that we had a two-valued truth domain, and without loss of generality order $\mathbf{f}$ below $\mathbf{t}$ in the truth ordering. In this case we cannot support negation. For suppose $s \sqsubseteq t$ with $g(s) = \mathbf{f}$ and $g(t) = \mathbf{t}$. g is monotonic. But, $\neg g$ cannot be monotonic. This implies that if we wish to support negation, we need to move beyond a two-valued logic.

3.3.3 Symbolic Version. Describing the properties of a system explicitly by a set of scalar formulas of TL is too tedious. Symbolic formulas allow a concise representation of a large set of scalar formulas. A symbolic formula represents the set of all possible instantiations of that symbolic formula.

TL is extended to symbolic domains by allowing boolean variables to appear in the formulas. Let $\mathcal{V}$ be a set of variable names $\{v_1, \ldots, v_n\}$. It would be possible to define the symbolic version of the logic by introducing quaternary variables. However, in practice, it is boolean variables which are needed, and introducing only boolean variables means that simpler and more efficient implementations of the logic can be accomplished. Furthermore, the effect of a quaternary variable can be created by introducing a pair of boolean variables. Note that this does not say that the quaternary values $\top$ and $\bot$ and the corresponding constant TL predicates are not needed – only that it does not seem that quaternary *variables* are needed.

Definition 3.7 (The Symbolic Logic — TL). *The syntax of the set of symbolic TL formulas, TL, is defined by:-*

$$\dot{T}L ::= G \mid \mathcal{V} \mid \dot{T}L \wedge \dot{T}L \mid \dot{\neg}\dot{T}L \mid \mathtt{Next}\,\dot{T}L \mid \dot{T}L\,\mathtt{Until}\,\dot{T}L$$

□

The derived operators are defined in a similar way to Definition 3.6. For convenience, where there is little chance of confusion, the dots on TL formulas are omitted.

The satisfaction relation is now determined by a sequence, a formula, *and* an *interpretation* of the variables.

Definition 3.8. *An interpretation, ϕ, is a mapping from variables to the set of constant predicates $\{\mathbf{f}, \mathbf{t}\}$. Let $\Phi = \{\phi : \phi\colon \mathcal{V} \to \{\mathbf{f}, \mathbf{t}\}\}$ be the set of all interpretations. Given an interpretation ϕ of the variables, there is a natural, inductively defined interpretation of TL formulas. For a given $\phi \in \Phi$, we extend the domain of ϕ from $\mathcal{V}$ to all of TL by defining:*

$$\phi(g) = g \text{ if } g \in G$$
$$\phi(\neg g) = \neg\phi(g)$$
$$\phi(g_1 \wedge g_2) = \phi(g_1) \wedge \phi(g_2)$$
$$\phi(\mathtt{Next}\, g) = \mathtt{Next}\,\phi(g)$$
$$\phi(g_1 \,\mathtt{Until}\, g_2) = \phi(g_1) \,\mathtt{Until}\, \phi(g_2)$$

This can be expressed syntactically: if $\phi(v_i) = b_i$, replace each occurrence of v_i with b_i, written as $\phi(g) = g[b_1/v_1, \ldots, b_n/v_n]$.

Given a sequence and a symbolic formula, the symbolic satisfaction relations, SAT_q, determine for which interpretations of variables the sequence satisfies the formula with which degree of truth. For each q, SAT_q is of type $S^\omega \times \mathrm{TL} \to \Phi$.

Definition 3.9 (Satisfaction relations for TL). *A number of satisfaction relations are defined.*

$-$ *For* $q = \mathbf{f}, \mathbf{t}, \top$,

$\quad SAT_q(\sigma, g) = \{\phi \in \Phi : q = \mathrm{Sat}(\sigma, \phi(g))\}$.

$-$ *For* $q = \mathbf{f}, \mathbf{t}, \top$,

$\quad SAT_{q\uparrow}(\sigma, g) = \{\phi \in \Phi : q \preceq \mathrm{Sat}(\sigma, \phi(g))\}$.   □

Note that if g is a (symbolic) formula and ϕ an interpretation, then $SAT_q(\sigma, g)$ $\subseteq \Phi$, while $Sat(\sigma, \phi(g)) \in \mathcal{Q}$. Informally,

$-$ $SAT_\top(\sigma, g)$ is the set of interpretations for which g and $\neg g$ hold. Such results are undesirable and verification algorithms should detect and flag them.

$\quad SAT_{\top\uparrow}(\sigma, g) = SAT_\top(\sigma, g)$.

$-$ $SAT_\mathbf{t}(\sigma, g)$ is the set of interpretations for which g is (sensibly) true.

$\quad SAT_{\mathbf{t}\uparrow}(\sigma, g) = SAT_\top(\sigma, g) \cup SAT_\mathbf{t}(\sigma, g)$.

$-$ $SAT_\mathbf{f}(\sigma, g)$ is the set of mappings for which g is (sensibly) false.

$\quad SAT_{\mathbf{f}\uparrow}(\sigma, g) = SAT_\top(\sigma, g) \cup SAT_\mathbf{f}(\sigma, g)$.

Thus each satisfaction relation defines a set of interpretations for which a desired relationship holds. Sets of interpretations can be represented efficiently using BDDs. For example, we may be interested in the interpretations of variables for which a sequence satisfies a formula with truth value $\mathbf{t}$, or the interpretations for which a sequence satisfies a formula with truth value at least $\mathbf{t}$. By being able to determine for which interpretations a property holds with a given degree of truth, we are able to construct appropriate verification conditions. Each satisfaction relation defines a set of interpretations for which a desired relationship holds.

4. Modelling Hardware

Section 4.1 describes types of circuits that can be modelled, and how accurate finite state machine (FSM) models can be constructed from circuit descriptions. Section 4.2 shows what a lattice space structure for these FSMs looks like, and also shows that a simple version of TL is appropriate for circuit models.

4.1 Modelling issues

The tool that we use is called the VossProver, based on the original Voss system [Sege93]. In the next section we discuss the front-end, the interface which the verifier sees. This section discusses the back-end $-$ how circuits are modelled and manipulated.

Since the basic engine in Voss is symbolic trajectory evaluation $-$ which essentially is a form of symbolic simulation $-$ the main requirement on any hardware description format is that it must have an operational semantics that can be made 'data oblivious'. In other words, we must be able to symbolically simulate the model. Fortunately, this is a fairly mild requirement and thus STE can deal with a range of circuit models.

At the lowest level, Voss can be used to verify properties of transistor netlists modelled as switch-level circuits. Similar to the COSMOS system from Carnegie-Mellon University, the Voss system uses symbolic switch-level analysis to compile a transistor netlist into a behavioural (multi-level) state machine description. Since it uses the same type of Gaussian elimination approach as COSMOS, there is essentially no size limitation for the circuits that can be compiled. In fact, except for some few (typically very 'analog') circuit structures, the compiler generates state machine descriptions that accurately reflect the behaviour of the transistor netlists. In order to avoid 'false positives' (i.e., the model checker saying that the a result is true when it is really false), the switch-level model is pessimistic. Finally, in order to deal with low-level timing issues, the switch-level models that the Voss system handles can be annotated with minimum and maximum rise and fall delays and various race analysis algorithms can be used in the modelling of the networks.

At the next level of abstraction, Voss can be used to reason about gate level circuits. Here the input format is one of EDIF (subset), structural VHDL (subset), or various proprietary gate-level formats. Again, the models can be annotated with delay values and the verification can be done using race analysis algorithms ranging from simple zero and unit-delay models up to complex bounded delay models.

Finally, at the most abstract level, Voss can also read in circuits described in (a subset) of behavioural VHDL. The restriction here is partly based on the need to be able to symbolically simulate the model, but there are also severe restrictions on the subset that simply stems from implementation artifacts.

4.2 Circuit Models as State Spaces

In practice, STE is applied to circuit models. The state space for a circuit model represents the values which the nodes in the circuit take on, and the next state function can be represented implicitly by symbolic simulation of the circuit. The nodes in a circuit take on high (H) and low (L) voltage values. It is useful, both computationally and mathematically, to allow nodes to take on unknown (U) and inconsistent or over-defined (Z) values. The set $C = \{U, L, H, Z\}$ forms the lattice defined in Figure 4.1. Thus if our circuit consists of n state-holding components, the state space will be C^n.

Q and C have a similar underlying mathematical structure, which is useful but can be confusing. However, it is important to distinguish between the two: Q represents truth values, and C represents state values.

The need for the U element is usually readily accepted, but the Z needs a little more motivation. We use the next state function $\mathbf{Y}$ to model the behaviour of the circuit through symbolic simulation. Essentially we use the

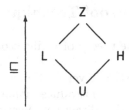

Fig. 4.1. Circuit node state space

information given by a TL formula to 'initialise' the state of the circuit and then use symbolic simulation to step through the state of the circuit over time. Given this, it is very easy for a user to provide an initialising TL formula which takes the circuit into an inconsistent state – for example one in which a node in the circuit must have both a high and low voltage. Whether this occurs may depend entirely on the next state function and the formula may not be invalid; in general in a situation like this we shall find that there are many sequences that satisfy the formula, but that no real run of the system (i.e. no realisable trajectory) satisfies it. Having the Z value is technically very convenient as we get a pleasant mathematical structure to compute in, and from a user's point of view practically useful since it allows our verification methodology to detect inconsistent specifications and report them to the user.

In terms of the logic that we use to describe a circuit model, the special case of the state space being a cross-product of quaternary sets need not be treated differently than the general case (when the state space is an arbitrary lattice) as all the above definitions apply. However, it is convenient to establish a specialised form of TL for circuit models called TL_n.

TL_n is the same as TL except that we only allow as our base predicates the constant predicates and the simple predicates that express whether individual nodes have given voltages, viz. formulas of the form

$$[N] = x$$

which ask whether the node N has the symbolic value x. We have shown that this is only a syntactic restriction since all simple predicates can still be expressed as TL_n formulas [Haze96].

It is useful here to pause to reemphasise the distinction between the truth domain and the circuit domain. Let s be a state of the circuit (i.e. an element of $\mathcal{C}^n$). Then $s[i]$ represents the value on a particular node in the circuit (and is an element of $\mathcal{C}$). The predicate $p \overset{\text{def}}{=} [i] = x$ asks whether node i has the value x ($x \in \mathcal{C}$). $p(s) \in \mathcal{Q}$ specifies the truth of the statement that in state s node i has the value x.

5. Specification and Proof Techniques

The section first presents the theory of verification, and then discusses practical issues.

Section 5.1 explores the style of verification adopted and shows what verification conditions look like; this introduces some useful notation and definitions and guides the rest of this discussion. Section 5.2 describes the theory of symbolic trajectory evaluation. This shows how verification conditions can be verified. Section 5.3 presents STE's compositional theory.

Our experience of verification using STE and its associated theory is heavily influenced by our experience in constructing and using verification tools, and so this section is centred around the use of Voss. This section describes how specifications are given and how verification is performed. Section 5.4 discusses the features and use of the Voss system. Section 5.5 explains how properties are expressed and verification is performed.

5.1 Verification assertions

Let the model structure of the system be $\mathcal{M} = (\langle \mathcal{S}, \sqsubseteq \rangle, \mathcal{R}, \mathbf{Y})$. $\mathcal{S}^\omega$ is the set of sequences of the state space. The partial order on $\mathcal{S}$ is extended point-wise to sequences ($\sigma^1 \sqsubseteq \sigma^2$ if for all i, $\sigma_i^1 \sqsubseteq \sigma_i^2$). Informally, the *trajectories* are all the possible runs of the system; formally, a *trajectory*, σ, is a sequence compatible with the next state function:

$$\forall i \geq 0, \mathbf{Y}(\sigma_i) \sqsubseteq \sigma_{i+1}.$$

The style of verification used in symbolic trajectory evaluation (STE) is to ask questions of the form:

Do all trajectories that satisfy g also satisfy h?

The formula g is known as the *antecedent*, and the formula h is known as the *consequent*. 'Satisfy' is a broad term — there are a number of satisfaction relations that can be used. Which one matches our notion of correctness? There are a number of possible ways of modelling correctness, and the key issue is how to deal with inconsistent information. How correctness is modelled depends on choices made in the specification and verification process — although guided by technical considerations, there is considerable flexibility. There are two obvious ways to formalise the notion of 'trajectory σ satisfies g'.

(5.1) $t \quad = \quad Sat(\sigma, g)$

(5.2) $t \quad \preceq \quad Sat(\sigma, g)$

Relation 5.1 captures a more precise notion — where inconsistency has not caused a predicate to be true of a trajectory. Intuitively, it is a better model of satisfaction than relation 5.2. However, the latter definition has some advantages. Firstly, it can be implemented very efficiently as will be seen when trajectory evaluation is described later. Secondly, it does capture some useful information — either the property 'really' is true, or the specification is inconsistent.

Corresponding to these two definitions, there are two ways of asserting correctness with respect to an antecedent, consequent pair.

Definition 5.1. $\models \langle g \Rightarrow h \rangle$ *if and only if* $\forall \sigma \in \mathcal{R}_T, \mathbf{t} = \mathrm{Sat}(\sigma, g)$ *implies that* $\mathbf{t} = \mathrm{Sat}(\sigma, h)$.

and

Definition 5.2. $\models \langle g \Rrightarrow h \rangle$ *if and only if* $\forall \sigma \in \mathcal{S}_T, \mathbf{t} \preceq \mathrm{Sat}(\sigma, \phi(g))$ *implies that* $\mathbf{t} \preceq \mathrm{Sat}(\sigma, \phi(h))$.

The first definition takes a very precise view of realisability, considering only realisable trajectories (if there are unrealisable trajectories with strange behaviour, then these are ignored) and requiring that trajectories satisfy formulas with value exactly $\mathbf{t}$. The second definition takes a more relaxed view of inconsistent behaviour. We consider the behaviour of all trajectories, whether realisable or not, and treat the truth values $\mathbf{t}$ and $\top$ as satisfying the notion of correctness. Therefore, for pragmatic reasons, in this chapter we concentrate on Definition 5.2, which will be central in the development of the theory. The alternative definition is interesting both theoretically and practically and has been developed elsewhere [Haze96].

5.2 Trajectory Evaluation

Only the scalar version of TL is examined here. Extension to the symbolic case is straightforward; however, there is enough extra notation and detail to make an exposition of the scalar case clearer.

Section 5.2.1 shows that a formula of TL can be characterised by the sets of minimal trajectories that satisfy it, and furthermore shows that these sets can be used to accomplish verification. The computation of such sets is not directly possible, but Section 5.2.2 shows that computing approximations of the sets *is* feasible (and as later experimental evidence will show, forms a good basis for practical verification).

5.2.1 Minimal Sequences and Verification. This section first formalises the notion of the sets of minimal trajectories satisfying formulas, and then shows how these sets can be used in verification.

The first definition is an auxiliary one: given a subset of a partially-ordered set, it is useful to be able to determine the minimal elements of the set. If B is a subset of A, then $b \in B$ is a minimal element of B if no other element in B is smaller than b (i.e. all elements of A smaller than b do not lie in B).

Definition 5.3. *If A is a set, $B \subset A$, and $\sqsubseteq$ a partial order on A, then*

$$\min B = \{b \in B : if \exists a \in A \text{ such that } a \sqsubseteq b, \text{ either } a = b \text{ or } a \notin B\}. \square$$

Definition 5.4. *If g is a (scalar) TL formula, then $\min g$ is the set of minimal trajectories satisfying g, where $\min g$ is defined by:* $\min g = \min\{\sigma \in S_T : t \preceq \text{Sat}(\sigma, g)\}$ $\qquad\qquad \square$

We now see a useful property of min.

Lemma 5.1. *If $\min g \subseteq \min h$, then every trajectory that satisfies g also satisfies h.*

Proof. Suppose σ satisfies g: then there must exist $\sigma' \in \min g$ such that $t \preceq Sat(\sigma', g)$ and $\sigma' \sqsubseteq \sigma$; but since $\min g \subseteq \min h$, $\sigma' \in \min h$ and hence $t \preceq Sat(\sigma', h)$; hence by monotonicity $t \preceq Sat(\sigma, h)$. $\qquad\qquad \square$

This gives some indication that manipulating and comparing the sets of minimal trajectories that satisfy formulas can be useful in verification.

Although we shall be comparing sets of sequences, containment is too restrictive, motivating a more general method of set comparison. The statement 'every trajectory that satisfies g also satisfies h' implies that the requirements for g to hold are stricter than the requirements for h to hold. Thus, if σ is a minimal trajectory satisfying g, σ must satisfy h. Since the requirements for g are stricter than the requirements on h, σ need not be a *minimal* trajectory satisfying h, but there must be a minimal trajectory, σ', satisfying h where $\sigma' \sqsubseteq \sigma$. This is the intuition behind the following definition, which defines a relation over $\mathcal{P}(S)$, the power set of S.

Definition 5.5. *If S is a lattice with a partial order $\sqsubseteq$ and $A, B \subseteq S$, then $A \sqsubseteq_P B$ if $\forall b \in B, \exists a \in A$ such that $a \sqsubseteq b$.* $\qquad\qquad \square$

To illustrate this definition, consider the example of Figure 5.1. Assume A and B are subsets of some partially ordered set S. Note that in this example that both A and B are upward closed. Although the definitions given here do not require this, we shall be dealing with upward closed sets.[3] Figure 5.1(a) depicts A. Let $A_m = \min A = \{\alpha, \beta, \gamma, \zeta\}$ be the set of minimal elements of A. Then A consists of all the elements above the dotted line. Similarly, Figure 5.1(b), depicts B. Let $B_m = \min B = \{\eta, \gamma\}$ be the set of minimal elements of B. Figure 5.1(c) is the superposition of Figures 5.1(a) and (b).

Note that $A_m \sqsubseteq_P B_m$. For each element of B_m there is an element of A_m less than or equal to it: $\alpha \sqsubseteq \eta$ and $\gamma \sqsubseteq \gamma$.

Suppose A is the set of elements with property g, and that B is the set of elements with property h. Then $\min g = A_m$ and $\min h = B_m$. By examining the figure it is easy to see that all elements of S that have property h also

---

[3] We shall be manipulating sets of trajectories and sequences that satisfy formulas; that these are upward closed follows from the monotonicity of the satisfaction relation.

have property g (h implies g). But, note that $\min h \not\sqsubseteq \min g$. However, it is the case that $\min g \sqsubseteq_{\mathcal{P}} \min h$, which motivates exploring the $\sqsubseteq_{\mathcal{P}}$ relation further.

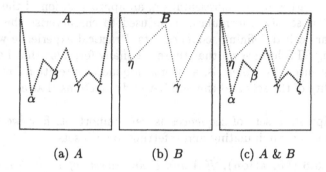

(a) A　　　　　(b) B　　　　　(c) A & B

Fig. 5.1. The Preorder $\sqsubseteq_{\mathcal{P}}$

Theorem 5.1. *If g and h are TL formulas, then $g \Longrightarrow h$ if and only if $\min h \sqsubseteq_{\mathcal{P}} \min g$.*

Although computing the minimal sets directly is often not practical, it is possible to find approximations of the minimal sets (they are approximations because they may contain some redundant sequences, i.e. there may be two sequences that are ordered by the partial order in the set when only the lesser of the two need be in the set). The next section shows how to construct two types of approximations to the minimal sets. $\Delta^{\mathbf{t}}(h)$ is an approximation of the set of minimal *sequences* that satisfy h, and $T^{\mathbf{t}}(g)$ is an approximation of the minimal trajectories that satisfy g. The importance of these approximations are that (i) $\Delta^{\mathbf{t}}(h) \sqsubseteq_{\mathcal{P}} T^{\mathbf{t}}(g)$ exactly when $g \Longrightarrow h$ (an analogue of Theorem 5.1), and (ii) there is an efficient method for computing these approximations, which we now turn to.

5.2.2 Scalar Trajectory Evaluation. The method of computing the approximations to the minimal sets of formulas is based on symbolic trajectory evaluation (STE), a model checking algorithm for checking partially-ordered state spaces. The original version of STE was first presented in [BrSe91] and a full description of STE can be found in [SeBr95]. In these presentations, the algorithm is applied only to trajectory formulas, a restricted, two-valued temporal logic. This chapter generalises earlier work in two important respects:

1. It presents the theory for applying STE to the quaternary logic.
2. It presents the theory for the full class of TL. In particular it deals with disjunction and negation.

This section examines the scalar version of TL and shows how, given a TL formula, a set of sequences characterising the formula can be constructed. Recall the definition of defining pair and defining set from Section 3.3.1. The defining set of a simple predicate characterises that predicate; this can be used as a building block to find a characterisation of all temporal predicates. By using the partial order representation, an approximation of the minimal sequences that satisfy a formula can be used to characterise the formula. These sets are called defining sequence sets. Practical experience with verification using STE has shown that there are many formulas that have small defining sequence sets – there is, at worst, a quadratic relationship between the cardinality of the sets and the number of disjunctions that appear in the formula.

As manipulating sets of sequences is very important, first we build up some notation for manipulating and referring to such sets.

Definition 5.6 (Notation). *If A and B are subsets of a lattice $\mathcal{L}$ on which a partial order $\sqsubseteq$ is defined, then $A \amalg B = \{a \sqcup b : a \in A, b \in B\}$. If $g : \mathcal{L} \to \mathcal{L}$, $g(A)$ continues to represent the range of g, and similarly, $g(\langle A, B \rangle) = \langle g(A), g(B) \rangle$.*   □

Note that we write $A \amalg B$ rather than $A \sqcup B$ since although $A \amalg B$ is a least upper bound (with respect to $\sqsubseteq_{\mathcal{P}}$) of A and B it is not *the* least upper-bound (this reflects the fact that $\sqsubseteq_{\mathcal{P}}$ is a preorder and not a partial order).

Given a formula, g, we can construct a defining sequence set, $\Delta^t(g)$, which is a good approximation of the set of minimal sequences (it is a superset thereof). There is also an counterpart, $\Delta^f(g)$, which is an approximation of the set of minimal sequences that satisfy g with truth value at least $\mathbf{f}$. The details of the construction of these sets can be found elsewhere [Haze96].

Definition 5.7 (Defining sequence set). *Let $g \in$ TL. Define the defining sequence sets of g as $\Delta(g) \stackrel{\text{def}}{=} \langle \Delta^t(g), \Delta^f(g) \rangle$, where the $\Delta(g)$ is defined recursively by:*

1. *If g is simple, $\langle \Delta^t(g), \Delta^f(g) \rangle$, where $\Delta^q(g) \stackrel{\text{def}}{=} \{s\mathsf{XX}\ldots : (\mathsf{s}, \mathsf{q}) \in D_g,$ or $(s, \top) \in D_g\}$. This says that provided a sequence has as its first element a value at least as big as s then it will satisfy g with truth value at least q. Note that $\Delta^q(g)$ could be empty.*

2. *$\Delta(g_1 \vee g_2) \stackrel{\text{def}}{=} \langle \Delta^t(g_1) \cup \Delta^t(g_2), \Delta^f(g_1) \amalg \Delta^f(g_2) \rangle$*
   Informally, if a sequence satisfies $g \vee h$ with a truth value at least $\mathbf{t}$ then it must satisfy either g or h with truth value at least $\mathbf{t}$. Similarly if it satisfies $g \vee h$ with a truth value at least $\mathbf{f}$ then it must satisfy both g and h with a truth value at least $\mathbf{f}$.

3. *$\Delta(g_1 \wedge g_2) \stackrel{\text{def}}{=} \langle \Delta^t(g_1) \amalg \Delta^t(g_2), \Delta^f(g_1) \cup \Delta^f(g_2) \rangle$*
   This case is symmetric to the preceding one.

4. *$\Delta(\neg g) \stackrel{\text{def}}{=} \langle \Delta^f(g), \Delta^t(g) \rangle$*

This is motivated by the fact that for $q = \mathbf{f}, \mathbf{t}$, σ satisfies g with truth value at least q if and only if it satisfies $\neg g$ with truth value at least $\neg q$.

5. $\Delta(\text{Next}\, g) \stackrel{\text{def}}{=} \text{shift}\,\Delta(g)$, *where* $\text{shift}(s_0 s_1 \ldots) = \mathsf{X}s_0 s_1 \ldots$
    $s_0 s_1 s_2 \ldots$ *satisfies* $\text{Next}\, g$ *with truth value at least q if and only if $s_1 s_2 \ldots$ satisfies g with at least value q.*

6. $\Delta(g_1\, \text{Until}\, g_2) \stackrel{\text{def}}{=} \langle \Delta^{\mathbf{t}}(g_1\, \text{Until}\, g_2), \Delta^{\mathbf{f}}(g_1\, \text{Until}\, g_2)\rangle$, *where*

    $\Delta^{\mathbf{t}}(\,g_1\, \text{Until}\, g_2) \stackrel{\text{def}}{=}$
    $\bigcup_{i=0}^{\infty} (\Delta^{\mathbf{t}}(\text{Next}^0 g_1)\, \amalg \ldots \amalg\, \Delta^{\mathbf{t}}(\text{Next}^{(i-1)} g_1)\, \amalg\, \Delta^{\mathbf{t}}(\text{Next}^i g_2))$

    $\Delta^{\mathbf{f}}(\,g_1\, \text{Until}\, g_2) \stackrel{\text{def}}{=}$
    $\amalg_{i=0}^{\infty} (\Delta^{\mathbf{f}}(\text{Next}^0 g_1)\, \cup \ldots \cup\, \Delta^{\mathbf{f}}(\text{Next}^{(i-1)} g_1)\, \cup\, \Delta^{\mathbf{f}}(\text{Next}^i g_2))$

Recall that $\text{Next}^k g = g$ if $k = 0$ and $\text{Next}^k g = \text{Next}\,\text{Next}^{k-1} g$ otherwise. Here we consider the until operator as a series of disjunctions and conjunctions and apply the motivation above when constructing the defining sequence sets.

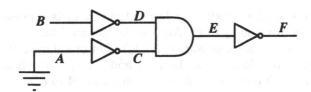

Fig. 5.2. Example Circuit

Example 5.1. Let us consider the simple circuit in Figure 5.2. The state space of this circuit is C^6. Each node is represented by one component of the tuple in the obvious way: the first component of the tuple represents A, the second B and so on. Consider the TL_n formula.

$$g \stackrel{\text{def}}{=} ([B] = L)\wedge([D] = L)\wedge\text{Next}\,([B] = H).$$

The defining sequence set $\Delta^{\mathbf{t}}(g)$ contains exactly one sequence:
$\Delta^{\mathbf{t}}(g) = \{(U,L,U,L,U,U)(U,H,U,U,U,U)(U,U,U,U,U,U)\ldots\}$. This is the weakest sequence that satisfies g.

The following lemma gives a key property of these defining sets: a sequence, σ, satisfies a formula g at least with degree q if and only if there is an element of the defining set $\Delta^q(g)$ that is less than or equal to σ. This proof and proofs of other results stated in this section can be found in [Haze96].

Lemma 5.2. *Let $g \in TL$, and let $\sigma \in S^\omega$. For $q = \mathbf{t},\mathbf{f}$, $q \preceq \text{Sat}(\sigma, g)$ iff $\exists \delta^g \in \Delta^q(g)$ with $\delta^g \sqsubseteq \sigma$.*   □

5.2.3 Defining Trajectory Sets. The defining sequence sets contain the set of the minimal sequences that satisfy the formula. It is possible to find the analogous structures for trajectories — we can find an approximation of the set of minimal trajectories that satisfy a formula. This section first shows how, given an arbitrary sequence, to find the weakest trajectory larger than it. Using this, the *defining trajectory sets* of a formula are defined. Finally, Theorem 5.2 is presented, which provides the basis for using defining sequence sets and defining trajectory sets to accomplish verification based on Definition 5.2.

The defining trajectory sets are computed by finding for each sequence in the defining sequence sets the smallest trajectory bigger than the sequence. Note that the use of the join operation requires that we are computing in a lattice with a top element (hence, the need for Z in a circuit model).

Definition 5.8. *Let* $\sigma = s_0 s_1 s_2 \ldots$ *. Define* τ *by* $\tau(\sigma) = t_0 t_1 t_2 \ldots$ *where:*

$$
t_i = \begin{cases} s_0 & \text{when } i = 0 \\ \mathbf{Y}(t_{i-1}) \sqcup s_i & \text{otherwise} \end{cases}
$$

□

$t_0 t_1 t_2 \ldots$ is the smallest trajectory larger than σ. s_0 is a possible starting point of a trajectory, so $t_0 = s_0$. Any run of the machine that starts in s_0 must be in a state at least as large as $\mathbf{Y}(s_0)$ after one time unit. So t_1 must be the smallest state larger than both s_1 and $\mathbf{Y}(s_0)$. By definition of join, $t_1 = \mathbf{Y}(s_0) \sqcup s_1 = \mathbf{Y}(t_0) \sqcup s_1$. This can be generalised to $t_i = \mathbf{Y}(t_{i-1}) \sqcup s_i$.

Definition 5.9 (Defining trajectory set). $T(g) = \langle T^{\mathbf{t}}(g), T^{\mathbf{f}}(g) \rangle$, *where* $T^q(g) = \{\tau(\sigma) : \sigma \in \Delta^q(g)\}$. □

By construction, if $\tau^g \in T^q(g)$ then there is a $\delta^g \in \Delta^q(g)$ with $\delta^g \sqsubseteq \tau^g$. $T(g)$ characterises g by characterising the trajectories that satisfy g. This is formalised in the following lemma.

Lemma 5.3. *Let* $g \in \text{TL}$, *and let* σ *be a trajectory. For* $q = \mathbf{t}, \mathbf{f}$, $q \preceq \text{Sat}(\sigma, g)$ *if and only if* $\exists \tau^g \in T^q(g)$ *with* $\tau^g \sqsubseteq \sigma$. □

Example 5.2. Consider again the simple circuit in Figure 5.2 and the formula

$$
g \stackrel{\text{def}}{=} ([B] = \text{L}) \wedge ([D] = \text{L}) \wedge \text{Next} \, ([B] = \text{H}).
$$

We saw that $\Delta^{\mathbf{t}}(g) = \{(\text{U}, \text{L}, \text{U}, \text{L}, \text{U}, \text{U})(\text{U}, \text{H}, \text{U}, \text{U}, \text{U}, \text{U})(\text{U}, \text{U}, \text{U}, \text{U}, \text{U}, \text{U}) \ldots \}$. We can now compute $T^{\mathbf{t}}(g)$, which contains exactly one element, the weakest trajectory satisfying g. Now we take into account the behaviour of the circuit.

$T^t(g)$ =
$$\{ \quad (U,L,U,L,U,U)$$
$$(L,H,U,H,L,U)$$
$$(L,U,H,L,U,H)$$
$$(L,U,H,U,L,U)$$
$$(L,U,H,U,U,H)$$
$$(L,U,H,U,U,U)\ldots$$
$$\ldots\}$$

The existence of defining sequence sets and defining trajectory sets provides a potentially efficient method for verification of assertions such as $g \Longrightarrow h$. The formula g, the antecedent, can be used to describe initial conditions or 'input' to the system. The consequent, h, describes the 'output'. This method is particularly efficient when the cardinalities of the defining sets are small. This verification approach is formalised in Theorem 5.2 which shows that comparing the appropriate defining sets tells us when an assertion (e.g. $g \Longrightarrow h$)) is true.

Theorem 5.2. *If g and h are TL formulas, then $\Delta^t(h) \sqsubseteq_P T^t(g)$ if and only if $g \Longrightarrow h$.*   □

In our experience, most useful formulas have small defining sequence sets. Indeed the computational limit is not so much the size of the sets, but the complexity of the BDDs needed to represent them.

Definition 5.10. *If $g \in$ TL, and $\exists \delta^g \in \Delta^t(g)$ such that $\forall \delta \in \Delta^t(g), \delta^g \sqsubseteq \delta$, then δ^g is known as the defining sequence of g. If the δ^g is the defining sequence of g, then $\tau^g = \tau(\delta^g)$ is known as the defining trajectory of g.*   □

Finite formulas with defining sequences are known as trajectory formulas. Seger and Bryant characterised these syntactically [SeBr95].

Two special cases of Theorem 5.2 should be noted because they can often be used. First, if A is a formula of TL with a defining sequence δ^A, and $h \in$ TL, then $\forall \delta \in \Delta^t(h), \delta \sqsubseteq \tau^A$ if and only if, for every trajectory σ for which $t \preceq Sat(\sigma, A)$ it is the case that $t \preceq Sat(\sigma, h)$.

Second, let A and C be formulas of TL with unique defining sequences δ^A and δ^C. Then $\delta^C \sqsubseteq \tau^A$ if and only if for every trajectory σ for which $q \preceq Sat(\sigma, A)$ it is the case that $q \preceq Sat(\sigma, C)$. This is essentially the result of Seger and Bryant [SeBr95] generalised to the four valued logic.

Example 5.3. In examples 5.1 and 5.2, $\Delta^t(g)$ and $T^t(g)$ both have exactly one element. So, we can refer to their elements as δ^g and τ^g respectively.

Example 5.4. Let $h \stackrel{\text{def}}{=} \text{Next}^3([F] = \text{L})$. $\Delta^t(h)$ consists of exactly one element, which we may refer to as δ^h.

$$\delta^h \stackrel{\text{def}}{=}$$

$$(\text{U}, \text{U}, \text{U}, \text{U}, \text{U}, \text{U})$$
$$(\text{U}, \text{U}, \text{U}, \text{U}, \text{U}, \text{U})$$
$$(\text{U}, \text{U}, \text{U}, \text{U}, \text{U}, \text{U})$$
$$(\text{U}, \text{U}, \text{U}, \text{U}, \text{U}, \text{L})$$
$$(\text{U}, \text{U}, \text{U}, \text{U}, \text{U}, \text{U}) \ldots$$

Example 5.5. Consider again the simple circuit in Figure 5.2 and the formula

$$g \stackrel{\text{def}}{=} ([B] = \text{L}) \wedge ([D] = \text{L}) \wedge \text{Next} ([B] = \text{H}),$$

and h above.

The assertion $\models \langle\!\langle g \Longrightarrow\!\!\!\gg h \rangle\!\rangle$ says that in every run of the system when at time 0 the nodes B and D have the value L and at time 1 the node B has the value H, then it is also the case that at time 3 the node F has the value L.

We can verify this assertion by checking to see whether $\delta^h \sqsubseteq \tau^g$. It is, and so the assertion holds.

All of these sequences can easily be represented with trivial BDDs.

The results of this section can easily be generalised to the symbolic version of TL. The constructs used in the previous section such as defining set all have symbolic extensions. Each symbolic TL formula is a concise encoding of a number of scalar formulas; each interpretation of the variables yields a (possibly) different scalar formula. To extend the theory of trajectory evaluation, symbolic sets are introduced; these can be considered as concise encodings of a number of scalar sets and can be represented compactly using BDDs. Symbolic sets can be manipulated in an analogous way to scalar sets. Using this approach, the key results presented above extend to the symbolic case. For full details of this see [Haze96].

5.3 Compositional Theory

The fundamental result of STE, Theorem 5.2, gives an efficient way of inferring whether an assertion $\models \langle\!\langle g \Longrightarrow\!\!\!\gg h \rangle\!\rangle$ is valid. This is an important result as the algorithm that puts the theorem into practical effect is capable of dealing with large problems. STE supports abstraction by allowing the human verifier to choose the level of information needed to prove the result. This suggests that a compositional approach would complement STE very well since the cost of verification is sensitive to the properties being checked rather than the size of the circuit. Instead of verifying an assertion directly using STE, we prove some 'smaller' properties first using STE, and then use a compositional theory to combine these smaller proofs into our desired goal. This approach also has the advantage that we can use appropriate data structures at different levels of reasoning: binary decision diagrams (BDDs) for STE; symbolic representations for using the compositional theory.

The compositional theory is outlined below. An earlier version of the compositional theory appeared in [HaSe95] and the version presented here is described in detail (with proofs) in [Haze96]. Many of the rules described below have analogies in logic. Some of these are worth stating as an introduction to our rules.

Example 5.6. If $a_1 \Rightarrow b_1$ and $a_2 \Rightarrow b_2$, then $a_1 \wedge a_2 \Rightarrow b_1 \wedge b_2$. Another is transitivity: suppose $a_1 \Rightarrow b_1$ and $a_2 \Rightarrow b_2$ then if $b_1 \Rightarrow a_2$ (or more weakly if $a_1 \vee b_1 \Rightarrow a_2$), $a_1 \Rightarrow b_2$.

Another good example is that suppose we know that a predicate $\forall x \in S.P(x)$ is true; then we know that $\forall x \in S.P(\xi(x))$, where ξ is a function mapping from S to S. Essentially we infer from a more general result (P is true for all x) a more specialised one (P is true for all y in the range of ξ).

The rest of this section describes some of the inference rules with simple examples. Although what follows is not an exhaustive list, the rules given are the major ones. One way of extending the list is to package some of the rules together in various ways. There is also an 'antecedent truncation' rule which uses causality (events at time t cannot affect events at time $t - 1$) to allow precondition weakening in some situations. For some discussion of the deductive power of an earlier version of this compositional theory see [ZhSe94].

5.3.1 Voss. This is the rule of Symbolic Trajectory Evaluation:

$$\frac{\Delta^t(h) \sqsubseteq_{\mathcal{P}} T^t(g)}{\models \langle\!\langle g \Longrightarrow h \rangle\!\rangle}$$

5.3.2 Identity. This trivial rule says:

$$\overline{\models \langle\!\langle g \Longrightarrow g \rangle\!\rangle} \ .$$

Although it is very simple, it does have occasional use in a proof. In much the same way a piece of code summing variables might initialise a sum variable to 0, we might use the identity rule to get a simple theorem that is subsequently transformed.

5.3.3 Time Shift. The Time Shift Rule and the Specialisation Rule allow us to transform one assertion into another – from the more general to the more particular. There are two reasons for doing this: first, in general, it is cheaper to use STE for more general results since the BDDs needed are simpler; second, we may wish to use an assertion in a number of slightly different contexts. Our goal will be to compose assertions, and it turns out that being able to transform assertions in sound ways increases the range of compositions that we can make. It also makes the use of trajectory evaluation less costly since we can now use the Voss rule on simpler assertions which are then transformed into more complex assertions, rather than having to use STE directly on the complex assertions.

The Time Shift rule is a very useful rule in proofs. Informally, an assertion says something like: *If g is true at time t_1, then h is true at time t_2.* The Time Shift rule says that what is important about timing in the assertion is not so much the actual times, but the relationship between the times, and it will allow us to infer from this assertion that for all $t \geq 0$: *If g is true at time $t_1 + t$, then h is true at time $t_2 + t$.* Note that we cannot Time Shift backwards as circuit behaviour also depends on the propagation of constant values (determined by some nodes being tied to ground or high voltage). Formally, the Time Shift rule says:

$$\frac{\models \langle\!\langle g \Longrightarrow h \rangle\!\rangle \quad t \geq 0}{\models \langle\!\langle \text{Next}^t g \Longrightarrow \text{Next}^t h \rangle\!\rangle}.$$

This rule has two main uses. First, it allows the verifier to abstract away from actual times – in verifying a particular property, it may not be useful to worry about particular times. The actual timings can be determined at a later stage in the proof (and the process of finding time shifts can be partially automated in our tool). More important is that often we may wish to use the property of a circuit at many different times. A component of a circuit may be reused hundreds of times in a computation; however, we need only verify the appropriate properties once using STE, and then use Time Shift (which can be implemented very cheaply) for all the other times. This yields simpler and much more efficient proofs.

5.3.4 Specialisation. Time Shift allows us to abstract from time; Specialisation allows us to abstract from actual values. We shall consider a simple example that not only illustrates the usefulness of this, but helps explain why the compositional theory can be used successfully. Consider the simple circuit of Figure 5.3, which takes three inputs, say a, b and c, and then computes $a \times b + c$. Suppose we wish to show that this adder adds. (We use the convention that capital letters A, B, ... label nodes, while lower case letters a, b, ... refer to values.)

Fig. 5.3. A Simple Adder and Multiplier

It is not possible to prove directly using STE that this adder, given inputs $a \times b$ and c produces output $a \times b + c$ because multiplication of two bit vectors in general does not have an efficient BDD representation. But, what we can easily show is that if the adder is given inputs x and y, it will produce output $x + y$ since the BDDs needed here are compact. Once we have proved this, the result can be stored *symbolically*. This result says that the adder works for all possible values of x and y. when $x = a \times b$ and $y = c$ (i.e. when E has the value $a \times b$ and C has the value c) since these are just special cases of the proved result. Since multiplication has an efficient symbolic representation, there is no impediment in doing a symbolic substitution of $a \times b$ for x and c for y. We shall come back to this example later after we have introducing the Transitivity Rule.

Another reason to specialise is that a particular component of the circuit may be used many times during execution, each time with different input values. By proving one general result and then specialising this result appropriately, very significant improvements in performance can be made.

We now formally define a specialisation. Let $\mathcal{E}$ be the set of boolean expressions over the variables.

Definition 5.11.
1. *The function $\xi: \mathcal{V} \rightarrow \mathcal{E}$ is a* substitution.
2. *A substitution $\xi: \mathcal{V} \rightarrow \mathcal{E}$ can be extended to map from TL to TL:*
- $\xi(g_1 \wedge g_2) = \xi(g_1) \wedge \xi(g_2)$
- $\xi(\neg g) = \neg \xi(g)$
- $\xi(\text{Next } g) = \text{Next } (\xi(g))$
- $\xi(g \, \text{Until} \, h) = \xi(g) \, \text{Until} \, \xi(h)$
- *Otherwise, if g is not a variable, $\xi(g) = g$*
- *If T is the assertion*

$$\models \langle g \Longrightarrow h \rangle$$

then $\xi(T)$ is the assertion

$$\models \langle \xi(g) \Longrightarrow \xi(h) \rangle.$$

Lemma 5.4 (Substitution Lemma). *Suppose $\models \langle g \Longrightarrow h \rangle$ and let ξ be a substitution.*
Then $\models \langle \xi(g) \Longrightarrow \xi(h) \rangle$.

Although substitution is useful, in practice sometimes a more sophisticated transformation is also desirable. A *specialisation* is a conjunction of conditional substitutions which allows us to perform different substitutions in different circumstances.

Definition 5.12. *Let $\Xi = [(e_1, \xi_1), \dots, (e_n, \xi_n)]$ where each ξ is a substitution and each e_i is a boolean expression, is a* specialisation. *If $g \in$ TL, then $\Xi(g) = \wedge_{i=1}^{n}(e_i \Rightarrow \xi_i(g))$.*

This allows us to define the Specialisation Rule.

$$\frac{\models \langle\!\langle g \Longrightarrow\!\!\!\!\Rightarrow h \rangle\!\rangle \qquad \Xi \stackrel{\text{def}}{=} [(e_1, \xi_1), \ldots, (e_n, \xi_n)], \text{ a specialisation}}{\models \langle\!\langle \Xi(g) \Longrightarrow\!\!\!\!\Rightarrow \Xi(h) \rangle\!\rangle}$$

Example 5.7. Suppose that we have proved:

$$\models \langle\!\langle [D] = d \Longrightarrow\!\!\!\!\Rightarrow \mathtt{Next}\,[F] = 2 \times d \rangle\!\rangle$$

We can specialise the result to:

$$\models \langle\!\langle ((a > b) \Rightarrow ([D] = a - b)) \wedge ((a \le b) \Rightarrow [D]a)$$
$$\Longrightarrow\!\!\!\!\Rightarrow$$
$$\mathtt{Next}\,(((a > b) \Rightarrow ([F] = 2 \times (a - b))) \wedge ((a \le b) \Rightarrow [F] = 2 \times a)) \rangle\!\rangle$$

In practice, substitutions are used far more frequently than specialisations, but specialisations do have their use.

5.3.5 Conjunction and Disjunction. We have pointed out already that verifying 'smaller' properties is cheaper than verifying 'larger' properties. By keeping the antecedent as weak and as possible, the level of abstraction is raised; and the weaker and simpler both antecedent and consequent, in general the fewer BDD variables will be needed when using STE. But, this means that we need to glue results together; the *Conjunction* rule is one of the most important rules for doing this gluing. Informally, it says that if we have established two assertions then if we conjunct the two antecedents together and conjunct the two consequents together we get a valid antecedent and consequent for a new assertion.

Formally, the Conjunction Rule is

$$\frac{\models \langle\!\langle g_1 \Longrightarrow\!\!\!\!\Rightarrow h_1 \rangle\!\rangle \quad \models \langle\!\langle g_2 \Longrightarrow\!\!\!\!\Rightarrow h_2 \rangle\!\rangle}{\models \langle\!\langle g_1 \wedge g_2 \Longrightarrow\!\!\!\!\Rightarrow h_1 \wedge h_2 \rangle\!\rangle}.$$

The simplistic example below illustrates this.

Example 5.8. We have a simple two cell memory, depicted in Figure 5.4. The input to the circuit is an address, and the output is the contents of memory. Suppose that we prove, using STE, that

$$\models \langle\!\langle [\text{AddIn}] = a) \wedge ([\text{Mem}[1]] = m_1) \Longrightarrow\!\!\!\!\Rightarrow \mathtt{Next}\,((a = 1) \Rightarrow [\text{Out}] = m_1) \rangle\!\rangle$$

and

$$\models \langle\!\langle [\text{AddIn}] = a) \wedge ([\text{Mem}[2]] = m_2) \Longrightarrow\!\!\!\!\Rightarrow \mathtt{Next}\,((a = 2) \Rightarrow [\text{Out}] = m_2) \rangle\!\rangle.$$

Now using conjunction we get:
$$\models \langle\!\langle ([\text{AddIn}] = a) \wedge ([\text{Mem}[1]] = m_1) \wedge ([\text{Mem}[2]] = m_2)$$
$$\Longrightarrow\!\!\!\!\Rightarrow \mathtt{Next}\,((a = 1) \Rightarrow [\text{Out}] = m_1 \wedge (a = 2) \Rightarrow [\text{Out}] = m_2) \rangle\!\rangle$$

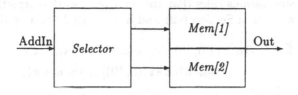

Fig. 5.4. A Simple Memory

There is also a Disjunction Rule:

$$\frac{\models \langle g_1 \Longrightarrow h_1 \rangle \quad \models \langle g_2 \Longrightarrow h_2 \rangle}{\models \langle g_1 \lor g_2 \Longrightarrow h_1 \lor h_2 \rangle}.$$

5.3.6 Transitivity. The Conjunction and Disjunction rules allow us to glue two 'parallel' results together. The Transitivity Rule allows us to glue two sequenced results together. As with Conjunction and Disjunction, the reason for using Transitivity is typically that while the individual results can be verified efficiently using STE (and hence have efficient BDD representations), the combined result does not have an efficient BDD representation, but can be represented symbolically.

Informally, suppose that we want to show for a given circuit that given some inputs i we get desired outputs o. One way of doing this is to show that given i, some intermediate nodes in the circuit achieve the value j, and that when the intermediate nodes achieve the value j then the output achieves o. Thus, the Transitivity Rule is very similar to transitivity rule of logic described on page 33, except that we need to take timing into account. Formally, the Transitivity Rule is:

$$\frac{\models \langle g_1 \Longrightarrow h_1 \rangle \quad \models \langle g_2 \Longrightarrow h_2 \rangle \quad \Delta^t(g_2) \sqsubseteq_{\mathcal{P}} \Delta^t(g_1) \sqcup \Delta^t(h_1)}{\models \langle g_1 \Longrightarrow h_2 \rangle}.$$

The side condition for this rule is that the defining sequence set of g_2 must be less than the join of the defining sequence sets of g_1 and h_1 (note that this is a weaker requirement than saying that the defining sequence set of g_2 must be less than the defining sequence set of h_1). Essentially, this condition says that every sequence that satisfies both g_1 and h_1 also satisfies g_2. Since the premises of the rule state that every trajectory that satisfies g_1 also satisfies h_1 and that every trajectory that satisfies g_2 satisfies h_2, the conclusion of the rule follows.

Example 5.9. Now that we have introduced the major rules, we can use Figure 5.3 to give an extended example. As the circuit is very simple, the example is somewhat contrived, but it illustrates the general ideas. Suppose that we have proved using various rules that the multiplier works correctly (how this is done is the subject of Section 6.2) and have proved the assertion:

$$T_1 \stackrel{\text{def}}{=} \models (\text{Global}\,[(0,10)]\,[A] = a \wedge [B] = b \wedge [C] = c$$
$$\Longrightarrow \text{Global}\,[(5,10)]\,[E] = a \times b),$$

and suppose that using trajectory evaluation we show that:

$$T_2 \stackrel{\text{def}}{=} \models (\text{Global}\,[(0,5)]\,[E] = e \wedge [C] = c$$
$$\Longrightarrow \text{Global}\,[(1,5)]\,[D] = e + c).$$

It is easy to see that T_1 and T_2 are related to each other, but that there are mismatches in T_2 so that we cannot apply the Transitivity Rule directly. Instead, we first time shift T_2 forward to get T_2',

$$T_2' \stackrel{\text{def}}{=} \models (\text{Global}\,[(5,10)]\,[E] = e \wedge [C] = c$$
$$\Longrightarrow \text{Global}\,[(6,10)]\,[D] = e + c).$$

and then specialise T_2' by substituting $a \times b$ for e to get:

$$T_2'' \stackrel{\text{def}}{=} \models (\text{Global}\,[(5,10)]\,[E] = a \times b \wedge [C] = c$$
$$\Longrightarrow \text{Global}\,[(6,10)]\,[D] = a \times b + c).$$

Inspecting T_1 and T_2'' shows that the Rule of Transitivity can be applied between the two to yield:

$$T \stackrel{\text{def}}{=} \models (\text{Global}\,[(0,10)]\,[A] = a \wedge [B] = b \wedge [C] = c$$
$$\Longrightarrow \text{Global}\,[(6,10)]\,[D] = a \times b + c).$$

5.3.7 The Until Rule. The Until Rule allows us to take two assertions and generalise their effect over time in much the same way as the Time Shift Rule does. Formally, Until Rule is:

$$\frac{\models (g_1 \Longrightarrow h_1) \quad \models (g_2 \Longrightarrow h_2)}{\models (g_1 \text{ Until } g_2 \Longrightarrow h_1 \text{ Until } h_2)}$$

Its use is best described by giving and explaining a corollary

Corollary 5.1.

$$\frac{\models (g \Longrightarrow h)}{\models (\text{Exists}\,g \Longrightarrow \text{Exists}\,h) \qquad \models (\text{Global}\,g \Longrightarrow \text{Global}\,h)}$$

Informally, $\models \langle g \Longrightarrow h \rangle$ says that if g is true, then so is h. The rule above now says we can infer from this assertion that if g is always true, then h is always true, and if g will be true some time in the future, then h will be true some time in the future. An example of this type of reasoning will be given in Section 6.7.1.

Another corollary will be of use later. This says that if we know that $\models \langle g \Longrightarrow h \rangle$, then if g holds over a given time interval, then so does h. The proof of this is by induction over t.

Corollary 5.2.
If $\models \langle g \Longrightarrow h \rangle$, *then for all* t, $\models \langle \texttt{Global}\,[(0,t)]\ g \Longrightarrow \texttt{Global}\,[(0,t)]\ h \rangle$.

5.4 The Voss Verification System: The Front End

In Section 4.1 we discussed how Voss could be used to generate accurate finite state machine (FSM) models from a variety of input formats. Given an FSM of a circuit, the user interacts with the Voss system to analyse and verify the circuit. Interaction is done by using FL, a lazy functional language, which has the following features [Sege93]:

– Boolean data can be represented and manipulated using an efficient BDD package. This is very important because circuits are modelled at gate or switch level, and so efficient methods of manipulating bit-level data is critical.
– Abstract data types can be declared in FL. This means that data can be represented and manipulated at a high level. This is important for making specifications understandable, and critical for overcoming the limitations of BDDs.
– FL provides built-in functions that are the interface to the STE engine.

The use of a fully programmable script language is a key factor in implementing our verification methodology. It means that our tool can be simple, but through the use of a flexible interface, a user can verify a wide range of problems. On top of Voss's facilities, we have implemented a simple theorem prover to implement the compositional theory presented in Section 5.3 – we have called this augmented system called VossProver.[4]

The VossProver provides a set of abstract data types (ADT) for the representation of the data the circuit is manipulating, for example boolean data, bit vectors, and integers. The abstract data type consists of the data declarations, functions to manipulate the data and to transform data into BDD form. It is worth emphasising that all data is by default represented symbolically. Boolean data which is represented symbolically is automatically converted to BDD form when necessary.

---

[4] We have actually implemented a number of such tools, experimenting with style and functionality. The description presented here is a general description of one of the latest versions.

There is also an ADT for TL formulas and trajectory assertions. The user can use the TL type to specify formulas, and then construct assertions to be checked. Once the assertion has been constructed, a routine of the VossProver can be invoked to check whether the assertion is true.

The Voss system itself provides the basic trajectory evaluation facilities. The VossProver extends these facilities so that richer temporal formulas (including expressions involving ADTs) can be checked. What gets provided by the Voss system itself, and what by the VossProver is hidden from the user.

5.5 Verification Style

Verification consists of proving that a set of assertions holds of the model. Each assertion, of the form $\langle\, g \Longrightarrow h \,\rangle$, says that each trajectory of the system that satisfies the antecedent temporal formula g also satisfies the consequent temporal formula h. For example, the assertion specifying an adder might be:

$$\langle\, ([I1] = i1) \wedge ([I2] = i2) \Longrightarrow \text{Next}^{10}([O] = i1 + i2) \,\rangle.$$

In our approach, verification is done on a completed design, which is done using traditional methods; given a design, the VossProver provides the facilities for verification. Part of the facilities are scalar and symbolic simulation which can be used to 'debug' the circuit. This is often useful as the first stage of verification, particularly for users who are relatively inexperienced with verification. Many straightforward errors can be caught this way, and Voss provides features such as timing diagrams that the user can inspect.

Formal verification is done by proving that the model of the design satisfies a given set of assertions. The proof that the model satisfies an assertion must be done using one of a set of verification routines provided by the VossProver. These routines provide an interface to the trajectory evaluation facilities of Voss, and implement the compositional theory discussed earlier. The basic philosophy of this approach is that trajectory evaluation is suitable for verifying low level properties of the circuit, properties which are often tedious for humans to prove. The compositional theory is then used to infer higher-level results, where human insight can be used productively. Trajectory evaluation is typically used to verify a property of a part of the circuit containing in the order of hundreds or thousands of state holding components.

In essence, the VossProver is a simple, specialised theorem prover, the rules of inference being packagings of the compositional theory. In our various implementations of the VossProver, we have implemented the compositional theory presented in Section 5.3 as a set of FL library routines. Some are direct implementations of the theory, others are packaged in various ways. A representative selection is listed here:

Voss: This takes an assertion and uses trajectory evaluation to determine whether it is valid.

TRANS: Implements the transitivity rule.

SPEC: Implements the specialisation rule.

SHIFT: Implements the time shift rule.

CONJ: Implements conjunction.

DISJ: Implements disjunction.

There are also some 'packaged rules'.

AUTOTRANS: Given two assertions T_1 ($\langle g_1 \Longrightarrow h_1 \rangle$) and T_2 ($\langle g_2 \Longrightarrow h_2 \rangle$) the VossProver tries to find an integer t so that either T_1 or T_2 can be time shifted in such a way that it is possible to combine T_1 (possibly shifted) with T_2 (possibly shifted) using transitivity. Heuristics are used to find the time shift. However, the safety of the system is not compromised because after time shifting one of the assertions, the VossProver invokes TRANS to do the combining.

SPTRANS: Given two assertions T_1 ($\langle g_1 \Longrightarrow h_1 \rangle$) and T_2 ($\langle g_2 \Longrightarrow h_2 \rangle$) the VossProver tries to find a specialisation ξ so that T_2 can be specialised in such a way that it is possible to combine T_1 with $\xi(T_2)$ using transitivity. Heuristics are used to find the specialisation. However, the safety of the system is not compromised because after specialising T_2, the VossProver invokes TRANS to do the combining.

The proof of an assertion is written as an FL program, invoking the proof rules as necessary. Typically, the program will prove a number of assertions using trajectory evaluation (the Voss rule), and then combine these results using the other inference rules in more complex ways. This is where the power of FL is particularly needed; with a relatively small program it is possible to write a proof containing a large number of steps.

6. Experimental Work

Symbolic trajectory evaluation has been used now for almost five years. Good examples of work done using STE can be found in [Beat93, Darw94]. In this section we report more recent work, particularly work that uses the compositional methodology, concentrating on the IFIP WG10.5 Benchmarks. Section 6.1 presents the verification of Benchmark 11, a circuit simulating a blackjack dealer. Section 6.2 presents the verification of the Benchmark 17, a simple multiplier circuit. We have also looked at other multiplier circuits, and the verification of an IEEE-compliant floating point multiplier is outlined in Section 6.3. The Benchmark 17 multiplier is used as a component in two more verification examples: Benchmarks 21 and 22 are systolic circuits containing a number of multipliers. Their verifications are described in Section 6.4 and 6.5. Most of the examples mentioned so far have had arithmetic as an important part; Section 6.6 presents the verification of a different type of circuit, an associative memory circuit which is Benchmark 20. Finally, Section 6.7 discusses Benchmarks 7 and 12, which are presently not well suited for our methods.

6.1 Benchmark 11: BlackJack Dealer

The first example simulates the behaviour of a blackjack dealer. It has a non-trivial control path, and a small data path. This example, can be verified using trajectory evaluation alone.

6.1.1 Implementation. For all the IFIP benchmarks that we verified, the circuit implementations were detailed gate-level descriptions based on the documentation given in the suite documentation, essentially a translation of the VHDL code given into the Voss's own 'EXE' format. (Voss is capable of taking as input the output of some VHDL compilers. However at an early stage in our work extending trajectory evaluation, we needed to manipulate the generated finite state machines directly. The practical effect of this was to force us to use the 'EXE' format. We no longer have this restriction, but we have not had time to redo the implementations.) For all gates, we use a unit-delay model – however, this was done for convenience, and Voss is capable of richer timing models.

A description of this circuit can be found in the Appendix. The EXE implementation of the circuit can be found in Appendix A.1.1, and the interested reader may wish to compare this with the VHDL code given in the benchmark description to assess the faithfulness of our translation. The translation of the structural VHDL is direct. Some of the latches in the circuit are described using behavioural VHDL. The only complication here is the modelling of clock events. A behavioural VHDL program can simply refer to Clk'EVENT, which has no correspondence in a gate-level description. To simulate this, we add, for each clock, a one-unit delay buffer: a clock event happens when the clock is high and the delay buffer is low. Direct implementations of these flip-flops would pose no problems. (In this particular circuit, we have made one addition to the circuit – this is explained later).

6.1.2 What we proved. We proved that when the circuit is dealt cards in sequence that each time a card is dealt the circuit computes the correct score and sets the Stand and Broke outputs appropriately. Both the timing and value of all the nodes is checked.

We saw two strategies for verifying this circuit. Our first strategy was to prove (1) that when the circuit is initialised and the first card is dealt that the correct output is produced, and then (2) that if the circuit has produced the correct output for n cards and another card is produced then the circuit produces the correct output for $n + 1$ cards. The problem with this strategy is that proving the second step requires knowing detailed information about the values that many of the internal nodes, including detailed information about timing. After attempting this, we concluded that while it might be a good strategy for someone who has intimate knowledge of the circuit, for those like us who didn't want intimate knowledge of the circuit, it was not the best approach.

The second strategy is a more direct approach, requiring more computational resources but less human intervention. First, we fix r, which denotes the maximum number of cards that we expect the dealer to hold. The verification strategy is to show that for all possible legal hands of up to and including r cards that after each card is dealt that the circuit responds correctly – we check that the correct score is computed, and that the Stand and Broke signals are set correctly.

We assume that if the dealer asks for r cards, that s/he will be given cards v_1 to v_r. A new card is dealt once a round (a round comprising four clock cycles). For round i we check that the Score output of the circuit is the sum of cards 1 to i, where the sum of the cards must be adjusted depending on whether one of the cards is an ace. We also check that the Stand and Broke outputs are set according to the Score value. We then verify the correctness of the circuit for round $i + 1$ for all cases where the cards already dealt do not complete the hand (when the dealer's score is less than 17). Note that the verification is symbolic in that we are not verifying the behaviour for particular cards v_1 to v_r, but *all* cards v_1 to v_r.

The complete and annotated verification script can be found in Section A.1.2.

6.1.3 Experience and cost of verification.
The verification process was straightforward, and we were able to verify both timing and functionality of the circuit. A number of (unintentional) errors had been made when translating to EXE format, and the output of the Voss system on the failed verification attempts provided the necessary debugging information.

Table 6.1 shows the computational costs for verifying the circuit with different values of r, that is the cost of a complete verification of the circuit when the dealer gets at most r cards. The time is shown in seconds. Note that the most cards a dealer can get in a multi-pack card is 11 (2, 2, 2, 2, 2, 2, 1, 1, 1, 1, 1), and, moreover, in many versions of blackjack a hand of five cards beats all hands except an ace and a face card. As can be seen the cost does increase as r increases, but the costs are quite reasonable. We expect the performance to be slightly worse than linear because as r increases, the time the circuit must be exercised increases linearly, but later iterations tend to take longer than earlier ones. The verification was completed on an SGI R4400 Indy.

In general, our methodology allows us to trade-off computation and human costs appropriately. In this example, verification can be done by trajectory evaluation alone, and no human intervention is necessary. As we shall see in subsequent examples, sometimes we need to reduce the computational costs by having more human intervention.

r	4	5	6	7	8	9	10	11	12
Time (s)	15	25	36	58	84	113	164	205	252

Table 6.1. Cost of verifying r rounds

Due to some timing problems, we had to make two changes to the circuit in order to complete the verification. First, we had to double the clock cycle to 40ns because for some inputs the critical path was so long that certain latches latched their values before the value being latched had stabilised. We also found that in some cases the clock went high before the input to the JK flip-flop storing the Acellflag value had attained its value. We rectified this by inserting a four-unit delay buffer between the clock and the clock input of that JK flip-flop. Although this problem may have occurred because of the particular delay values we assigned to gates, it is a good example of a common problem in circuit design.

The biggest problem we had in the verification is that we cannot print out some of the properties in a meaningful way – using FL it is possible to express properties that at the moment do not have concise string representations. For example, it is easy to define a function which given a hand of cards returns a TL expression describing whether the hand contains an ace. However, when this is printed out, the definition of the function is expanded out (a disjunct of boolean expressions). While it is useful to be able to do this, it is also important for the human verifier to provide a higher level description Although this is not a fundamental problem, it is not merely an aesthetic one – the human verifier must be able to inspect the output of the verification tool in a meaningful way.

6.2 Benchmark 17: Multiplier

6.2.1 Implementation. The circuit was implemented as a gate-level model, with the given VHDL model translated into EXE format.

6.2.2 Overview of proof. Dealing with multiplication is difficult since general decision procedures for arithmetic with multiplication do not exist. BDD-based approaches have some difficulty in dealing with multiplication as the representation of the multiplication of two integers is exponential in size [Brya91a]. Other data structures, such as newer types of decision diagrams and the symbolic representation method used in the VossProver, do not have this problem. However, if the circuit design is given at the bit-level, the most likely initial representation of the circuit behaviour is with BDDs.

Many different designs for integer multipliers exist and we have verified two other designs (a simplistic combinational circuit, and a Wallace tree multiplier). The typical design consists of a number of stages, each stage comprising an adder and additional circuitry. The obvious approach to verification is to verify each stage individually, and then to prove that the combination of the stages yields the correct approach. This falls quite naturally into the framework of our verification methodology. Trajectory evaluation can be used to verify the correctness of each stage (for example, using STE to verify a 64-bit adder is very cheap), and the compositional theory can be used to combine the individual verifications.

Note that this compositional approach requires that the human verifier identify structure in the proof; however, it does *not* require that the circuit be decomposed. This is effective because STE is relatively insensitive to the size of the circuit being verified – the properties being verified determine the computational cost of verification.

The reasons that this compositional strategy works are:

- BDDs (used to implement STE efficiently) are the right data structure to use for the verification of the individual stages, particularly as it matches the circuit design. The low-level of the properties being proved makes an automatic verification method appropriate.
- Once we have verified that each of the stages works correctly, the results are represented symbolically, which means that we are no longer bound by the limitations of BDDs. The VossProver is able to reason about the symbolic representation directly. In application of the compositional inference rules, it is this symbolic representation that is manipulated.

Figures 6.1–6.3 give an overview of the proof. In this discussion, timing is ignored; however, the proof itself takes timing into account. We assume that node A gets the input a and node B gets the inputs b, where both a and b are four bit numbers. We use the notation $b[i]$ to refer to the i-th bit of b, and $b[i-0]$ to refer to the $i+1$ lower order bits of b.

Figure 6.1(i) shows that we can prove using STE alone that the output of the first stage of the multiplier is $ab[0]$ (a multiplied by the zero-th bit of b). Figure 6.1(ii) shows that we can prove using STE alone that if the input to the second stage of the multiplier is the number c, then the output of the second stage is $c + 2ab[1]$. This is a more general result than we need, so Figure 6.1(iii) shows that we can specialise the previous result to: if we know the input of the second stage of the multiplier is $ab[0]$, then the output of the second stage is $ab[0] + 2ab[1]$ which is just $ab[1-0]$ (a multiplied by the two lower order bits of b). At this stage, all the results are being represented symbolically, so we do not have to worry about BDDs increasing in size.

Figure 6.2(i) shows that if the input to the third stage is c, then the output from the third stage is $c + 4ab[2]$. Figure 6.2(ii) specialises the result to: if the input to the third stage is $ab[2-0]$, then the output is $ab[2-0] + 4ab[2]$, which is $ab[3-0]$.

Figure 6.3(i) shows that if the input to the fourth stage is c, then the output from the fourth stage is $c + 8ab[3]$. Figure 6.3(ii) specialises the result to: if the input to the third stage is $ab[3-0]$, then the output is $ab[2-0] + 8ab[3]$, which is $ab[3-0]$, which is just ab. The final step in the proof is shown in Figure 6.3(iii). Here the results of Figures 6.1(i), 6.1(iii), 6.2(ii) and 6.3(ii) are combined using transitivity repeatedly (the consequent of the one matches the antecedent of the next). This allows us to conclude the final result.

The key steps in the proof are to prove that the i-th stage works for arbitrary input, and then to specialise to the actual inputs. Although it would

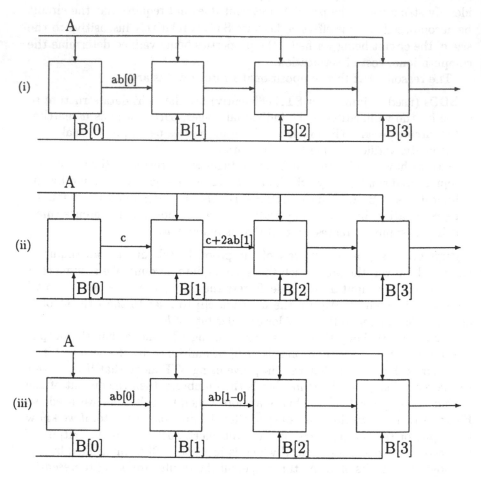

Fig. 6.1. First steps of multiplier proof

not be a computational problem for a four bit multiplier, in general it would not be possible to prove directly using STE that the i-th stage works.

6.2.3 What we proved. In the Benchmark 17 circuit description, there are two n-bit inputs, A and B, and a $2n$-bit output, P. We prove the following property of the circuit:

$$\models \langle\, \texttt{Global}\, [(0, 100)]\, ([A] = a \wedge [B] = b) \Longrightarrow \texttt{Global}\, [(22, 100)]\, ([P] = a * b) \,\rangle.$$

This says that if A and B have the values a and b respectively from time 0 to 100 inclusive, then from time 22 to 100 inclusive, P will have the value $a * b$. The time 22 is determined by the design of the circuit and the time-delay

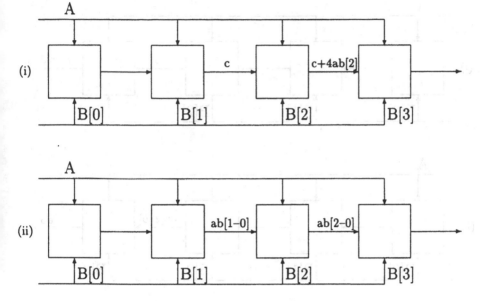

Fig. 6.2. Next steps of multiplier proof

model of its components. a and b are arbitrary n-bit integers. The unit of time is whatever unit of time the circuit model uses for computation.

In summary, both functional and timing properties of the circuit are verified.

6.2.4 Cost of verification. The computational cost of verification of circuits with different bit-widths is shown in Table 6.2. All times are given in seconds, and the verification was done on a DEC Alpha 3000.

Bit width	Size (gates)	Time (s)
4	135	3.9
8	473	9.8
16	1841	36.0
32	7265	168.7
64	28865	1081.9

Table 6.2. Verification Times for Benchmark 17 Multiplier

It is not easy to quantify the human cost of verification. To give some flavour for what a proof looks like, the complete FL code for verification is given in Appendix A.2. It is difficult to describe this properly without going into some level of detail about the syntax of the VossProver, which is inappropriate here. However, the proof is about 115 lines of code, including

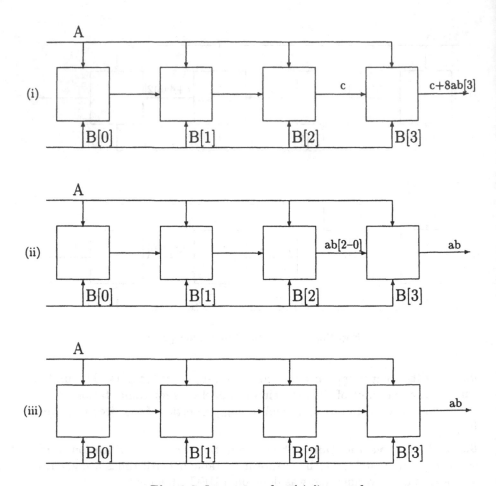

Fig. 6.3. Last steps of multiplier proof

comments and constant declarations. We estimate that given a *correct* circuit description, a circuit like this would take less than a day to verify.

Our verification of this took somewhat longer, since there were several errors in the circuit description due to our errors in translating the given description, and at the same time we were developing the VossProver. Our experience here was that the methodology was useful in discovering errors. When STE failed, a counter-example was generated. If necessary, we could then simulate the circuit with scalar values to find why the error occurred.

The only difficulty in the circuit verification was posed by an anomaly in the circuit design, which complicated the proof a little. The circuit design is very regular (this regularity is in itself neither helpful or unhelpful in the proof). The effect of this regularity is a little redundancy which gives the carry output of the most significant cell in each stage zero. In turn, this means that one of the inputs into the final *or*-gate is *always* zero; the correctness of the final result depends on this result. However, initially it was not clear that this result was needed, costing us a few hours extra work.

6.3 IEEE FP multiplier

One of the largest verifications done using the theory presented in this section is the verification of an IEEE compliant floating point multiplier by Aagaard and Seger [AaSe95]. The multiplier, implemented in structural VHDL, includes the following features:
– double precision floating point;
– radix eight multiplier array with carry-save adders;
– four stage pipeline; and
– three 56-bit carry-select adders.

The circuit verified is approximately 33 000 gates in size.

Aagaard and Seger estimate that verifying the circuit took approximately twenty days of work. The computational cost of the verification was reasonable (a few hours on a DEC Alpha 3000).

6.4 Benchmark 21: One Dimensional Systolic Array

Benchmark 21 of the IFIP WG10.5 suite is based on Kung's design of a systolic array that computes convolution. Given k weights, $w_1, \ldots, w_k$, and n inputs $x_1, \ldots, x_n$, it computes the sequence $y_1, \ldots, y_{n+1-k}$ where $y_i = w_1 x_i + w_2 x_{i+1} + \ldots + w_k x_{i+k-1}$.

6.4.1 Implementation. A brief description of the circuit is given here to aid presentation of the proof. Full details can be found in the Appendix.

For performance reasons, the circuit is implemented as a parallel architecture, using systolic array of cells, one cell for each weight. Each cell contains a multiplier and an adder. The overall architecture is shown in Figure 6.4 (in our examples, we assume an array of four cells, and an input bit width of 4). Input to the circuit is StreamIn, which is b bits wide, and output is ResultOut, which is $2b + k + 1$ bits wide. The Control lines contain a two clocks (clock 2 is the inverse of clock 1), a reset line and a number of other control lines.

In the first k clock cycles the weights are input to the circuit. In the next $2n$ clock cycles the data is input. The control lines are set appropriately in these periods so the cells know what to do with the input data. Data passes into the circuit from left to right, and as results are computed in the circuit they move from right to left. Examining the implementation of each cell, shown in Figure 6.5, this means that StreamOut of cell i is connected to StreamIn of cell $i + 1$ and ResultOut of cell i is connected to ResultIn of cell $i - 1$. In each clock cycle, the cell takes the new input x of StreamIn, the

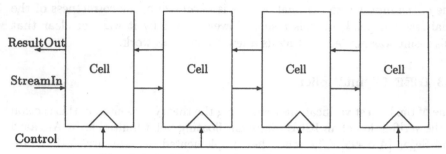

Fig. 6.4. Architecture of one dimensional array

weight w stored in the cell, the partial result y on ResultIn and computes $wx + y$, putting out the result on ResultOut.

The circuit implementation is given as a detailed gate-level design, essentially a direct translation of the VHDL code into EXE format.

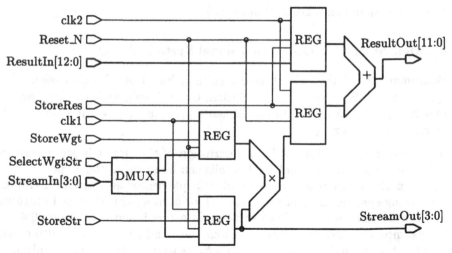

Fig. 6.5. Implementation of a cell

6.4.2 Overview of proof. Our proof approach is shown below:

1. Verify each cell. We need to show that if a weight w is stored in the cell and at the beginning of a cycle the cell is given inputs x and y, that at the beginning of the next cycle, the cell computes $wx + y$. This entails showing:

   a) the multiplier works correctly. The multiplier is Benchmark 17 of the IFIP verification examples;

   b) the adder – a carry-ripple adder – works correctly;

   c) the demux and latches all work correctly

   d) these are all connected correctly.

2. Show that if the circuit as a whole gets as its inputs $w_1, \ldots, w_k, x_1, \ldots, x_n$ and that the control signals are set accordingly, each cell stores the right weight and gets the x_i inputs at the right time.

3. Show using the previous results that as the circuit executes, the correct results are proved.

Verifying each cell. The verification of Benchmark 17 has been described previously so there is no need to elaborate this. The only modification to the proof script is taking into account the timings induced by the input latches (Benchmark 17 is not clocked) and the fact that the input and output nodes have different names. The verification of the other components in the cell are straightforward. The various proofs are put together using transitivity and specialisation. The assertion that the k-th cell is correct is labelled *StageProof k*.

Verifying inputs. The verification that the inputs to the circuit are stored correctly and pass through the systolic array at the right cycles can be proved by using trajectory evaluation alone. This example very nicely shows the abstracting power of the lattice state space — even though the state space is very large, it is possible to prove interesting results directly using STE.

    To make the result more readable here, the predicates which give clocking information and describe when and how the control lines are set are abbreviated as 'ClockingInfo' and 'ControlInfo' respectively. StreamIn.i is the StreamIn input of the i-th cell. The weight in cell i is stored in a latch which has output labelled Dpath2.i. This assertion, shown below, is called *DataAndWeighSetThm*.

```
always [(11200, 12799)]  StreamIn.0 = d[5]<|3--0|> and
always [(9600, 11199)]   StreamIn.0 = d[4]<|3--0|> and
always [(8000, 9599)]    StreamIn.0 = d[3]<|3--0|> and
always [(6400, 7999)]    StreamIn.0 = d[2]<|3--0|> and
always [(4800, 6399)]    StreamIn.0 = d[1]<|3--0|> and
always [(3200, 4799)]    StreamIn.0 = d[0]<|3--0|> and
ClockingInfo  and ControlInfo  and
always [(0, 799)]        StreamIn.0 = w[0]<|3--0|> and
always [(800, 1599)]     StreamIn.0 = w[1]<|3--0|> and
always [(1600, 2399)]    StreamIn.0 = w[2]<|3--0|> and
always [(2400, 3199)]  StreamIn.0 = w[3]<|3--0|>
   ==>>
    always [(3601, 4400)]  StreamIn.1 = d[0]<|3--0|> and
    always [(4401, 5200)]  StreamIn.2 = d[0]<|3--0|> and
    always [(5201, 6000)]
       StreamIn.1 = d[1]<|3--0|> and StreamIn.3 = d[0]<|3--0|> and
    always [(6801, 7600)]
       StreamIn.1 = d[2]<|3--0|> and StreamIn.3 = d[1]<|3--0|> and
    always [(8401, 9200)]
```

```
        StreamIn.1 = d[3]<|3--0|> and StreamIn.3 = d[2]<|3--0|> and
always [(10001, 10800)]
        StreamIn.1 = d[4]<|3--0|> and StreamIn.3 = d[3]<|3--0|> and
always [(11601, 12400)]
        StreamIn.1 = d[5]<|3--0|> and StreamIn.3 = d[4]<|3--0|> and
always [(13201, 14000)] StreamIn.3 = d[5]<|3--0|> and
always [(6001, 6800)]
        StreamIn.2 = d[1]<|3--0|> and  StreamIn.4 = d[0]<|3--0|> and
always [(7601, 8400)]
        StreamIn.2 = d[2]<|3--0|> and StreamIn.4 = d[1]<|3--0|> and
always [(9201, 10000)]
        StreamIn.2 = d[3]<|3--0|> and StreamIn.4 = d[2]<|3--0|> and
always [(10801, 11600)]
        StreamIn.2 = d[4]<|3--0|> and StreamIn.4 = d[3]<|3--0|> and
always [(12401, 13200)]
        StreamIn.2 = d[5]<|3--0|> and StreamIn.4 = d[4]<|3--0|> and
always [(14001, 14800)]
        StreamIn.4 = d[5]<|3--0|> and
always [(3200, 17599)]
        Dpath2.0 = w[3]<|3--0|> and Dpath2.1 = w[2]<|3--0|> and
        Dpath2.2 = w[1]<|3--0|> and Dpath2.3 = w[0]<|3--0|>
```

Verifying the circuit as a whole. The pseudo-code in Figure 6.6 shows the steps needed to prove the circuit as a whole. *Explanatory notes:* start is the first cycle in which the data is fed into the circuit. New data is fed into the circuit every second cycle. The effect of this is that when a cell computes a result, it keeps the same output for two cycles. Moreover, the odd numbered and even numbered cells produce new output on alternate cycles.

```
    prove DatAndWeightSetThm;
    for each cell k, prove stageProof k;
    current := DataAndWeightSetThm;
    for j=start to maxcycles do begin
        active_list := cells computing new output in cycle j;
        for k in active_list do begin
            this := stageProof k;
            this := TimeShift j this;
            current := GenTransThm current this;
        end;
    end;
```

Fig. 6.6. Overview of complete verification

The first step is to prove *DataAndWeightSetThm* and, for each cell, *stageProof k* as described above. For each cycle, we decide which cells will compute new results in that cycle (the active list). Suppose that *current* is the proof of the behaviour of the circuit up to the beginning of cycle j. Then, for each cell, k active in cycle j:

− We take the proof of correctness of cell k, *stageProof k*;

– Time-shift the proof forward by j cycles so that it can be used in cycle j;
– Combine *current* and this time shifted proof by specialising the latter and
  using transitivity. The packaged proof rule *GenTransThm* finds the specialisation automatically and then applies transitivity.

Once we have done this for each cell *current* is now the proof of the behaviour
of the circuit up to the beginning of cycle $j + 1$.

The final proof for a circuit of four cells with an input bit-width of four
is shown in Figure 6.7. To make the result slightly more readable, we have
abbreviated all the clocking information as 'ClockingInfo'.

```
always [(1400, 1599)] StreamIn.0 = d[5]<|3--0|> and
always [(1200, 1399)] StreamIn.0 = d[4]<|3--0|> and
always [(1000, 1199)] StreamIn.0 = d[3]<|3--0|> and
always [(800, 999)]   StreamIn.0 = d[2]<|3--0|> and
always [(600, 799)]   StreamIn.0 = d[1]<|3--0|> and
always [(400, 599)] StreamIn.0 = d[0]<|3--0|> and
ClockingInfo and
always [(0, 3), (10, 2200)] Reset_N = T and
always [(4, 9)] Reset_N = F and
always [(0, 299), (400, 2199)]
       StoreWgt = F and StoreStr = T and SelectWgtStr = T and
always [(0, 599)]      StoreRes = F and
always [(600, 2199)]   StoreRes = T and
always [(0, 99)]      StreamIn.0 = w[0]<|3--0|> and
always [(100, 199)]   StreamIn.0 = w[1]<|3--0|> and
always [(200, 299)]   StreamIn.0 = w[2]<|3--0|> and
always [(300, 399)]
    StoreWgt = T and  StoreStr = F and SelectWgtStr = F and
    StreamIn.0 = w[3]<|3--0|>
  ==>>
  always [(1550, 1600)]
    ResultOut =
        d[2]<|3--0|>*w[0]<|3--0|> + d[3]<|3--0|>*w[1]<|3--0|> +
        d[4]<|3--0|>*w[2]<|3--0|> + d[5]<|3--0|>*w[3]<|3--0|>and
  always [(1350, 1400)]
    ResultOut =
        d[1]<|3--0|>*w[0]<|3--0|> + d[2]<|3--0|>*w[1]<|3--0|> +
        d[3]<|3--0|>*w[2]<|3--0|> + d[4]<|3--0|>*w[3]<|3--0|>  and
  always [(1150, 1200)]
    ResultOut =
        d[0]<|3--0|>*w[0]<|3--0|> + d[1]<|3--0|>*w[1]<|3--0|> +
        d[2]<|3--0|>*w[2]<|3--0|> + d[3]<|3--0|> * w[3]<|3--0|>
```

Fig. 6.7. Proof of Benchmark 21

6.4.3 Cost of verification.
The cost of verifying a four cell version of the
circuit for various bit widths is shown in Table 6.3. The bit width given is the
bit width of the input data; the output of the circuit is just over double the

input bit width. These times include the cost of constructing the model and are measured in minutes. The total memory used is given in megabytes. All runs were on a 175MHz SGI R4400 Indy with 160M RAM. This verification shows the correctness of all the individual components and the entire design. As can be seen from the table, even large versions of the circuit can be verified with reasonable computational costs.

Bit width	Time (min)	Memory (M)
4	4	15
8	6	18
16	16	27
32	110	61
64	485	173

Table 6.3. Computational cost of verification

The human cost of verification was approximately three working days. This is a fairly realistic cost since the translation from the VHDL code to EXE code introduced a number of errors which had to be found and debugged. Furthermore, although the design is relatively straight-forward, the lack of familiarity with the detail of the design added to the cost.

6.5 Benchmark 22: Two Dimensional Systolic Array

A filter circuit based on a design of Mead and Conway is Benchmark 22 of the IFIP WG10.5 suite [MeCo80]. The filter is a matrix multiplication circuit for band matrices. A band matrix of band width w is a matrix in which zeros must be in certain positions (the matrices contain natural numbers), and the maximum number of non-zero items in a row or column is w.

6.5.1 Implementation. The filter uses a systolic array of cells, thereby gaining performance advantage through hardware parallelism. We analysed a 4×4 array of cells. Each cell contains buffers (including clocking circuitry), an integer multiplier, and an integer adder. A cell has three inputs: two inputs a and b which are two of the coefficients of the matrices being multiplied, and c, a partial result computed so far. The cell then computes $ab + c$, and produces three outputs: a and b are passed on to neighbours to process, and the new partial result $ab + c$ is passed on to another neighbour.

The entire circuit is clocked. In each clock cycle:

- Some of the coefficients of the arrays being multiplied are given as input to some of the inputs of the circuit; zeros are applied to the other inputs;
- Each cell takes its inputs (which may be the inputs of the circuit, or the outputs of other cells), computes its result for that cycle.

For details of the implementation, see the benchmark documentation.

6.5.2 Overview of proof. The structure of the circuit offers a natural decomposition of the proof. (Again, we stress that it is the proof, not the circuit that must be decomposed).

First the functionality (and timing properties) of each cell is verified. This requires verifying the multiplier (which is the multiplier of Benchmark 17), the adder and the buffer circuits of each cell. Even though the cells are all identical, each cell is verified in turn. This may be seen to be an advantage or disadvantage of our approach, depending on the requirements of the verification exercise, and is an issue that will be discussed later.

Once each cell has been verified, we can use the compositional rules to verify that when connected as designed, the overall circuit produces the correct result. The proof naturally follows the operation of the circuit.

An outline of the proof is:

1. Verify the behaviour of each cell in clock cycle 0 for arbitrary input ;
2. For each clock cycle, k, until all output has been produced:
   - For each cell, j:
     a) Time Shift the result for cell j proved in 1 forward by $k-1$ cycles;
     b) For cells which take inputs for the entire circuit, specialise this result for the actual inputs received in cycle k;
     c) Take the proof of the behaviour of the cell's neighbours in cycle $k-1$;
     d) Combine the proof of 2a or 2b (depending on the cell) and 2c by specialising the latter and the using transitivity. The VossProver can find this specialisation automatically.

6.5.3 What we proved. The benchmark documentation gives the specification of the circuit using a table which describes in which clock cycle what input data is given to which input lines, and on which output lines which output data can be found. Our specification is based on these tables; the FL code which contains this information is shown in Figure 6.8. For example, we see that in cycle 6 the input ports a0 and a3 get as input values the coefficients a_{12} and a_{31} of the A matrix, and the ports b0 and b3 get the coefficients b_{21} and b_{13} of the B matrix.

The FL program produces a table showing the outputs of the circuit at the appropriate times. However, it is also possible for the FL program to print out the assertions proved; Figure 6.9 gives one example, verifying that the circuit does compute c_{22}. It is clear why the table method is preferred.

6.5.4 Cost of Proof. The FL proof script uses STE and the inference rules to prove what the output of the circuit is at different stages. The first time the verification was attempted, a significant error was discovered in the benchmark (the specification gave wrong cycles for many outputs), which has subsequently been corrected as a result of this work. The proof script, including the proof of the correctness of all the multipliers and declarations, is approximately 650 lines long. The program itself is straightforward, although the use of a two dimensional array does not show off a functional, interpreted

```
//---------------- Input specifications
let    the_inputs =
       //   a0   a1   a2   a3     b0   b1   b2   b3
       [ ([ '0,  '0,  '0,  '0], [ '0,  '0,  '0,  '0]),   //0
         ([ '0,  '0,  '0,  '0], [ '0,  '0,  '0,  '0]),   //1
         ([ '0,  '0,  '0,  '0], [ '0,  '0,  '0,  '0]),   //2
         ([ '0,  '0,  '0,  '0], [ '0,  '0,  '0,  '0]),   //3
         ([ '0, a11,  '0,  '0], [ '0, b11,  '0,  '0]),   //4
         ([ '0,  '0, a21,  '0], [ '0,  '0, b12,  '0]),   //5
         ([a12,  '0,  '0, a31], [b21,  '0,  '0, b13]),   //6
         ([ '0, a22,  '0,  '0], [ '0, b22,  '0,  '0]),   //7
         ([ '0,  '0, a32,  '0], [ '0,  '0, b23,  '0]),   //8
         ([a23,  '0,  '0, a42], [b32,  '0,  '0, b24]),   //9
         ([ '0, a33,  '0,  '0], [ '0, b33,  '0,  '0]),   //10
         ([ '0,  '0, a43,  '0], [ '0,  '0, b34,  '0]),   //11
         ([a34,  '0,  '0,  '0], [b43,  '0,  '0,  '0]),   //12
         ([ '0, a44,  '0,  '0], [ '0, b44,  '0,  '0]),   //13
         ([ '0,  '0,  '0,  '0], [ '0,  '0,  '0,  '0]),   //14
         ([ '0,  '0,  '0,  '0], [ '0,  '0,  '0,  '0])];//15;


//---------------- Output specifications
let    timeForOutputs =
       //   1   2   3   4
       // ----------------
       [ [ 6,  7,  8,  9],        // 1
         [ 7,  9, 10, 11],        // 2
         [ 8, 10, 12, 13],        // 3
         [ 9, 11, 13, 15]        // 4
       ];
```

Fig. 6.8. FL Input and Output Specifications

language at its best. The complete verification of a 4 × 4 systolic array of 32 bit multipliers takes just under three hours on a DEC Alpha 3000 with 512M of memory (or about 11 hours on a Sun 10/51 with 64M of RAM).

The verification is of a detailed gate-level description of a particular circuit. Unlike other approaches we verify each multiplier of a given bit-width. Clearly, there are computational disadvantages to this as we cannot just verify one parameterised design. On the other hand, this approach means that we can deal with timing easily; this is important since the timing characteristics of a circuit can be very sensitive to the bit-width. Moreover, from a proof development point of view, this is not a problem since proofs can easily be reused (the proof script for a 4 bit multiplier and a 64 bit multiplier differs in one line only).

Nor do we exploit the regularity of design in order to reduce computational cost (i.e. we verify each multiplier in the systolic array). The reason for this is that we are directly verifying a gate or switch level model of the circuit in which there is no structure apparent. While the VHDL program

```
Global [(1800, 1999)] A0.0 = a23 and B0.0 = b32 and
Global [(1400, 1599)] A1.0 = a22 and B0.1 = b22 and
Global [(0, 199), (600, 799)]  A0.0 = 0 and B0.0 = 0 and
Global [(0, 199)] A3.0 = 0 and C_Out4.1 = 0 and B0.3 = 0 and
                 C_Out1.4 = 0 and A0.0 = 0 and B0.0 = 0 and
Global [(200, 399)]
  C_Out4.2 = 0 and C_Out2.4 = 0 and A1.0 = 0 and B0.1 = 0 and
Global [(400, 599)]
  A2.0 = 0 and C_Out4.3 = 0 and C_Out3.4 = 0 and B0.2 = 0 and
Global [(600, 799)] A0.0 = 0 and B0.0 = 0 and A3.0 = 0 and
                   C_Out4.1 = 0 and C_Out4.4 = 0 and
                   B0.3 = 0 and C_Out1.4 = 0 and
Global [(800, 999)]
  A1.0=a11 and B0.1=b11 and C_Out4.2=0 and C_Out2.4=0 and
Global [(1000, 1199)]
  A2.0=a21 and B0.2=b12 and C_Out4.3=0 and C_Out3.4=0 and
Global [(1200, 1399)] A0.0=a12 and B0.0=b21 and C_Out4.4=0 and
Global [(0, 99), (200, 299), (400, 499), (600, 699), (800, 899),
  (1000,1099), (1200,1299), (1400,1499), (1600,1699), (1800,1899),
  (2000,2099), (2200,2299), (2400,2499), (2600,2699), (2800,2899),
  (3000,3099), (3200,3299), (3400,3499), (3600,3699), (3800,3899)]
     CLK = F and
Global [(100,199), (300,399), (500,599), (700,799), (900,999),
  (1100,1199), (1300,1399), (1500,1599), (1700,1799), (1900,1999),
  (2100,2199), (2300,2399), (2500,2599), (2700,2799), (2900,2999),
  (3100,3199), (3300,3399), (3500,3599), (3700,3799)]
     CLK = T
       ==>>
       Global [(2000, 2100)]
          A0.1 = a23 and
          B1.0 = b32 and
          C_Out0.0 = ((a21*b12)+(a22*b22)+(a23*b32))<|11--0|>
```

Fig. 6.9. Assertion of c_{22}

that describes the circuit does have structure, the FSM generated does not. It would be possible to extend our tools to allow compositional reasoning on the structure of the model. However, this would need careful thought in how the front and back ends of the VossProver would be integrated as the compositional theory would be complicated. Moreover, taking this approach would also lose some accuracy in the modelling of the circuit. The advantage of our approach is that we are verifying the circuit design at the last stage before it is laid out on silicon – thus the effects of any circuit optimisation etc. are taken into account.

As this verification was done at the same time as circuit implementation and system development, it is difficult to estimate how much human time it took — our very rough estimate is that it took a week of work. In order to use our system effectively, it is important to be a competent FL programmer, and this will be the most important part of the learning curve for an engineer.

The theorem prover itself is a fairly simple with only a limited number of proof rules. It is also possible to use Voss in a graduated way, starting off using it as a symbolic simulator, then moving to verification using symbolic trajectory evaluation, and then starting to use the compositional theory in increasingly sophisticated ways.

6.6 Benchmark 20: Associative Memory

Previous examples illustrate how the data representation method together with the compositional theory are effective in dealing with arithmetic circuits. Benchmark 20 – an associative memory circuit – is a very different type of circuit.

6.6.1 Implementation. It is based on a design by Kohonen [Koho77]. The entire memory contains a memory of w words of n bits each, and some control circuitry. For the purpose of this exposition, a simplified version of the circuit description can be found in Figure 6.10.

There are two main inputs of n-bits each: *DataIn* and *Mask*, a number of control lines (including a clock – the details of the other control lines can be ignored for the purpose of this exposition). The circuit searches (in parallel) each word in the memory for a word that matches with *DataIn*. All bits of memory words that are not masked (a '1' in a mask bit indicates that that bit is masked) are compared with the corresponding bits of the *DataIn* word. A match with a memory word occurs if all non-masked bits of the memory word are identical with the corresponding bits of the *DataIn* word. Put more formally: suppose that *DataIn* is the bit-vector $d_1 d_2 \ldots d_n$, that *Mask* is $m_1 \ldots m_n$, and that the contents of the j-th memory cell is $w_1 w_2 \ldots w_n$. The j-th cell matches exactly when the following boolean expression is true.

$$((d_1 = w_1) \vee m_1) \wedge ((d_2 = w_2) \vee m_2) \wedge \ldots \wedge ((d_n = w_n) \vee m_n).$$

The major components of the circuit are AssocMemArray (actual memory), Prior (the selection circuitry to resolve multiple hits), AssocMemCtl (which controls output among other functions), and three input and one output registers. These registers latch their inputs when the clock goes high.

The internal nodes we must identify are:

– *Match*, a w-bit array, the j-th bit of which indicates whether the data input matched the j-th word (a number of bits may have a 1 value), and
– *Sel* a w-bit array, the j-th bit of which will be 1 if the j-th word is to be output (at most one bit of *Sel* will be 1).

The actual circuit verified was a detailed gate-level design – essentially a translation from the VHDL code given in the suite documentation (i.e. we verified the benchmark circuit, not the simplified description above).

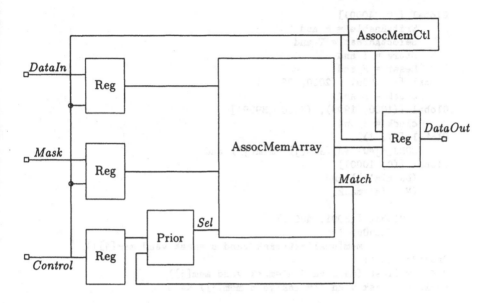

Fig. 6.10. Simplified Associative Memory Circuit

6.6.2 What we proved. There are several properties one might want to prove – whether the memory can be updated correctly, the response of the circuit to read requests given different control signals. We only verified one property of the circuit.

> Given arbitrary *DataIn* and *Mask* words, and given an arbitrary contents of the memory, does the circuit produce the correct output?

We assume a clock cycle of 2000ns. The result of our verification for a four word memory is shown in Figure 6.11. Informally this result says:

If:
- from times 0 to 4000, the *Reset* and *WriteEnable* control lines are held low, and the *SelectAdrMat* and *Store* control lines are held high; and
- the clock is low from times 0 through 999 and 2000 through 2999; and the clock is high from times 1000 through 1999 and 3000 through 3999; and
- at time 1, for $j = 0 \ldots , 3$, memory word j contains bit vector mem_j; and
- the *DataIn* port contains the bit vector d from times 0 through 1000; and
- the *MaskIn* port contains the bit vector $mask$ from times 0 through 1000.

Then:
- from times 3001 to 4000, *DataOut* contains the contents of the lowest memory word that matches the given input and mask.

Provided that:
- The index of lowest memory word that matches the given input and mask is between 0 and 3 (i.e. that such a word exists).

```
Global [(0, 4000)]
    WriteEnable = F and
    SelectAdrMat = T and
    Store = T and
    Reset = F and
Global [(0, 999), (2000, 2999)]
    clock = F and
Global [(1000, 1999), (3000, 3999)]
    clock = T and
Global [(1, 1)]
    CONJ_{j=0}^{3}(Mem[j]'=mem[j]) and
Global [(0, 1000)]
    (DataIn'=d) and
    (MaskIn'=mask)
    ==>>
      Global [(3001, 4000)]
        DataOut '=
            mem[smallest(mask vand d'=mask vand mem[j])]
Provided that:
0 <= smallest (mask vand d'=mask vand mem[j])
smallest (mask vand d'=mask vand mem[j]) <= 3
```

Fig. 6.11. Formal result of correctness proof for four word memory

6.6.3 Overview of Proof.

Memory array: The first part of the proof is to verify the behaviour of the memory array, AssocMemArray. Since the memory is large in general, it is not feasible to verify the whole array in one go using STE. What we do instead is to prove for each cell in the memory array, that given the appropriate control signals and arbitrary data input and mask, that the *Match* output of that cell is true exactly when the data input and mask matches the contents of the cell. STE is used for each proof. As the results are proved, they are conjuncted together, so that at the end of the process, we have a proof of what the entire *Match* array is set to with a given input. This proof is called *set_match_proof.*

This verification includes a verification of the behaviour of the various input registers with their clocking behaviour. It seems wasteful to repeat this for each cell of memory, as it would be possible to break down the proof into first proving the behaviour of the input registers, then separately proving AssocMemArray, and then combining the two proofs. There is a trade-off between computational cost and work required by human verifier; here we preferred to off-load work on to the computer. Even if it were the wrong choice in the end, it would probably be a good start to the verification process.

Selection Circuitry: The next part of the proof is to verify the behaviour of the selection circuitry, Prior, which can be done using only one trajectory evaluation. Given a bit vector v, the vector $\min v$ is the vector with a 1 in the first position that v has a 1, and has a 0 everywhere else (if v only contains zeros, then so does $\min v$).

The selection/priority circuitry chooses the first memory cell that matches with given input. We prove, using STE, that this priority circuitry takes a vector *match* on the *Match* nodes, and computes $\min match$ on the *Sel* nodes. Again timing is taken into account. This proof is called *priority_circuit_proof*.

We now combine *set_match_proof* and *priority_circuit_proof* by specialising the latter and using transitivity (the specialisation is found automatically by the VossProver). The new proof is called *set_sel_proof*.

Output circuitry We show that for each of the w possible inputs on *Sel* that *DataOut* produces the right output (i.e. this verifies AssocMemArray, AssocMemCtl and the output register). For each verification, STE is used and all the results combined in one proof called *assoc_out*. We now combine this using the rules of transitivity and specialisation with *set_sel_proof* to get the final result.

6.6.4 Cost of Proof. In human cost, the proof is relatively straight-forward – the FL code contains fewer than 200 lines of code, about 50 of which are constant declarations. However, this hides the fact that the VossProver is not stable and is undergoing development, a point which is discussed in the conclusion.

Table 6.4 shows the computational cost of verification for different values of n (bit-width) and w (number of cells). Each entry in the table gives the time (in seconds) and memory used (in MB). The verification was run on a 175MHz SGI R4400 Indy with 160M RAM.

Number of cells	Bit-width				
	8	16	32	64	128
4	11 12				
8	21 13	30 14	60 15	185 18	629 24
16	44 14	72 14	164 17	524 22	1857 33
32	109 15	204 17	505 22	1623 29	5977 50
64	329 18	644 23	1768 30	5389 48	20984 84
128	1225 28	2396 35	6388 52	21788 86	85193 153
256	5271 50	10021 72			

Table 6.4. Computational cost of associative memory verification: time (s)/memory usage (Mbytes)

As can be seen, verification is computationally intensive. Nevertheless, this methodology is capable of verifying fairly large circuits in a reasonable time. For example, the verification of the circuit containing 128 cells of 64 bits each, a circuit containing roughly 75 000 state holding components, takes roughly 6 hours for a complete verification of the functional and timing properties of the circuit.

It is difficult to determine the computational cost analytically since it requires estimating the cost of trajectory evaluation with respect to a given circuit and properties to be checked. However, a simple inspection of the proof and circuit shows that the cost must be $\Omega(nw^2)$. This lower limit can most easily be seen in the proof of $assoc\_out$. For each of the w possible inputs, the cost of checking that the right output is produced is nw, so the overall cost is nw^2. The other proofs have a complexity with a lower or similar lower bound. The experimental evidence shown above indicates that it is $\Theta(nw^2 + n^2)$, although more evidence would be needed to be sure.

Errors discovered We found $w+1$ errors in the benchmark implementation. In the computation of the Match array AssocMemArray, the circuitry misses a not gate for each word (there are alternative solutions, but this is the easiest). The remaining error is an RS flip-flop in AssocMemCtl which has its inputs swapped.

6.7 Liveness Properties: Benchmarks 7 and 12

These examples are one which our methodology is currently not well suited, and many of the other verification methodologies discussed in this book would be much better to use here. The fundamental problem is that the important properties to be proved are liveness properties, which we cannot handle well; this will be discussed in the conclusion of this chapter. However, we shall look at one of the examples to show how the fundamental compositional theory introduced earlier can be built on; particularly through the use of induction on time, composite, problem-specific inference rules can be developed.

6.7.1 Single Pulser. Johnson has used the Single Pulser – a textbook example circuit – to study different verification methods [JoMC94]. The original problem statement for the circuit is:

> We have a debounced pushbutton, on (true) in the down position, off (false) in the up position. Devise a circuit to sense the depression of the button and assert an output signal for one clock pulse. The system should not allow additional assertions of the output until after the output has released the button.

Johnson reformulates this into:
- the pulser emits a single unit-time pulse on its output for each pulse received on i,
- there is exactly one output pulse for every input pulse, and
- the output pulse is in the neighbourhood of the input.

Figure 6.12 illustrates the external interface of the pulser. The port In is the button to be pressed (if it has the value H, the button is pressed, if L then it is not), and Out is the output.

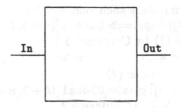

Fig. 6.12. Single Pulser

Johnson presents the verification of this circuit in a number of different systems. This section presents a paper verification using the compositional theory of STE. This attempt is not as general as some of Johnson's approaches since the specification is very specific about the timing of the output with relation to the button being pressed.

6.7.2 An Example Composite Compositional Rule. The motivation for the lemma below is that the essence of the behaviour of the pulser can be described by three assertions that show how the pulser reacts immediately to stimulation. By using induction over time, these results can be combined and generalised.

Lemma 6.1. *Let s, t, and u be arbitrary integers such that $0 \leq s \leq t < u$. Suppose:*

1. $\models \langle\!\langle \neg g_1 \Longrightarrow \text{Next } h_1 \rangle\!\rangle$,
2. $\models \langle\!\langle (\neg g_1 \land \text{Next } g_1) \Longrightarrow (\text{Next}^2 h_2) \rangle\!\rangle$, *and*
3. $\models \langle\!\langle g_1 \Longrightarrow \text{Next}^2 h_1 \rangle\!\rangle$;

then
1. $\models \langle\!\langle \text{Global}\,[(s,t)]\,(\neg g_1) \Longrightarrow \text{Global}\,[(s+1, t+1)]\,h_1 \rangle\!\rangle$.
2. $\models \langle\!\langle (\text{Global}\,[(s,t)]\,(\neg g_1) \land \text{Global}\,[(t+1, u)]\,g_1)$
$$\Longrightarrow$$
$\quad (\text{Global}\,[(s+1, t+1)]\,h_1) \land (\text{Next}^{(t+2)} h_2) \land (\text{Global}\,[(t+3, u+2)]\,h_1) \rangle\!\rangle$

Proof. The proof of 1 comes straight from Corollary 5.2. For 2, let s, t, and u be arbitrary natural numbers such that $s \leq t < u$.

(1) $\models \langle\!\langle\, \mathtt{Global}\,[(s,t)]\,(\neg g_1) \Longrightarrow \mathtt{Global}\,[(s+1, t+1)]\,h_1 \,\rangle\!\rangle$
  From hypothesis (1) by Corollary 5.2

(2) $\models \langle\!\langle\, \mathtt{Next}^t(\neg g_1) \wedge \mathtt{Next}^{(t+1)} g_1 \Longrightarrow \mathtt{Next}^{(t+2)} h_2 \,\rangle\!\rangle$
  Time-shifting hypothesis (2)

(3) $\models \langle\!\langle\, \mathtt{Global}\,[t+1, u)]\,g_1 \Longrightarrow \mathtt{Global}\,[(t+3, u+2)]\,h_1 \,\rangle\!\rangle$
  From hypothesis 3, by Corollary 5.2

(4) $\models \langle\!\langle(\mathtt{Global}\,[(s,t)]\,(\neg g_1) \wedge \mathtt{Global}\,[(t+1, u)]\,g_1)$
   $\Longrightarrow (\mathtt{Global}\,[(s+1, t+1)]\,h_1 \wedge \mathtt{Next}^{(t+2)} h_2 \wedge$
    $\mathtt{Global}\,[(t+3, u+2)]\,h_1)\rangle\!\rangle$
  Conjunction of (1), (2), (3)

6.7.3 Application to Single Pulser. Given a candidate circuit, we can use STE to verify the following three properties (since we know this is not a good example for STE, we have not actually verified the circuit; this discussion shows how we could go about it):

1. $\models \langle\!\langle\,(\neg[\mathtt{In}]) \Longrightarrow \mathtt{Next}\,(\neg[\mathtt{Out}])\,\rangle\!\rangle$;
2. $\models \langle\!\langle\,(\neg[\mathtt{In}] \wedge \mathtt{Next}\,[\mathtt{In}]) \Longrightarrow (\mathtt{Next}^2[\mathtt{Out}])\,\rangle\!\rangle$, and
3. $\models \langle\!\langle\,[\mathtt{In}] \Longrightarrow \mathtt{Next}^2(\neg[\mathtt{Out}])\,\rangle\!\rangle$.

Using these results, the above lemma can be invoked to show that

1. $\models \langle\!\langle\,\mathtt{Global}\,[(s,t)]\,(\neg[\mathtt{In}]) \Longrightarrow \mathtt{Global}\,[(s+1, t+1)]\,\neg[\mathtt{Out}]\,\rangle\!\rangle$, and
2. $\models \langle\!\langle\,(\mathtt{Global}\,[(s,t)]\,(\neg[\mathtt{In}]) \wedge \mathtt{Global}\,[(t+1, u)]\,[\mathtt{In}])$
  $\Longrightarrow$
  $\mathtt{Global}\,[(s+1, t+1)]\,(\neg[\mathtt{Out}]) \wedge$
  $\mathtt{Next}^{(t+2)}[\mathtt{Out}] \wedge$
  $\mathtt{Global}\,[(t+3, u+2)]\,(\neg[\mathtt{Out}]))\rangle\!\rangle$

The first result says that if the input does not go high (the button is not pushed), then the output does not go high. The second result says when the button is pushed (input goes from low to high), the output goes high for exactly one pulse and then goes low and stays low at least as long as the button is still pushed.

We argue that these two properties capture the intuitive specification of Johnson. However, the specification is more restrictive; there are valid implementations that satisfy Johnson's specification which would not pass this specification, showing the limitations of our current methods. It is possible to give a more general specification based on Johnson's SMV specification[5], but currently there are not efficient model checking algorithms for these specification.

One further point is that the pulser typically forms part of a bigger circuit. The important practical question is not whether we can verify abstract

---

[5] Note that although the timing constraints in Johnson's SMV specification are more general, this SMV specification is also implementation dependent — in particular, it requires some knowledge of the internal structure of the implementation, which this proof does not.

properties of the pulser in isolation, but whether we can verify the behaviour of the larger circuit. Benchmark 11 described above is a good example. It contains a pulser, but the properties we wish to prove of that circuit are all safety properties. Trajectory evaluation easily handles the essential behaviour of the pulser.

6.7.4 Arbiter. For similar reasons to the Single Pulser example, our methodology is not well suited to this benchmark example since the interesting properties to verify are liveness properties. We have not attempted to verify this.

7. Conclusion

7.1 Summary

We have shown that a fairly simple verification methodology can be used to verify large designs with an accurate model of time, using reasonable computational and human resources. The specification is given as a set of temporal logic assertions, and the model is extracted automatically from a gate or switch-level design. The components of the methodology are:
– a powerful model checking algorithm with implicit support for abstraction;
– a simple compositional theory;
– the use of a specialised theorem prover that implements the compositional theory, and allows the methodology to use different representation schemes for data at different levels;
– a flexible script language for writing proofs.

7.2 Limitations of specification

Our approach is suitable for verification of models with very large state spaces – typical of detailed circuit designs. It is particularly suitable for verifications which need accurate timing models.

There are two weaknesses of STE. First, the models must be deterministic (although this limitation is somewhat obviated by the use of the partial order state space, and the fact that real circuits are usually deterministic). Second, the temporal logic that STE supports is limited, and that not all TL formulas can be model checked efficiently. The major limitations are that we cannot efficiently use STE to model check formulas with infinite or indeterminate times, and that assertions which have antecedents with many disjunctions also have efficiency problems.

For example, we cannot efficiently check whether at some indeterminate time in the future a property becomes true (we can efficiently check whether at some specific time in the future a property becomes true). Depending on the application, this may or may not be a problem – for many circuit specifications, we want to know that a given signal happens between two times close together rather than that a signal happens at some time in the future.

There are examples of real circuits that have specifications that do require temporal formulas that cannot be checked efficiently using STE, but can be checked efficiently using other model checking approaches (such as SMV).

Example 7.1. One example of this was the partial verification of a B8ZS encoder and decoder[6] where an obvious specification is 'no matter how many bits in the input, the circuit produces the correct output'. Since the state space of the circuit is relatively small, we were able to verify this easily using SMV. However, the TL formula corresponding to the general case cannot be checked efficiently. Instead we can only check approximations, such as 'for n bits of input, the circuit produces the correct output'. We are able to do this for $n = 100$ quite easily.

Another example of this problem can be seen with the verification of the Single Pulser (Benchmark 7), which is discussed in Section 6.7.1. Other approaches are therefore more suitable for the verification of high-level designs, or properties that need to be checked deal with liveness properties.

7.3 Tool-building experience

Where a methodology consists of combining different approaches, an integrated tool is essential. While it may be necessary for user-given assumptions to be used in a proof, these steps must be carefully isolated. Using one consistent notation for expressing properties at all levels in a proof is usually desirable. Our tool is essentially a combined model checker and theorem prover. The model checking algorithm is symbolic trajectory evaluation; the theorem proving implements the compositional theory.

A powerful model checker is important. STE has proved to be a very successful basis for the tool. It is a powerful model checking algorithm capable of verifying very large state spaces, with detailed timing models. Its weakness is that the TL formulas that can be model checked efficiently are not as expressive as other temporal logics.

Good tools are necessary. The use of FL and Voss as the basis of the tool was very important. Voss gave us good models and an efficient bases of model checking. FL gives the user and tool builder tremendous flexibility and expressiveness, which means we can keep our basic tool simple without depriving the user of power. It is true that there are some performance limitations in using an interpreted language (although a compiled version of

---

[6] Bipolar with eight zero substitution (B8ZS) encoding is used in some ISDN systems. The details of the encoding are not important here — the essence is that circuit must take an unbounded bit-stream of input and produce the corresponding encoded output.

FL is certainly possible). However, key parts of the system are written in C (for example the core of trajectory evaluation and the BDD package), and automatic caching of results in FL also improves performance considerably.

A simple compositional theory is a good basis for a theorem prover The experience with the HOL-Voss system showed the potential of combining STE and theorem-proving [JoSe93]. However, the use of a full-blown theorem prover was unnecessary, and in some environments might be a practical problem as the learning curve for such systems is much higher than for model checkers. More importantly, the compositional theory is critical for extending the power of the verification system to overcome the limits of model checking, and it is the implementation of this proof system that is key.

Assistance for users is important. From a verifier's point of view, using the proof system requires the most effort. We found that it was important to provide help to the user to make the task reasonable. One aspect of this is to provide methods of reasoning about the ADTs being manipulated (in particular, those representing the data components of the circuit). We experimented with several ways of doing this. In our view the best approach is to have:

- Methods of introducing ADTs;
- Powerful decision procedures for reasoning about ADTs (e.g. integers, bit-vectors);
- Facilities for the verifier to provide facts about the data domain;
- A flexible interface to the VossProver to allow other tools to communicate with the VossProver. This would allow simple, special purpose, auxiliary theorem prover or even a powerful tool such as HOL or PVS to be used when really necessary.

This approach gives the user the ability to decide exactly how much they are prepared to pay (in terms of level of skill needed) for power provided.

The other aspect of user assistance is providing heuristics in the Voss-Prover to 'spot' possible specialisations or time shifts that could be used with transitivity. While not critical, these heuristics proved useful in many proofs in reducing the level of detail which the human verifier has to deal with.

While there is a trade-off between power and ease of use, where the trade-off is should be left to the user rather than the tool designer. We believe that our first attempts have shown the feasibility of designing such a tool. We claim for our methodology the elements that support this methodology: a good automatic model checking algorithm, a simple compositional theory, a general and powerful scripting language. This allows a tool that is extensible and one that a user can get value out of with different levels of competence.

7.4 Future work

Two immediate challenges are how to provide good debugging information and how to provide the tool with theories about the domain of the circuit (e.g. integer, bit vectors, floating point numbers etc).

- Automatic model checking has the great advantage that it provides counter-examples when a verification attempt fails. The use of a compositional theory and the raising of the level of abstractions makes this more difficult to do.
- The use of domain knowledge is critical; we have had some success with different approaches but a satisfactory way of doing this cleanly and generally has so far eluded us – perhaps not surprising given the underlying computational complexity of the problem.

There are a number of interesting scientific questions that are worth exploring, which have both theoretical and practical importance.

- Improving our trajectory evaluation algorithms;
- Extending the algorithms to support a richer logic;
- Supporting non-determinism more fully;
- The use of inductive methods to improve performance for regular architectures.
- Applying the idea of lattice data structures to other model checking algorithms;
- The synthesis of a model from a set of assertions (theoretically this relates to the question of the completeness of the compositional logic; practically this may be useful for specification validation and multi-level trajectory evaluation.

Acknowledgement. Much of the work reported here was done while the authors were with the Integrated Systems Design Laboratory of the Department of Computer Science at the University of British Columbia, Vancouver, Canada. We wish to thank all in the ISD Lab for their support over the years, especially Mark Greenstreet, Mark Aagaard and Andy Martin. The work at UBC was supported by a University Graduate Fellowship from the University of British Columbia for the first author, operating grant OGPO 109688 from the Natural Sciences and Engineering Research Council of Canada, a fellowship from the B.C. Advanced Systems Institute for the second author, and by equipment grants from Sun Microsystems Canada Ltd. and Digital Equipment of Canada Ltd. The work has also been supported by grants by the University of the Witwatersrand Research Committee. Thomas Kropf and other members of the HVG group gave very detailed comments on drafts of this chapter, which improved the chapter considerably. We also wish to thank Thomas Kropf for his help with the benchmarks. We thank Conrad Mueller, Amitha Perera and Adi Attar for reading and commenting on a draft of this chapter.

A. Appendix

A.1 Benchmark 17

A.1.1 Implementation.

```
// 'library' units
let nMUX_four A B C D S0 S1  Y =
     let notS0 = Not (Val S0) in
     let notS1 = Not (Val S1) in
     if notS1 then
        (if notS0 then
              Y <=== A
        else
              Y <=== B)
     else
        (if notS0 then
              Y <=== C
          else
              Y <=== D);


let  REG_5_clr intclock DataIn Clear Load Reset Clk DataOut =
     let clock_event = (Val Clk) And (Not (Val intclock)) in
     let ResetLowActive    = Not (Val Reset) in
     (if ResetLowActive then
         DataOut <=== [Zero, Zero, Zero, Zero, Zero]
       else
        (if clock_event then
              (if (Val Clear) then
                DataOut <=== [Zero, Zero, Zero, Zero, Zero]
              else
                (if (Val Load) then
                     DataOut <=== Valv DataIn
                  else
                     DataOut <=== Valv DataOut))
           else
              DataOut <=== Valv DataOut));

//Architecture Structure of BlackJack_CTRL

let clockEvent1 = fakeClockEvent oldClk Clk;
let clockEvent2 = fakeClockEvent oldnClk nClk;

let BlackJack_CTRL =
        (high <== One) |_|
        (nCard_rdy_d <== Not (Val Card_rdy_d)) |_|
        (nScoreGT21   <== Not (Val ScoreGT21))  |_|
        (nStateA     <== Not (Val StateA))      |_|
        (nStateB     <== Not (Val StateB))      |_|
        (nAce11flag <== Not (Val Ace11flag)) |_|
        (nCard_rdy_s <== Not (Val Card_rdy_s)) |_|
        (S_Get <==      (Val nStateA) And (Val nStateB)) |_| // 0 0
```

```
        (S_Add <==      (Val StateA)  And (Val nStateB)) |_| // 1 0
        (S_Use <==      (Val nStateA) And (Val StateB)) |_| // 0 1
        (S_Test<==      (Val StateA)  And (Val StateB))  |_|
        (MuxB0 <== Zero) |_|
        (MuxB1 <== One)  |_|
        (MuxB2 <== One)  |_\vert
        (MuxB3 <== Val Test_3)  \vert_\vert
        (MuxA0 <== Val Get_2)    |_|
        (MuxA1 <== Not((Val Acecard) And (Val nAce11flag))) |_|
        (MuxA2 <== One) |_|
        (MuxA3 <== Val Test_3) |_|
        (nMUX_four [Val MuxA0] [Val MuxA1]
                   [Val MuxA2] [Val MuxA3]
                   StateA StateB [MuxAout]) |_|
        (nMUX_four [Val MuxB0] [Val MuxB1]
                   [Val MuxB2] [Val MuxB3]
                   StateA StateB [MuxBout]) |_|
        (DFFsr oldClk (Val MuxAout) high Reset Clk StateA)  |_|
        (DFFsr oldClk (Val MuxBout) high Reset Clk StateB)  |_|
        (Get_1 <== (Val S_Get) And (Val nCard_rdy_s)) |_|
        (Get_2 <== (Val S_Get) And (Val Card_rdy_s)
                   And (Val nCard_rdy_d)) |_|
        (Get_3 <== (Val Get_2) And (Val S_or_B)) |_|
        (Test_1<== (Val S_Test) And (Val ScoreGT16)
                   And (Val nScoreGT21)) |_|
        (Test_2<== (Val S_Test) And (Val ScoreGT21)
                   And (Val nAce11flag))|_|
        (Test_3<== (Val S_Test) And (Val ScoreGT21)
                   And (Val Ace11flag)) |_|
        (S_or_B<== (Val Stand)  Or (Val Broke)) |_|
        (Clr_Ace11flag
                <== (Val Get_3) Or (Val Test_3)) |_|
        (Ld_Score
               <== (Val S_Add) Or (Val S_Use) Or (Val Test_3)) |_|
        (Hit <== Val Get_1) |_| (Set_Stand <== Val Test_1) |_|
        (Clr_Stand <== Val Get_2) |_|
        (Set_Broke <== Val Test_2) |_|
        (Clr_Broke <== Val Get_2) |_|
        (Set_Ace11flag <== Val S_Use) |_|
        (Adder_S1  <== Val S_Add) |_|
        (Adder_S0  <== Val Test_3) |_|
        (Clr_Score <== Val Get_3);

let Delay =
    if (Val Reset) then
        (("d1" <== Val Clk) |_| ("d2" <== Val "d1") |_|
         ("d3" <== Val "d2") |_|
         ("d4" <== Val "d3")|_| ("d5" <== Val "d4") )
    else
        (("d1" <== Zero) |_| ("d2" <== Zero) |_|
         ("d3" <== Zero) |_|
         ("d4" <== Zero)|_| ("d5" <== Zero) );
```

```
let BlackJack_DP =
    let N  = length internal_value in
    // Debounce Card Ready Button
    (DFFsr oldClk (Val Card_Ready) high Reset Clk Crs) |_|
    (DFFsr oldClk (Val Crs) high Reset Clk Card_rdy_d) |_|
    // instantiate various FF for status info
    (JK_FF_R
        oldClk Set_Stand Clr_Stand Clk Reset Stnd open1) |_|
    (JK_FF_R
        oldClk Set_Broke Clr_Broke Clk Reset Brke open2) |_|
    Delay |_|
    (JK_FF_R
        "d4" Set_Ace11flag Clr_Ace11flag "d5" Reset AcFlag open3)|_|
    // data path for black jack
    (nCMPN "AF" (Valv Card) ACEVALUE Acecard open4) |_|
    (expand Card brCard) |_|
    (nMUX_four Ten_plus Ten_minus (Valv brCard)
                low Adder_S0 Adder_S1
        internal_value) |_|
    (CRA internal_value  internal_Score (el 1 low)
        internal_Sum open5 intnames (N+1) "dpcra")) |_|
    (nClk <== Not (Val Clk)) |_|
    (REG_5_clr oldnClk internal_Sum Clr_Score Ld_Score Reset
        nClk internal_Score )            |_|
    (nCMPN "G16" Sixteen (Valv internal_Score)
            open6 ScoreGT16)    |_|
    (nCMPN "G21" Twentyone (Valv internal_Score)
            open7 ScoreGT21) |_|
    (Score <=== Valv internal_Score) |_|
    (Card_rdy_s <== Val Crs) |_|
    (Stand <== Val Stnd) |_|
    (Broke <== Val Brke) |_|
    (Ace11flag <== Val AcFlag);

let circ = clockEvent1 |_| clockEvent2 |_|
            BlackJack_CTRL |_| BlackJack_DP;

//make_fsm -- library call to Voss: turns EXE file into FSM
let M = make_fsm circ;
```

A.1.2 Proof script for Benchmark 11. This section presents the complete and annotated FL proof script for the verification of Benchmark 11 for the case of $r = 5$. Those readers familiar with functional languages should be able to understand the code, and the explanations should help those who do not.

The proof script starts by declaring some basic constants, and then loading a standard library.

```
//===========================================================
// Proof script for BlackJack dealer
//-------------- constants
```

```
let     clock_phase = 20;
let     clock_cycle = 2*clock_phase;
let     ace = '1;

lib_load "vlib.fl";
```

The next step declares the node values that we wish to check. We use the convention that a name ending with 'N' is an integer node, and a name ending with 'B' is a boolean node. All of these are nodes that exist in the circuit.

```
//-------     nodes
let     ResetB =   bnode   Reset;
let     ClkB    =  bnode   Clk;
let     ScoreN =   nnode   NScore;
let     StandB =   bnode   Stand;
let     HitB     = bnode   Hit;
let     BrokeB =   bnode   Broke;
let     CardN   =  nnode   NCard;
let     Card_ReadyB = bnode Card_Ready;
```

The next declarations are of integer variables that we wish to use in the specification. We declare v_1 to v_5 as four-bit integers. Note this makes the specification more general than it needs since all possible values of the v_i are verified (including zero and greater than ten).

```
//--------   variables ------
let v1 = (nvar "v1")<|3--0|>;
let v2 = (nvar "v2")<|3--0|>;
let v3 = (nvar "v3")<|3--0|>;
let v4 = (nvar "v4")<|3--0|>;
let v5 = (nvar "v5")<|3--0|>;
```

We define two simple predicates to indicate whether the reset is active or not. The circuit uses active low reset.

```
//-------- predicates ---------
let     ActiveReset  = LoSignal ResetB;
let     InactiveReset = HiSignal ResetB;
```

The next part of the script defines the antecedent. This is the temporal logic formula that specifies the expected behaviour of the environment – the stimulus to the circuit. The first function defines the behaviour of the Ready signal for a given round (recall that a round comprises four cycles): this shows that for a given round that the ready line is low for about 30ns, high for about 20ns, and then low for the rest of the round.

```
let ReadySignalRound {r :: int} =
    let base = (r * clock_cycle * 4) in
    let pt1  = 'base in
    let pt2  = '(base+clock_cycle+clock_phase/2) in
    let pt2b = '(base+clock_cycle+clock_phase/2+1) in
    let pt3a = '(base+clock_cycle+clock_phase+clock_phase/2-1) in
```

```
        let pt3  = '(base+clock_cycle+clock_phase+clock_phase/2) in
        let pt4  = '(base+4*clock_cycle-1) in
          (always [(pt1,pt2),(pt3,pt4)] (LoSignal Card_ReadyB)) and
          (always [(pt2b,pt3a)] (HiSignal Card_ReadyB));
```

The next function, GeneralAnt, defines the main antecedent. First, the clocking information is given – the clock ClkB is set to run for 40 cycles, with the first part of the cycle being low. Second, the behaviour of the reset line is specified – it is low (active) for 4 ns, and then high for a very long time. Third, for the next five rounds, we first specify the behaviour of the ready line, and then specify what the card input is for each round.

```
let GeneralAnt =
        ClockInfo ClkB F 40    and
        (always [('0,'3)] ActiveReset) and
        (always [('4,'10000)] InactiveReset) and
        (ReadySignalRound 0) and
        (always (cyclerange 0 4) (CardN '= v1)) and
        (ReadySignalRound 1) and
        (always (cyclerange 4 8) (CardN '= v2)) and
        (ReadySignalRound 2) and
        (always (cyclerange 8 12) (CardN '= v3)) and
        (ReadySignalRound 3) and
        (always (cyclerange 12 16) (CardN '= v4)) and
        (ReadySignalRound 4) and
        (always (cyclerange 16 20) (CardN '= v5)) and
        (ReadySignalRound 5);
```

Now the script starts to define the consequent – the temporal logic formula that specifies the behaviour of the circuit when exposed to the given environment. The first part of this is two functions which give the times at which the score should be computed and stable, and when the Stand and Broke control lines should be computed and stable.

```
let scoreoutputround {x :: int} =
    let start = (x+1)*4*clock_cycle+clock_cycle+clock_phase+3 in
    let finish = start +clock_phase-1  in
      [('start,'finish)];

let cntloutputround {x :: int} =
    let start = (x+1)*4*clock_cycle+3+clock_cycle+clock_phase in
    let finish = start + clock_phase-1  in
    [('start,'finish)];
```

Now some auxiliary functions are defined. AceInHand takes as its argument a list of cards and checks to see whether at least one of them is an ace. AdjustScore takes two arguments, a raw score and a list of cards. If at least one of the cards is an ace, and the raw score is strictly less than twelve, then the score should be adjusted (by adding ten). The compute_score function takes a list of cards. It computes the raw score (adding up the face value of the cards). If the raw score should be adjusted, the function returns ten pus the raw score, if not the raw score is returned. HandComplete is a function that takes a list of cards and checks to see whether the score of the cards is greater than 16. StandWithHand is a function that checks to see whether the score of a hand of cards is greater than sixteen but no more than 21. BrokeHand checks to see whether the score of the cards is greater than 21.

Note that the functions that return boolean values do not return constant true or false – they return a symbolic boolean expression which depends on the symbolic values given as parameters. Similarly compute_score does not return a constant integer value – it returns a symbolic integer expression.

```
cletrec AceInHand [{c :: N}] = c '==' ace
/\      AceInHand ({c :: N}:rest) =
               (c '==' ace) '| (AceInHand rest);


clet AdjustScore {raw_score :: N} cards  =
          (AceInHand cards) & (raw_score '<' 12);


clet compute_score cards =
     let raw_score = itlist (\x.\y.{x :: N} + {y :: N})
                         (tl cards) (hd cards) in
          if_c (AdjustScore raw_score cards)
              then_c (raw_score + 10)
              else_c  raw_score;


clet     HandComplete cards =
         let score = compute_score cards in
              Predicate (score '>' 16);


clet     StandWithHand cards =
         let score = compute_score cards in
             (score '>' 16) & (score '<=' 21);

clet     BrokeHand  cards =
         let score = compute_score cards in
             score '>' 21;
```

ResultForCards is the heart of the specification of the consequent. It takes two arguments: an integer specifying which round is being considered, and the list of cards dealt. It computes the score of the cards, and the timings when output should be seen, and returns as its value the temporal logic formula that states that the circuit node ScoreN should have the value cardscore during time interval 1, and that the StandB and BrokeB nodes should have appropriate values in during time interval 2.

```
let  ResultForCards {round :: int} cards =
     let cardscore = compute_score cards in
         let timing1   = scoreoutputround round in
```

```
let timing2    = cntloutputround round in
(always timing1 (ScoreN '= cardscore)) and
(always timing2 ((StandB '= (StandWithHand cards)) and
                 (BrokeB '= (BrokeHand cards)))));
```

GeneralCon is a recursive function that returns the temporal logic formula making the consequent. It takes two arguments: cards is the complete list of cards that could be dealt, and i is number of the current round (which varies from 0 to $r - 1$). The function first chooses the first $i + 1$ cards as the cards that have been dealt.

- If $i = r$ then we have gone through all the rounds, and so the TL formula, t (expressed in FL as const_true) is returned.
- Otherwise, we return the TL formula $\phi_i \wedge \psi_{i+1}$, where

$$\phi_i = \text{ResultForCards } i \text{ dealt}$$

(which expresses the correctness condition for the circuit's output for the current round) and

$$\psi_{i+1} = \neg(\text{HandComplete dealt}) \Rightarrow \text{GeneralCon cards } (i + 1)$$

(which expresses that if the current hand does not make the dealer stand or go broke that output for subsequent rounds is correct too).

```
letrec  GeneralCon cards {i :: int} =
        let dealt = take (i+1) cards in
        r = length cards
            => const_true
            | (ResultForCards i dealt   and
              if_c  (not (HandComplete dealt))
                then_c
                  {GeneralCon cards (i+1)  :: TL}
                endif);
```

The consequent Con is defined by calling GeneralCon with two arguments – a set of arbitrary five cards and 0, the round from which verification should start.

```
let    Con =    GeneralCon [v1, v2,v3,v4,v5] 0;
```

The final part of the proof script is the definition of the theorem that we want proved – that whenever the environment satisfies the TL formula given by GeneralAnt, the circuit satisfies the TL formula given by Con. The prove_voss_fsm is a library function that invoked STE to perform the verification. The first argument is the BDD order given. Here we give an empty list, and leave it to the VossProver to pick a good ordering (the VossProver also provide a set of library routines for finding good orderings for common expressions). The argument 'M' is the finite state machine created from the circuit description.

```
let     ProofThm = prove_voss_fsm [] M GeneralAnt Con;
```

It is worth describing here the verification process of this example. First we set $r = 0$ so that the behaviour of the circuit with one card could be examined. Also, instead of using an arbitrary variable v_1, we used a constant. The first attempt at verification picked up a number of errors introduced in the translation. The next step was to generalise the constant to the variable v_1, which picked up a number of other errors. Then we incremented r and went through a similar process, first using constants as inputs to pick up 'easy' errors and then generalising to variables to do a complete verification – it was here we realised the need to increase the clock cycle. We then incremented r and went though the same process. The remaining timing problems were discovered. We then continued incrementing r and doing a complete verification – no more errors were discovered.

A.2 FL Code for Proof of Multiplier

```
// miscellaneous
let high_bit = entry_width - 1; // 0..entry_width-1
let max_time = 800;
let out_time = 3;

//--------------- Node, variable declarations

let     A = Nnode AINP;
let     B = Nnode BINP;
let     RS i = Nnode (R_S i);
let     RC i = Nnode (R_C i)<<(high_bit-1)--0>>;
let     TopBit i = Nnode (R_C i)<<high_bit>>;

let     a   = (Nvar "a")<<(entry_width-1)--0>>;
let     b   = (Nvar "b")<<(entry_width-1)--0>>;
let     c   = Nvar "c";
let     d   = (Nvar "d")<<(high_bit-1)--0>>;

let     partial {n :: int} = c  <<(n+high_bit)--0>>;

// BDD variable ordering for each stage of multiplier

let     m_bdd_order {n::int}  =
          n = 0
            => order_int_1 [b, a]
            | n=entry_width
                => order_int_1 [partial n, d]
                  | order_int_1 [b<<n>>, a, partial n, d];

let     zero_cond i = ((TopBit i)==('0))??;

let     interval  n =
          n <= entry_width
            => [('(n*out_time), 'max_time)]
```

```
        | [('(n*out_time+2*entry_width), 'max_time)];

let     InputAnts = Always (interval 0)
                              (( (A == a) ?? ) and  ( (B == b) ??));
let     OutputCons =
          let lhs = RS entry_width in
          let rhs = (a * b)<<(2*entry_width-1)--0>> in
             Always (interval (entry_width+1)) ((lhs==rhs)??);

// Antecedent for row n of the multiplier
let     MAnt {n::int}   =
          n = 0
           => Always (interval 0)
                     ( ( (A == a)??) and
                       ( (B<<n>> == b<<n>>)?? )
                     )
              | Always (interval n)
                     (( (A == a)??) and
                      ( (B<<n>> == b<<n>>)?? ) and
                      ( (RS (n-1) == (partial(n-1)))??)and
                      ( (RC (n-1) == d)??)     and
                      ( zero_cond (n-1)) );

// Consequent of row n of the multiplier
let res_of_row n =
    let power n = Npow ('2) ('n)  in
    let lhs = (RS n) + (power (n+1))*(RC n) in
    let rhs =
      n=0
       => a * b <<0>>
        | ((partial (n-1))+(power n)* d) +
          (power n)*a *(b <<n>>) in
      ((lhs == rhs)??);

let     Con_of_stage n =
          let power n = Npow ('2) ('n)  in
          let lhs = (RS n) + (power (n+1))*(RC n) in
          let rhs =  a * b<<n--0>> in
             Always (interval (n+1))
                    ((lhs == rhs)?? and (zero_cond n));


let     MCon {n::int}  =
             Always  (interval (n+1))
                     ((res_of_row n ) and (zero_cond n));

let     Mthm n =
           let bdd_order = (m_bdd_order n) in
           let ant      = MAnt n in
           let con      = MCon n in
             prove_voss bdd_order multiplier ant con;

let     preamble_thm =
```

```
            let start = Mthm 0 in
              Precondition InputAnts start;

    letrec  do_proof_main_stage  n m previous_step  =
    let curr  = Mthm n in
    let curr' = GenTransThm previous_step curr in
    let current  = Postcondition (Con_of_stage n) curr' in
        n = m
            => current
    | do_proof_main_stage (n+1) m current;

    let    main_stage  =
              do_proof_main_stage 1 high_bit preamble_thm;

    let    adder_proof  =
            let post_ant_cond =
      (( (RS high_bit)      == (partial high_bit))??) and
      (( (RC high_bit)      == d)??) and
      (( (TopBit high_bit) == ('0))??)
         in
           let post_ant = Always (interval entry_width) post_ant_cond
             in
           let power  = Npow ('2) ('entry_width)  in
           let rhs    =
               ((partial high_bit) + power * d)<<(bit_width-1)--0>> in
           let post_con_cond = ((RS entry_width) == rhs)?? in
           let post_con      = Always (interval (entry_width+1))
                                    post_con_cond in
          prove_voss (m_bdd_order entry_width)
                multiplier post_ant post_con;

    let    proof = GenTransThm main_stage adder_proof;
```

Automated Verification with Abstract State Machines Using Multiway Decision Graphs

E. Cerny, F. Corella, M. Langevin, X. Song, S. Tahar, and Z. Zhou

1. Introduction

1.1 Motivations

Formal verification methods can be classified into two main categories: interactive verification using a theorem prover and automated finite state machine (FSM) verification based on state enumeration [Gupt92].

The most general approach to verification is to state the correctness condition for a system as a theorem in a mathematical logic and to generate a proof of this theorem that is verified using a general-purpose theorem-prover. Theorem provers use powerful formalisms such as higher-order logic [GoMe93] that allow the verification problem to be stated at many levels of abstraction. This approach has attained significant success in verifying microprocessor designs, for example [Hunt85, Joyc90, SrMi95, TaKu95]. However, theorem-proving-based verification has a drawback, viz. the user is responsible for coming up with the proof of correctness and for feeding it to the theorem prover, which can be quite difficult and time consuming.

At the other extreme of the spectrum lies state space exploration of finite state machines. State enumeration techniques permit automatic behavioral comparison and model checking [TSLB90, BCLM94]. They are effective for detecting design errors in finite-state systems. The major problem with these methods is that the size of the state space may grow very rapidly with the size of the model. This is known as the *state explosion problem*.

Many strategies have been proposed to alleviate the state explosion problem [BoFi89a, BrBS91, BCLM94, CHJP90, CPVM91, CoMa90, TSLB90]. They exploit Bryant's Reduced Ordered Binary Decision Diagrams (ROB-DDs) [Brya86] to encode sets of states and to perform an implicit enumeration of the state space, making it possible to verify FSMs with a large number of states. For some specific circuits with datapath, these methods achieve linear complexity with respect to the data width. However, these methods are not adequate in general for verifying circuits with large and complex datapaths, still leading to the state explosion problem. Even the ROBDD encoding cannot resolve the problem because of the binary representation of the circuit. More specifically, every individual bit of every data signal must be represented by a separate Boolean variable, while the size of an ROBDD grows, sometimes exponentially, with the number of variables. This means that ROBDD-based verification methods often take too much time, or run out of memory, when applied to circuits having a complex data path. Furthermore, these methods do not permit an abstract representation of the circuit, in contrast to the approaches based on theorem proving.

To overcome some of the above drawbacks, we present here a new verification approach based on *abstract descriptions of state machines* (ASM) which are encoded by a new class of decision graphs, called *Multiway Decision Graphs* (MDGs) [CZSL94], of which ROBDDs are a special case. The essential contribution of MDGs is that they make it possible to integrate two verification techniques that have been very successful: *implicit state enumeration* on one hand, and *the use of abstract sorts and uninterpreted function symbols* on the other. MDGs are decision graphs that can represent relations as well as sets of states. They allow sharing of isomorphic subgraphs which decreases the size of the graphs. MDGs incorporate variables of abstract types to denote data values and uninterpreted function symbols to denote data operations. This means that sequential circuits can be verified with a runtime that is independent of the width of the datapath. In MDG-based verification, abstract descriptions of state machines (ASM) are used to model the systems. Note that the ASMs are not a new kind of a state machine, but rather a new way of describing state machines at a higher level of abstraction. While the state machines that we want to verify are ordinary finite state machines (FSM), the abstract descriptions admit non-finite state machines as models in addition to their intended finite interpretations. The motivation for such abstract descriptions is eminently practical: it is possible to verify a circuit at the register transfer (RT) level without getting bogged down in the details of a gate-level implementation. Thus, we can raise the level of abstraction of automated verification methods to approach those of interactive methods, without sacrificing automation.

1.2 Limitations of the approach

Our approach, on the other hand, has its own significant limitations. First, the fact that function symbols denoting data operations are uninterpreted means that correctness must not depend on their intended denotation. That is, the implementation and the specification must be stated in terms of the same uninterpreted function symbols, and the correctness statement to be verified must hold for any allowable interpretation of those function symbols. For example, a circuit that computes the GCD of two numbers by repeated subtraction cannot be compared against a specification where the GCD is computed by repeated division, since the implementation and the specification use different function symbols in this case, and correctness depends on the arithmetic meaning of those symbols.

This limitation can be alleviated by the use of *term rewriting* or other automated deduction techniques. A conditional term rewriting algorithm for MDGs is provided with the MDG package, but will be described elsewhere. Rewrite rules can be viewed as axioms that limit the range of allowable interpretations of the function symbols that denote data operations. (Some authors say that the function symbols become *partially interpreted*.) The use of rewrite rules extends the class of verification problems that can be solved but reduces the degree of automation, since the user has to provide a problem-specific set of rules. The possibility of combining rewriting and other automated deduction techniques with state exploration opens up exciting possibilities for further research.

A second limitation is the fact that the computation of the set of reachable states does not always terminate. This is discussed in Section 4.2.3. A third limitation is the fact that we have not implemented algorithms for the verification of liveness properties. We expect to be able to do this in the future.

1.3 Related Work

Interactive verification by theorem proving does not require a Boolean representation of the circuit: it is usually carried out at a higher level of abstraction. Indeed, part of the inspiration for our work comes from prior work on interactive verification, and in particular from the fact that Joyce verified the Tamarack-3 microprocessor at such a high level of abstraction that he did not even mention the width of the datapath [Joyc90]. This is in striking contrast with ROBDD-based methods, where an increase in the width of the datapath often makes verification impossible.

In the automated verification community, the difficulties faced by Boolean methods when verifying circuits with a substantial datapath are well known, and have been tackled by many researchers.

Clarke, Grumberg, and Long [ClGL92, Long93] have shown examples where verifications problems involving circuits with wide datapaths can be reduced by a *data abstraction* technique to simpler problems involving *parameterized circuit descriptions* where the datapath is only a few bits wide. These simpler problems can then be solved by ROBDD-based model checking. However, the fact that correctness of the simpler circuit implies correctness of the original circuit is not always obvious, and is not verified mechanically. Also, the data abstraction function has to be provided by the user; this may require considerable ingenuity, and has to be done anew for each verification problem. Similarly, Kurshan [Kurs89] has proposed the use of *homomorphic reductions* to simplify verification problems stated as tests of ω-language containment. Again, each problem requires its own homomorphism, which has to be provided by the user.

Wolper [Wolp86] has shown that *data independent systems* can be verified by reducing the domain of data values to a very small set. But data independent systems are essentially systems that transfer data without observing it or performing any computation on it, and thus the method is not widely applicable.

In some cases it is possible to restate a verification problem concerning a circuit that consists of a datapath and a controller in terms of the controller only [VLAD92, Fuji92]. In this case the CPU time needed for verification is of course independent of the width of the datapath, which is not the case for the data abstraction method of [ClGL92]. However this is practically feasible only when the interface between the datapath and the control circuitry is easy to specify, and the equivalence of the original problem to the restated one is not verified mechanically.

In contrast to these *problem reduction* techniques, new representation tools have been developed which expand the range of circuits that can be verified directly, without recourse to ingenious problem transformations. Recently, a number of ROBDD extensions such as BMDs [BrCh95], HDDs [ClFZ95] and K*BMDs [DrBR95] have been developed to represent functions that map Boolean variables to integer values. They are mainly useful for verifying arithmetic circuits.

Our approach has its roots in the work of Langevin and Cerny [LaCe91, LaCe91a, LaCe94] and Corella [Core93, Core94], who have independently developed similar techniques based on the use of variables of abstract type to denote data values and uninterpreted function symbols to denote data operations. These approaches are well-suited for verifying simple microprocessors, as well as circuits produced by high-level synthesis, since in both cases data operations are viewed as black boxes. However, explicit control state enumeration was used, and this is not adequate for circuits containing complex controllers.

The immediate precursors of MDGs are Langevin and Cerny's EOB-DDs [LaCe94]. EOBDDs were used to represent the transition and output relation of a sequential circuit. However, they were not used to represent *sets of states*. With MDGs we go one step further: we are able to represent sets of abstract states, just like ROBDDs can be used to represent sets of states in the Boolean domain. We are thus able to lift the technique of *implicit state enumeration* from the Boolean domain (where ROBDDs are used) to the domain of abstract types (where MDGs are used); we call the lifted technique *abstract implicit enumeration.*

MDGs are similar in name and structure to the *Multivalued Decision Diagrams* (MDDs) of [SKMB90], but the similarity is superficial. MDDs and MDGs have in common that any number of edges can issue from a given node. In MDDs, however, those edges are labeled by constants that denote pairwise distinct values comprising the entire range of values for the node. In MDGs the labels of the edges can be first-order terms, need not be mutually exclusive, and need not denote all the values in a given range. This makes it possible to use variables of abstract type and uninterpreted function symbols in MDGs, which is not possible in MDDs.

More recently, a number of automatic verification methods emerged which are also based on the use of abstract sorts and uninterpreted function symbols. Burch and Dill [BuDi94, JoDi95] used a validity checking algorithm for instruction-set processor verification. A logic expression representing the correctness statement is generated using symbolic simulation. The algorithm is then used to check its validity. The authors verified a subset of the RISC pipeline processor DLX [BuDi94] and a protocol processor (PP) [JoDi95], using problem-specific heuristics.

Galter [Galt94] also presented a similar symbolic approach for the verification of processors. Two IF-expressions (If-Then-Else) which represent the functions of the specification and the implementation are derived using symbolic execution. They are then compared for syntactic equivalence. As IF-expressions may grow exponentially, a technique called IF-algebra was developed to simplify the expressions. The benchmark Tamarack-3 microprocessor was verified using the method.

Barringer [Barr95] proposed a verification methodology which can be characterized as symbolic simulation plus theorem proving. The symbolic simulation is performed on the implementation and the specification for a finite number (system-dependent) of steps, generating a pair of logical expressions which represent the circuit behaviors. These two expressions are further analyzed automatically and decomposed into sets of smaller expressions called *equivalent verification conditions* which are then checked by the theorem prover PVS.

Cyrluk and Narendran [CRSS94] defined a first-order temporal logic – *Ground Temporal Logic* (GTL) which also uses uninterpreted function symbols. Using a decidable fragment of GTL, they are able to automate part of the verification at a higher level of abstraction in the PVS theorem-proving system.

All the above methods are in fact validity checking procedures of logic formulas. Therefore, they are not applicable to state exploration-based verification such as model checking or behavioral equivalence checking. In contrast, MDGs are capable of both validity checking and verification based on state-space exploration.

1.4 Outline

We describe the theoretical foundations of our approach in Section 2. In particular, we define the formal logic used and the structure of MDGs, and briefly describe the basic MDG manipulation algorithms. Furthermore, we formulate the abstract description of a state machine and show how abstract state enumeration proceeds using MDGs. In Section 3, we describe hardware modeling using our approach, i.e., how to describe circuit components using MDGs. In Section 4, we present the application of our method to hardware verification. In particular, several techniques for combinational and sequential circuits are discussed. In Section 5, we report experimental results on a number of the IFIP benchmarks. In Section 6, we present a case study of formal verification of the Fairisle 4×4 ATM (Asynchronous Transfer Mode) switch fabric using MDGs, including experimental results. In Section 7, we summarize the contributions of the paper and point out the direction of further work.

2. Foundations of the Methodology

While Boolean logic is sufficient to represent circuits at the bit level, to represent and reason about circuits using abstract types and uninterpreted function symbols we need a first-order logic. We use a many-sorted first-order logic with a distinction between abstract and concrete sorts that mirrors the hardware distinction between data path and control. Multiway Decision Graphs are canonical representations of a certain class of quantifier-free formulas of the logic, which we call *Directed Formulas* (DFs). DFs can represent the transition and output relations of a state machine, as well as the set of possible initial states and the sets of states that arise during reachability analysis. We refer to state machines whose transition relation, output relation, and the set of initial states are given by DFs, or equivalently by MDGs, as *Abstract State Machines* (ASMs).

2.1 Logic

2.1.1 Syntax. As in ordinary many-sorted first-order logic, the vocabulary consists of *sorts, constants, variables,* and *function symbols* (or *operators*). Constants and variables have sorts. An n-ary function symbol ($n > 0$) has a type $\alpha_1 \times \ldots \times \alpha_n \to \alpha_{n+1}$, where $\alpha_1 \ldots \alpha_{n+1}$ are sorts. We deviate from standard many-sorted first-order logic by introducing a distinction between *concrete* (or *enumerated*) sorts, and *abstract sorts*; the difference is that concrete sorts have *enumerations*, while abstract sorts do not. The enumeration of a concrete sort α is a set of distinct constants of sort α. We refer to constants occurring in enumerations as *individual constants*, and to other constants as *generic constants*. An individual constant can appear in the enumeration of more than one sort α, and is said to be of sort α for each of them. Variables and generic constants, on the other hand, have unique sorts.

The distinction between abstract and concrete sorts leads to a distinction between three kinds of function symbols. Let f be a function symbol of type $\alpha_1 \times \ldots \times \alpha_n \to \alpha_{n+1}$. If α_{n+1} is an abstract sort then f is an *abstract function symbol*. If all the $\alpha_1 \ldots \alpha_{n+1}$ are concrete, f is a *concrete* function symbol. If α_{n+1} is concrete while at least one of $\alpha_1 \ldots \alpha_n$ is abstract, then we refer to f as a *cross-operator*. While abstract function symbols are used to denote data operations, cross-operators are used to denote feedback from the data path to the control circuitry. Both abstract function symbols and cross-operators are *uninterpreted*, i.e. their intended interpretation is not specified. However, information about them can be provided by axioms such as conditional equations which can be used as conditional rewrite rules. Such axioms limit the range of allowable interpretations.

The terms and their types (sorts) are defined inductively as follows: a constant or variable of sort α is a term of type α; and if f is a function symbol of type $\alpha_1 \times \ldots \times \alpha_n \to \alpha_{n+1}$, $n \geq 1$, and $A_1 \ldots A_n$ are terms of types $\alpha_1 \ldots \alpha_n$, then $f(A_1, \ldots, A_n)$ is a term of type α_{n+1}. A term consisting of a single occurrence of an individual constant has multiple types (the sorts of the constant) but every other term has a unique type. The *top symbol* of a term is defined as follows: the top symbol of $f(A_1, \ldots, A_n)$ is f, and the top symbol of a term consisting of a single occurrence of a variable or a constant is that variable or constant.

We say that a term, variable or constant is concrete (resp. abstract) to indicate that it is of concrete (resp. abstract) sort. A term is *concretely reduced* if and only if it contains no concrete terms other than individual constants. Thus a concretely reduced term can contain abstract function symbols, abstract variables, abstract generic constants and individual constants, but it can contain no cross-operators, concrete function symbols, concrete generic constants, or concrete variables; and a concretely reduced term that is itself concrete must be an individual constant. A term of the form "$f(A_1, \ldots, A_n)$" where f is a cross-operator and $A_1 \ldots A_n$ are concretely-reduced terms is called a *cross-term*. For example, if f is an abstract function symbol, c is an individual constant, x is a variable of concrete sort, and y is a variable of abstract sort, then $f(c, y)$ is a concretely-reduced term (assuming that it is well-typed) while $f(x, y)$ is not. And if g is a cross-operator, then $g(c, y)$ is a cross-term (again, assuming that it is well typed) but $g(x, y)$ is not.

A (well-typed) *equation* is an expression "$A_1 = A_2$" where the left-hand side (LHS) A_1 and the right-hand side (RHS) A_2 are terms of same type α. The *atomic formulas* are the equations, plus T (truth) and F (falsity). The *formulas* are defined inductively as follows: an atomic formula is a formula; if P and Q are formulas, then $\neg P$, $P \wedge Q$ and $P \vee Q$ are formulas; if P is a formula and x is a variable, then $(\exists x)P$ is a formula (with x bound in P). We use the abbreviation $P \Leftrightarrow Q$ for $(P \Rightarrow Q) \wedge (Q \Rightarrow P)$.

2.1.2 Semantics. An *interpretation* is a mapping ψ that assigns a denotation to each sort, constant and function symbol, and satisfies the following conditions:

1. The denotation $\psi(\alpha)$ of an abstract sort α is a non-empty set.
2. If α is a concrete sort with enumeration $\{a_1, \ldots, a_n\}$ then $\psi(\alpha) = \{\psi(a_1), \ldots, \psi(a_n)\}$ and $\psi(a_i) \neq \psi(a_j)$ for $1 \leq i < j \leq n$.
3. If c is a generic constant of sort α, then $\psi(c) \in \psi(\alpha)$. If f is a function symbol of type $\alpha_1 \times \ldots \times \alpha_n \to \alpha_{n+1}$, then $\psi(f)$ is a function from the cartesian product $\psi(\alpha_1) \times \ldots \times \psi(\alpha_n)$ into the set $\psi(\alpha_{n+1})$.

V being a set of variables, a *variable assignment* with domain V compatible with an interpretation ψ is a function ϕ that maps every variable $v \in V$ of sort α to an element $\phi(v)$ of $\psi(\alpha)$. We write Φ_V^ψ for the set of ψ-compatible assignments to the variables in V.

The denotation of a term under an interpretation ψ and a ψ-compatible variable assignment ϕ whose domain contains all the variables that occur in the term is defined by induction as follows: a constant c denotes $\psi(c)$; a variable x denotes $\phi(x)$; and if $A_1 \ldots A_n$ denote $\nu_1 \ldots \nu_n$, then $f(A_1, \ldots, A_n)$ denotes $(\phi(f))(\nu_1, \ldots, \nu_n)$. The truth of a formula P under an interpretation ψ and a ψ-compatible variable assignment ϕ whose domain contains the variables that occur free in P, written $\psi, \phi \models P$, is also defined by induction: $\psi, \phi \models A_1 = A_2$ iff A_1 and A_2 have same denotation; $\psi, \phi \models \neg P$ iff it is not the case that $\psi, \phi \models P$; $\psi, \phi \models P \wedge Q$ iff $\psi, \phi \models P$ and $\psi, \phi \models Q$; $\psi, \phi \models P \vee Q$ iff $\psi, \phi \models P$ or $\psi, \phi \models Q$; and $\psi, \phi \models (\exists x)P$ iff $\psi, \phi' \models P$ for some ϕ' that assigns an arbitrary value to x and otherwise coincides with ϕ.

We write $\psi \models P$ when $\psi, \phi \models P$ for every ψ-compatible assignment ϕ to the variables that occur free in P, and $\models P$ when $\psi \models P$ for all ψ. Two formulas P and Q are *logically equivalent* iff $\models P \Leftrightarrow Q$. A formula P *logically implies* a formula Q iff $\models P \Rightarrow Q$.

2.1.3 Directed Formulas. Given two disjoint sets of variables U and V, a directed formula of type $U \to V$ is a formula in disjunctive normal form (DNF) such that

1. Each disjunct is a conjunction of equations of the form
   $A = a$, where A is a term of concrete sort α of the form "$f(B_1, \ldots, B_n)$" (f is thus a cross-operator) that contains no variables other than elements of U, and a is an individual constant in the enumeration of α, or
   $u = a$, where $u \in U$ is a variable of concrete sort α and a is an individual constant in the enumeration of α, or
   $v = a$, where $v \in V$ is a variable of concrete sort α and a is an individual constant in the enumeration of α, or
   $v = A$, where $v \in V$ is a variable of abstract sort α and A is a term of type α containing no variables other than elements of U;
2. In each disjunct, the LHSs of the equations are pairwise distinct; and
3. Every abstract variable $v \in V$ appears as the LHS of an equation $v = A$ in each of the disjuncts. (Note that there need not be an equation $v = a$ for every concrete variable $v \in V$.)

Intuitively, in a DF of type $U \to V$, the U variables play the role of independent variables, the V variables play the role of dependent variables, and the disjuncts enumerate possible cases. In each disjunct, the equations of the form $u = a$ and $A = a$ specify a case in terms of the U variables, while the other equations specify the values of (some of the) V variables in that case. The cases need not be mutually exclusive, nor exhaustive. The condition that every abstract variable $v \in V$ must appear in every disjunct is less stringent than it seems. In practice, one can introduce an additional dependent variable u and add an equation $v = u$ to a disjunct where v is missing.

A DF is said to be *concretely reduced* iff every A in an equation $A = a$ is a cross-term, and every A in an equation $v = A$ is a concretely reduced term. It is easy to see that every DF is logically equivalent to a concretely reduced DF, given complete specifications of the concrete function symbols and concrete generic constants; the reduction can be accomplished by case splitting.

A concretely reduced DF contains no concrete function symbols and no concrete generic constants; and, in a concretely reduced DF of type $U \to V$, if A is the cross-term in the LHS of an equation $A = a$, or the concretely reduced term in the RHS of an equation $v = A$, then every variable that occurs in A is an abstract variable $u \in U$. We refer to such an occurrence of a variable as a *secondary occurrence* in the DF. A *primary occurrence* of a variable, on the other hand, is an occurrence as the LHS of an equation. From now on, by *DF* we shall mean *concretely reduced DF*.

For example, suppose that $U = \{u_1, u_2\}$ and $V = \{v_1, v_2\}$, where u_1 and v_1 are variables of a concrete sort *bool* with enumeration $\{0, 1\}$ while u_2 and v_2 are variables of an abstract sort *wordn*. Suppose that f is an abstract function symbol of type *wordn* $\to$ *wordn* and g is a cross-operator of type *wordn* $\to$ *bool*. Then the formula

$$(2.1) \qquad \begin{aligned} &((f(u_2) = 0) \wedge (v_2 = u_2)) \; \vee \\ &((f(u_2) = 1) \wedge (v_1 = u_1) \wedge (v_2 = g(u_2))) \end{aligned}$$

is a DF of type $U \to V$. In the case $f(u_2) = 0$ it assigns the symbolic value u_2 to v_2. In the case $f(u_2) = 1$ it assigns the symbolic values u_1 to v_1 and $g(u_2)$ to v_2. Note that, in the case $f(u_2) = 0$, the value of v_1 is left unspecified and thus is arbitrary.

The above DF (2.1) is *not* concretely reduced. This is because the right-hand side term u_1 in the second conjunct of the second disjunct is not concretely reduced. A concretly reduced DF logically equivalent to (2.1) can be obtained by further distinguishing the cases $u_1 = 0$ and $u_1 = 1$ in the case where $f(u_2) = 1$:

$$(2.2) \qquad \begin{aligned} &((f(u_2) = 0) \wedge (v_2 = u_2)) \; \vee \\ &((f(u_2) = 1) \wedge (u_1 = 0) \wedge (v_1 = 0) \wedge (v_2 = g(u_2))) \\ &((f(u_2) = 1) \wedge (u_1 = 1) \wedge (v_1 = 1) \wedge (v_2 = g(u_2))) \end{aligned}$$

Note that, in the absence of abstract sorts, a DF contains only equations of the form $u = a$ or $v = a$, and the sets of variables U and V play symmetrical roles. If there is only one sort, and that sort is concrete with enumeration $\{0, 1\}$, then a DF is a simply a Boolean formula in DNF.

2.2 Multiway Decision Graphs

2.2.1 Structure. An ROBDD is usually viewed as the representation of a function, with the leaf nodes labeled by values (0 or 1). But it can also be viewed as representing an assertion, with the leaf nodes labeled by propositions (truth or falsity). This latter view is the one that can be generalized to accommodate abstract types.

Let G be a finite directed acyclic graph with one root, whose internal nodes are labeled by terms and whose leaves are labeled by formulas of the logic. Then G can be viewed as representing a formula defined inductively as follows: (i) if G consists of a single leaf node labeled by a formula P, then G represents P; (ii) if G has a root node labeled A with edges labeled $B_1 \ldots B_n$ leading to subgraphs $G'_1 \ldots G'_n$, and if each G'_i represents a formula P_i, then G represents the formula $\bigvee_{1 \leq i \leq n}((A = B_i) \wedge P_i)$.

The concept of MDG is relative to two orderings, the *standard term ordering* and the *custom symbol ordering*. The standard term ordering is a total ordering of all the terms of the logic. The custom symbol ordering is a total ordering on a set of symbols C that includes the cross-operators, the concrete variables, and some, but not necessarily all, of the abstract variables. The custom symbol ordering need not be compatible with the standard term ordering. Variables that are elements of C are said to participate in the custom symbol ordering.

Let U and V be disjoint sets of variables, such that all the abstract variables in V participate in the custom symbol ordering. An *MDG of type $U \to V$* is a directed acyclic graph (DAG) G with one root and ordered edges, such that:

1. Every leaf node is labeled by the formula T, except if G has a single node, which may be labeled T or F.
2. For every internal node N, either
   a) N is labeled by a cross-term A of type α with variables in U, and the edges that issue from N are labeled by individual constants in the enumeration of α, or
   b) N is labeled by a variable $u \in U$ of concrete sort α and the edges that issue from N are labeled by individual constants in the enumeration of α, or
   c) N is labeled by a variable $v \in V$ of concrete sort α and the edges that issue from N are labeled by individual constants in the enumeration of α, or
   d) N is labeled by a variable $v \in V$ of abstract sort α and the edges that issue from N are labeled by concretely reduced terms of sort α with variables in U.
3. Along every path, every abstract variable $v \in V$ appears as a node label, there are no duplicate node labels, the top symbols of the node labels appear in the custom symbol order, and nodes labeled by cross-terms with same cross-operator appear in the standard term order.
4. The edges issuing from a given node are arranged in the standard term order.

5. There are no distinct isomorphic subgraphs, and no redundant nodes, a node being *redundant* iff it is labeled by a concrete variable or cross-term of sort α whose edges are labeled by all the individual constants in the enumeration of α, and all lead to the same subgraph.

6. If a node N is labeled by an abstract variable x, and an abstract variable y participating in the custom symbol order occurs in a term A that labels one of the edges that issue form N, then y comes before x in the custom symbol order. Similarly, if N is labeled by a cross-term A with cross-operator f, and y is an abstract variable that occurs in A and participates in the custom symbol order, then y comes before f in the custom symbol order.

The *primary occurrences* and *secondary occurrences* of variables are defined in the same manner for MDGs as for DFs. Note that, given an MDG G, if U is the set of variables having secondary occurrences in G, and V the set of variables having primary occurrences, then G is of type $U \rightarrow V$.

When we say that an MDG is of type $U \rightarrow V$, it will always be understood that U and V are disjoint sets of variables, and that all the abstract variables in V participate in the custom symbol order.

An MDG is a graph representation of a formula as defined above. The formula represented by an MDG of type $U \rightarrow V$ is usually not in DNF. However, it can be put in DNF by distributing $\wedge$ over $\vee$. It is easy to see that the resulting formula is a concretely reduced DF of type $U \rightarrow V$, whose disjuncts correspond to the paths of the MDG. In this sense, we say that an MDG is a representation of a concretely reduced DF. As an example, the MDG shown in Figure 2.1 represents the DF (2.2).

Conversely, given a concretely reduced DF P of type $U \rightarrow V$, a standard term order, and a custom symbol order comprising all the variables in V and all the cross-operators in P, it is easy to construct an MDG representing a DF that coincides with P up the ordering of the disjuncts in each conjunct and the ordering of the conjuncts themselves.

The following theorem states that MDGs are a *canonical representation*:

Theorem 2.1. *For a given custom symbol order and a given standard term order, if G and G' are MDGs representing formulas P and P' respectively, and $\models P \Leftrightarrow P'$, then G and G' are isomorphic graphs.*

Although MDGs represent DFs, which are first-order formulas, this result is not surprising, because DFs are a *restricted class* of first-order formulas. The proof of the theorem can be found in [CZSL94]. The proof uses a notion of Herbrand model suitable for our logic. If G and G' are not isomorphic, a Herbrand model can be constructed that satisfies one of the formulas P or P', but not the other.

2.2.2 Basic algorithms. We have implemented the following basic MDG algorithms, which are the building blocks of the procedures for combinational

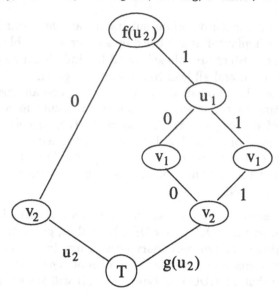

Fig. 2.1. MDG representing (2.2)

verification and reachability analysis. To simplify the description of the algorithms we shall identify an MDG with the formula that it represents.

Disjunction. Given two MDGs P_1 and P_2, there does not always exist an MDG R such that $\models R \Leftrightarrow (P_1 \vee P_2)$. For example, let x and y be distinct abstract variables, and a and b distinct abstract generic constants. Let P_1 be $x = a$ (i.e. an MDG with a root node labeled x and a single edge labeled a leading to T) and let P_2 be $y = b$. Then it can be shown that there exists no MDG R such that $\models R \Leftrightarrow (P_1 \vee P_2)$. But in the case where P_1 and P_2 have the same set of abstract primary variables, it is possible to compute an MDG R logically equivalent to $P_1 \vee P_2$.

Our disjunction algorithm is n-ary. It takes as inputs a set of MDGs P_i, $1 \leq i \leq n$, of types $U_i \to V$, and produces an MDG $R = \mathsf{Disj}(\{P_i\}_{1 \leq i \leq n})$ of type $(\bigcup_{1 \leq i \leq n} U_i) \to V$ such that

$$\models R \Leftrightarrow (\bigvee_{1 \leq i \leq n} P_i).$$

The algorithm computes the disjunction of its n inputs in one pass.

Relational product. As in the case of disjunction, given two MDGs P_1 and P_2, there does not always exist an MDG R such that $\models R \Leftrightarrow (P_1 \wedge P_2)$. For example, let x be an abstract variable, and let a and b be distinct abstract generic constants. Let P_1 be $x = a$ (i.e. an MDG with a root node labeled x and a single edge labeled a leading to T) and let P_2 be $x = b$. Then it can be shown that there exists no MDG R such that $\models R \Leftrightarrow (P_1 \wedge P_2)$.

But if P_1 and P_2 have no abstract primary variables in common, then it is possible to compute an MDG R logically equivalent to $P_1 \wedge P_2$. The abstract primary variables of R are those of P_1 and P_2. A secondary variable of R is a secondary variable of at least one of P_1, P_2 without being a primary variable of the other. (If a variable has secondary occurrences in one graph and primary occurrences in the other, the secondary occurrences are eliminated by substitution.)

Instead of implementing a conjunction algorithm, we have implemented a relational product algorithm that combines conjunction, existential quantification, and renaming. As in the case of disjunction, we have implemented an n-ary version of the algorithm. It takes as inputs a set of MDGs P_i, $1 \leq i \leq n$, of types $U_i \rightarrow V_i$, a set of variables E to be existentially quantified, and a renaming substitution η, and produces an MDG $R = \mathsf{RelP}(\{P_i\}_{1 \leq i \leq n}, E, \eta)$ such that

$$\models R \Leftrightarrow \left(\left((\exists E)(\bigwedge_{1 \leq i \leq n} P_i) \right) \cdot \eta \right).$$

The algorithm computes the conjunction of the P_i, existentially quantifies the variables in E, and applies the renaming substitution η, all in one pass. For $1 \leq i < j \leq n$, V_i and V_j must not have any abstract variables in common, otherwise the conjunction cannot be computed, because, in general, there is no MDG logically equivalent to the conjunction.

Let us determine the type of the MDG R computed by the algorithm. (It will be useful in Section 2.3.3.) The result of only computing the conjunction would be an MDG of type

$$((\bigcup_{1 \leq i \leq n} U_i) \setminus (\bigcup_{1 \leq i \leq n} V_i)) \rightarrow (\bigcup_{1 \leq i \leq n} V_i).$$

The set E of variables to be existentially quantified must be a subset of $(\bigcup_{1 \leq i \leq n} V_i)$. The result of only computing conjunction and existential quantification would be an MDG of type

$$((\bigcup_{1 \leq i \leq n} U_i) \setminus (\bigcup_{1 \leq i \leq n} V_i)) \rightarrow ((\bigcup_{1 \leq i \leq n} V_i) \setminus E).$$

The domain of η must be a subset of $((\bigcup_{1 \leq i \leq n} V_i) \setminus E)$, and η must preserve the custom symbol order when applied to the set

$$((\bigcup_{1 \leq i \leq n} U_i) \setminus (\bigcup_{1 \leq i \leq n} V_i)) \cup ((\bigcup_{1 \leq i \leq n} V_i) \setminus E).$$

The type of the result R is

$$((\bigcup_{1 \leq i \leq n} U_i) \setminus (\bigcup_{1 \leq i \leq n} V_i)) \rightarrow (((\bigcup_{1 \leq i \leq n} V_i) \setminus E) \cdot \eta).$$

Pruning by subsumption. It takes as inputs two MDGs P and Q of types $U \to V_1$ and $U \to V_2$ respectively, where U contains only abstract variables that do not participate in the custom symbol ordering, and produces an MDG $R = \mathsf{PbyS}(P, Q)$ of type $U \to V_1$ derivable from P by *pruning* (i.e. by removing some of the paths and reducing the resulting graph to satisfy the well-formedness conditions) such that

$$(2.3) \qquad \models R \lor (\exists U)Q \Leftrightarrow P \lor (\exists U)Q.$$

The paths that are removed from P are *subsumed* by Q [CZSL94], hence the name of the algorithm.

Since R is derivable from P by pruning, after the formulas represented by R and P have been converted to DNF, the disjuncts in the DNF of R are a subset of those in the DNF of P. Hence $\models R \Rightarrow P$. And, from (2.3)), it follows tautologically that $\models P \land \neg(\exists U)Q \Rightarrow R$. Thus we have

$$\models (P \land \neg(\exists U)Q \Rightarrow R) \land (R \Rightarrow P).$$

We can then view R as approximating the logical difference of P and $(\exists U)Q$. In general, there is no MDG logically equivalent to $P \land \neg(\exists U)Q$. If R is F, then it follows tautologically from (2.3) that $\models P \Rightarrow (\exists U)Q$.

2.2.3 Implementation. As in ROBDD packages, we use a *reduction table* (also called *unique table*) to maintain MDG canonicity during the computations, and a *results table* (also called *computed table*) to ensure that each distinct computation is performed only once.

Disjunction is straightforward to implement, given that all the arguments have the same set of abstract variables.

The relational product algorithm is more complex. If an abstract variable x has primary occurrences in one of the MDGs to which RelP is applied, and secondary occurrences in another, then the secondary occurrences are replaced with labels of edges that issue from nodes labeled by the primary occurrences. These substitutions are facilitated by condition 6 in the definition of MDG given in Section 2.2.1. Since terms appearing as edge and node labels can be very large in some cases, we implement them as DAGs, using a reduction table to maximize sharing and assign unique identifiers to all the terms and subterms. We use a results table for substitution. Also, with each MDG node, we keep a list of the abstract variables that participate in the custom symbol ordering and occur in the subgraph rooted at the node. Reordering of cross-terms is necessary after substitution, but is localized, since cross-terms with same cross-operator are consecutive along every path.

The PbyS algorithm is also quite complex. As the algorithm is recursively invoked in a top-down traversal, the edges labels and cross-terms of P are matched against those of Q in order to instantiate the secondary variables of Q. The algorithm must take into account the omission of redundant nodes from P. A path π of P is pruned if there exists an MDG M obtained from (the single-path MDG) π by addition of zero or more redundant nodes, such that, for every path π' of M, there exists an instantiation π'' of a path of Q such that every node-edge pair of π'' is a node-edge pair of π'.

Detailed descriptions of these three algorithms can be found in [CZSL94].

2.2.4 Other algorithms. Given an MDG P, there does not always exist an MDG R such that $\models R \Leftrightarrow (\neg P)$. For example, there exists no such R if P is $x = a$, where x is an abstract variable and a is an abstract generic constant. However, it is straightforward to compute R in the case where all the nodes in P are labeled by concrete variables or cross-terms. We refer to this special case as *concrete negation*. We shall implement a concrete negation algorithm when the need arises.

We have implemented an algorithm that simplifies an MDG by applying a set of conditional rewrite rules involving the abstract function symbols and cross-operators in the vocabulary of the logic. This algorithm will be described elsewhere.

2.3 Abstract State Machines

The presence of uninterpreted symbols in the logic means that we must distinguish between a state machine M and its abstract description D in the logic. A given abstract description D will determine a machine M for every interpretation ψ. For the purpose of hardware verification we are interested only in finite state machines (FSMs). However, an abstract description will represent infinite as well as finite state machines, since abstract sorts admit infinite interpretations. We call *Abstract State Machine* a state machine given by an abstract description in terms of MDGs, or equivalently DFs, as explained below.

2.3.1 Representing sets using MDGs. Let P be an MDG of type $U \rightarrow V$. Then, for a given interpretation ψ, P can be used to represent the set of vectors

$$Set_V^\psi(P) = \{\phi \in \Phi_V^\psi \mid \psi, \phi \models (\exists U)P\}.$$

In the next section, MDGs will thus be used in this fashion to represent sets of states and sets of output vectors. We shall also see how MDGs can be used to represent relations.

2.3.2 Describing state machines with MDGs. An abstract description of a state machine M is a tuple $D = (X, Y, Z, F_I, F_T, F_O)$, where

X, Y, Z are disjoint sets of variables, viz. the input, state, and output variables respectively. Let η be a one-to-one function that maps each variable y to a distinct variable $\eta(y)$ obtained, for example, by adorning y with a prime. The variables in $Y' = \eta(Y)$ are used as the next-state variables. X, Y and Z must be disjoint from Y'.

Given an interpretation ψ, an input vector of the state machine M represented by D is a ψ-compatible assignment to the set of input variables

X; thus the set of input vectors, or input alphabet, is Φ_X^ψ. Similarly, Φ_Z^ψ is the output alphabet. A state is a ψ-compatible assignment to the set of state variables Y; hence the state space is Φ_Y^ψ. A state ϕ can also be described by an assignment $\phi' = \phi \circ \eta^{-1} \in \Phi_{Y'}^\psi$, to the next state variables.

F_I is an MDG representing the set of initial states, of type $U \rightarrow Y$, where U is a set of abstract variables disjoint from $X \cup Y \cup Y' \cup Z$. Typically, F_I is a one-path MDG where each internal node N is labeled by a variable $y \in Y$, and the edge that issues from N is labeled by the symbolic initial value of y, which can be an individual constant, an abstract generic constant, or an abstract variable $u \in U$. It is possible to specify that two data registers have the same value, but that this common value is arbitrary, by using the same u as symbolic initial value of the abstract state variables representing the two registers.

Given an interpretation ψ, a state $\phi \in \Phi_Y^\psi$ is an initial state iff $\psi, \phi \models (\exists U)F_I$. Thus the set of initial states of the state machine M represented by D is

$$S_I = \{\phi \in \Phi_Y^\psi \mid \psi, \phi \models (\exists U)F_I\} = Set_Y^\psi(F_I).$$

F_T is an MDG of type $(X \cup Y) \rightarrow Y'$ representing the transition relation. Given an interpretation ψ, an input vector $\phi \in \Phi_X^\psi$ and a state $\phi' \in \Phi_Y^\psi$, a state $\phi'' \in \Phi_Y^\psi$ is a possible next state iff $\psi, \phi \cup \phi' \cup \phi'' \circ \eta^{-1} \models F_T$. Thus the transition relation of the state machine M represented by D is

$$R_T = \{(\phi, \phi', \phi'') \in \Phi_X^\psi \times \Phi_Y^\psi \times \Phi_Y^\psi \mid \psi, \phi \cup \phi' \cup (\phi'' \circ \eta^{-1}) \models F_T\}.$$

F_O is an MDG of type $(X \cup Y) \rightarrow Z$ representing the output relation. Given an interpretation ψ, the output relation of the state machine M represented by D is

$$R_O = \{(\phi, \phi', \phi'') \in \Phi_X^\psi \times \Phi_Y^\psi \times \Phi_Z^\psi \mid \psi, \phi \cup \phi' \cup \phi'' \models F_O\}.$$

To recapitulate, for every interpretation ψ of the sorts, constants and function symbols of the logic, the abstract description $D = (X, Y, Z, F_I, F_T, F_O)$ represents the state machine $M = (\Phi_X^\psi, \Phi_Y^\psi, \Phi_Z^\psi, S_I, R_T, R_O)$ with input alphabet Φ_X^ψ, state space Φ_Y^ψ, output alphabet Φ_Z^ψ, set of initial states S_I, transition relation R_T, and output relation R_O.

2.3.3 State exploration. Given an abstract state machine description $D = (X, Y, Z, F_I, F_T, F_O)$ we can compute the set of reachable states of a state machine $M = (\Phi_X^\psi, \Phi_Y^\psi, \Phi_Z^\psi, S_I, R_T, R_O)$ represented by D, for any ψ, using the MDG algorithms mentioned above, while at the same time checking that a given condition on the outputs of the machine, the *invariant*, holds in all the reachable states. The invariant is represented by an MDG C of type $W \rightarrow Z$, where W is a set of abstract variables disjoint from X, Y, Y', Z and U. (Recall that F_I is of type $V \rightarrow Y$.) For a given ψ, an output vector is deemed to satisfy the invariant iff $\psi, \phi \models (\exists W)C$; thus $Set_Z^\psi(C)$ is the set of output vectors that satisfy the invariant.

The procedure, called ReAn for *Reachability Analysis*, is the result of lifting the algorithm given in [CoBM89b] to the realm of abstract types and MDGs. It can be described by the following pseudo-code:

```
1.      ReAn(D, C)
2.              R := F_I; Q := F_I; K := 0;
3.              loop
4.                      K := K + 1;
5.                      I := Fresh(X, K);
6.                      O := RelP({I, Q, F_O}, X ∪ Y, ∅);
7.                      P := PbyS(O, C);
8.                      if P ≠ F then return failure;
9.                      N := RelP({I, Q, F_T}, X ∪ Y, η);
10.                     Q := PbyS(N, R);
11.                     if Q = F then return success;
12.                     R := PbyS(R, Q);
13.                     R := Disj(R, Q);
14.             end loop;
15.     end ReAn;
```

In this pseudo-code, I, N, P, Q and R are program variables that take as values MDGs representing sets of states, and O takes as values MDGs representing sets of output vectors. We will identify the program variables and their values in the following explanations when there is no risk of confusion.

Before each loop iteration, R represents the set of reachable states found so far, while Q represents the frontier set, i.e., a subset of $Set_Y^\psi(R)$ containing at least all those states that entered $Set_Y^\psi(R)$ for the first time in the previous iteration.

In line 5, $Fresh(X, K)$ constructs a one-path MDG representing a conjunction of equations $x = u$, one for each abstract input variable $x \in X$, where u is a fresh variable from the set of auxiliary abstract variables U. The value of the loop counter K is used to generate the fresh variables. This one-path MDG is assigned to I, which represents the set of input vectors.

In line 6, the relational product operation is used to compute the MDG representing the set of output vectors produced by the states in the frontier set. The resulting MDG is assigned to O. Then, in line 7, the pruning-by-subsumption operation is used to remove from O paths representing output vectors that satisfy the invariant C. The resulting MDG is assigned to P. In line 8, if P is not F, then the procedure stops and reports failure. We have implemented a counterexample facility that can then be invoked to produce a most general symbolic trace leading to a state for which the outputs do not satisfy the invariant. Examples of such a trace can be found in [ZSTC96]. If P is F, then $Set_Z^\psi(O) \subseteq Set_Z^\psi(C)$, i.e. every output vector produced by a state in the frontier set satisfies the invariant, and the verification procedure continues.

In line 9, the relational product operation is used again, this time to compute the MDG representing the set of states that can be reached in one state from the frontier set. Note that the MDG Q representing the frontier set is of type $U \to Y$, the MDG I representing the set of input vectors is of type $U \to X$, and the MDG F_T representing the transition relation is of type $(X \cup Y) \to Y'$. The result of taking the conjunction of these three MDGs would be of type $U \to (X \cup Y \cup Y')$, the result of subsequently removing the variables in $X \cup Y$ by existential quantification would be of type $U \to Y'$, and the result of subsequently applying the renaming substitution η would be of type $U \to Y$. The RelP operation performs these three operations in one pass, and assigns the resulting MDG of type $U \to Y$ to N.

Lines 10 and 11 check whether $Set_Y^\psi(N) \subseteq Set_Y^\psi(R)$ by the same method used in lines 7 and 8 to check whether $Set_Z^\psi(O) \subseteq Set_Z^\psi(C)$. If this is indeed the case, then every state reachable from the frontier set was already in $Set_Y^\psi(R)$. The fixpoint has been reached and R represents all the reachable states. Therefore, the procedure terminates and reports success. Otherwise the MDG assigned to Q in line 10 represents the new frontier set.

Line 12 simplifies R by removing from it any paths that are subsumed by Q, using PbyS. There may be such paths because Q was not computed earlier as an exact difference. Then line 13 computes the new value of R by taking the disjunction of R and Q, which represents the set of states $Set_Y^\psi(R) \cup Set_Y^\psi(Q)$, and assigning it to R.

In the general case, this procedure may not terminate and may produce false negatives. These limitations are discusses below, in Section 4.2.3 and Section 4.2.4 respectively.

3. Modeling Hardware with MDGs

A circuit is described at the RT level as a collection of components interconnected by nets that carry signals. Each signal is represented by a variable. Variables denoting control signals have concrete sorts, while variables denoting data values have abstract sorts. We show how various kinds of components can be represented by MDGs through the following examples. The parser in our MDG tools automatically transforms a component predefined in our Prolog-style MDG–HDL [ZhBo95] into its MDG representation.

- Gates: For gates, the input and output signals are always of Boolean sort. Figure 3.1(a) and Figure 3.1(b) show an OR gate and its MDG representation for a particular ordering of the variables. Boolean MDGs are essentially the same as ROBDDs.
- Multiplexer: For a two-way multiplexer as shown in Figure 3.2(a), we may have different MDGs depending on the signals being multiplexed. There is a very compact MDG (Figure 3.2(b)) if x_1, x_2 and y are all of an abstract

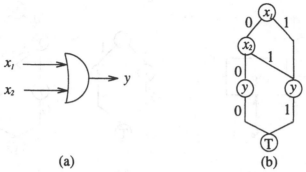

(a)   (b)

Fig. 3.1. The MDG for an OR gate.

sort. If x_1, x_2 and y are of a concrete sort with enumeration $\{c_i\}_{1 \le i \le m}$, then c_i are enumerated in the MDG as shown in Figure 3.2(c).

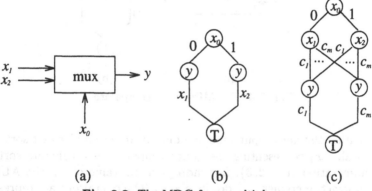

(a)   (b)   (c)

Fig. 3.2. The MDG for a multiplexer.

- Registers: Figure 3.3(a) and Figure 3.3(b) show a register r and its MDG when x and y are of an abstract sort. The variable y' denotes the next state of the register. If x and y are of a concrete sort with enumeration $\{c_i\}_{1 \le i \le m}$, we also have to enumerate c_i in the MDG as shown in Figure 3.3(c).
- Control operation: Figure 3.4(a) shows a comparator that produces a control signal y from two data inputs x_1 and x_2. Both x_1 and x_2 are variables of abstract sort while y is a Boolean variable. An uninterpreted cross-operator eq is used to denote the functionality of the comparator. If the meaning of eq matters, rewrite rules, such as $eq(x, x) \rightarrow 1$ should be used. An MDG of the comparator is shown in Figure 3.4(b).
- Data operation: Data operations are viewed as black boxes and are represented by uninterpreted function symbols. Figure 3.5(a) shows the ALU of the Tamarack-3 microprocessor [Joyc90]. The variables x_1, x_2 and y

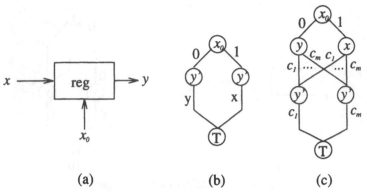

Fig. 3.3. The MDG for a register.

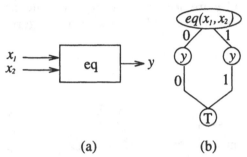

Fig. 3.4. The MDG for a comparator.

representing the data inputs and the output are of an abstract sort, while the variable x_0 representing the control input is of a concrete sort with the enumeration $\{0, 1, 2, 3\}$. Depending on the value of x_0, the ALU can add, subtract, increment, or produce *zero*. The operations are represented by symbols *add*, *sub* and *inc*. The symbol *zero* is a generic constant. The corresponding MDG shown in Figure 3.5(b) is quite compact.

Generally speaking, the behavior of a functional block involving data operations can be described by a directed formula (DF). The DF can then be transformed into an MDG by (i) creating an MDG for each atomic formula; (ii) for a disjunct of DF, conjuncting all the MDGs of its atomic formulas; and (iii) disjuncting all the MDGs representing the disjuncts.

Besides structural descriptions, MDG-HDL can also be used for the description of behavioral specifications. A behavioral description is given by high-level constructs as ITE (If-Then-Else) formulas, CASE formulas or tabular representations. The tabular construct is similar to a truth table but allows first-order terms in rows.

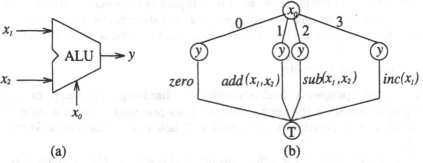

(a)   (b)

Fig. 3.5. The MDG for an ALU.

4. MDG-based Verification Techniques

We implemented in Prolog an MDG package including algorithms for disjunction, relational product (image computation), pruning by subsumption, and rewriting. We developed a reachability analysis algorithm (abstract implicit enumeration), and provided applications for hardware verification such as combinational circuits verification, safety property checking and equivalence checking of two abstract state machines. The latter two are based on the reachability analysis.

In the following sections, we detail the above applications to hardware verification.

4.1 Combinational Circuits

For *combinational verification*, we take advantage of the fact that MDGs are a canonical representation; we can thus lift the corresponding OBDD technique. Given two combinational circuits to be compared, we compute for each of them an MDG representing its input-output relation by combining the MDGs of the components of the circuit using the relational product operation. The canonicity of MDGs tells us that comparing the functionality of two combinational circuits reduces to computing the MDGs representing their input/output relations. If the two circuits have the same functionality, the two MDGs must represent logically equivalent formulas, and hence they must be isomorphic. By the use of a reduction table in the MDG package, this amounts to checking whether the two MDGs have the same Identification number (ID), a constant-time operation.

Functional comparison of two combinational circuits can also be accomplished using partitioned input/output relations. Instead of computing a single MDG for each circuit it is possible to compute a separate MDG for each output of the circuit. These separate MDGs may be much smaller than a monolithic MDG involving all the outputs. We then check whether the corresponding individual MDGs in the two partitioned relations have the same IDs.

The same technique can be used to compare two sequential circuits when a one-to-one correspondence between their registers exists and is known: it then suffices to compare the combinational parts of the sequential circuits.

4.2 Sequential Circuits

4.2.1 Safety property and equivalence checking.

The safety property checking is based on the reachability analysis procedure. Given a state machine M and an invariant C, we check if C holds in all the reachable states of M.

One application of the safety property checking is the *behavioral equivalence* (or input-output equivalence) checking of two sequential circuits. To verify that two machines produce the same sequence of outputs for every sequence of inputs, we feed the same inputs to the two circuits, i.e., we form the product state machine. Then, we perform reachability analysis on the parallel composition using an invariant that asserts the equality of the corresponding outputs in all the reachable states. For machines at different time scales, it is possible to synchronize them first if they have cyclic behavior. Thereafter we can perform reachability analysis on the product machine as usual. This technique can be used for the verification of non-pipelined microprocessor implementations against their instruction-set architecture specifications.

An invariant condition is specified by a combinational circuit whose output signals are named by the variables that occur in the condition. By convention, an assignment of values to those variables satisfies the condition if and only if the outputs of the combinational circuit take those values for *some* assignment of values to the inputs. An MDG representing the invariant is obtained from the MDG representing the functionality of the combinational circuit by existentially quantifying the concrete inputs. The variables representing abstract inputs are left in the graph as implicitly quantified secondary variables. For example, the combinational circuit of Figure 4.1(a), a simple fork, may yield different MDGs depending on the sort of the signals. If x and y are of *bool* sort, then u is existentially quantified and we get the MDG as shown in Figure 4.1(b) which simply represents $x = y$. If x and y are of an abstract sort, then we get an MDG as shown in Figure 4.1(c) which represents the formula $x = u \land y = u$. Taking the secondary variable u to be existentially quantified, the invariant becomes $(\exists u)(x = u \land y = u)$ which is logically equivalent to $x = y$.

Pruning-by-subsumption is used to check that the invariant is satisfied for the states in each frontier set. If we want to check the equality of two outputs x and y in an output MDG O, we just prune O against Inv which is the same MDG as Figure 4.1(b). This technique makes it possible to state the equality of two abstract signals without having recourse to a cross-operator eq and the rewrite rule $eq(x, x) \rightarrow 1$.

4.2.2 Simple microprocessors.

The *instruction set architecture* of a microprocessor is the specification of the effect that each instruction is intended

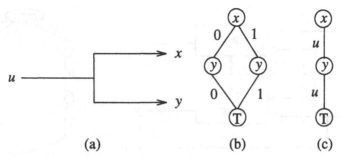

Fig. 4.1. Representation of the invariant $x = y$.

to have on the programmer's model which consists of the visible registers and memory. To verify a microprocessor against its instruction set architecture is to verify that the execution of every instruction has the intended effect.

The control FSM of a microprocessor has a distinguished *ready* state that is the starting point of instruction execution. When the control state is kept in a microprogram counter, the *ready* state is typically $mpc = 0$. We say that the microprocessor itself is in a ready state when the control FSM is in its *ready* state. Precisely stated, the problem is to verify two properties of the circuit C consisting of processor and memory, i.e., that (i) if s_1 and s_2 are consecutive reachable ready states, the visible portion of state s_1 is related to the visible portion of state s_2 as prescribed by the architecture for the executed instruction, and (ii) from every reachable ready state, a ready state is eventually reached again. Currently, we can verify the safety property (i), while property (ii) can be verified if the maximum number of clock cycles before a ready state to be reached from any reachable state is known (i.e., the liveness property is thus converted to a safety property).

To verify (i) we compare C with an ideal state machine C' whose state is the visible state of C and where each transition corresponds to the execution of an instruction as specified by the architecture. We refer to C and C' as the *implementation* and *specification*, respectively. C' is synchronized with C by a *ready* signal extracted from C: when *ready*=1 the specified transition takes place, otherwise C' remains in the same state. We perform reachability analysis on the synchronized composition of the implementation and the specification, checking an invariant that asserts the equality of the visible state in C and C' when *ready*=1. This amounts to verifying (i).

Figure 4.2(a) shows a circuit representing an invariant that asserts the equality of x and y, but only when $mpc = 0$. It is assumed that mpc and u_mpc have a concrete sort with enumeration $\{0, \dots, m\}$. The MDG of Figure 4.2(b) is obtained from the circuit by existentially quantifying the concrete input u_mpc. The formula which it represents after existentially quantifying the secondary variables u, u_1, u_2 is logically equivalent to

$$mpc = 0 \Rightarrow x = y.$$

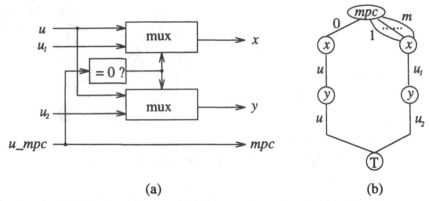

(a)   (b)

Fig. 4.2. Representation of the invariant "$x = y$ if $mpc = 0$"

4.2.3 Non-termination Problem and Initial State Generalization.

There are cases where the set of reachable states is not representable by a finite MDG of type $W \rightarrow Y$, and in such cases the reachability analysis procedures will not terminate. For example, consider a microprocessor having a program counter whose initial value is 0, denoted by a generic constant *zero* of abstract sort. An instruction that does not change the flow of control increments the program counter; assume that an abstract function symbol *inc* is used to represent this. An MDG P_k of type $W \rightarrow Y$ representing the set of states reachable in up to k steps must have at least k disjuncts (state descriptions), containing the equations $y_{pc} = zero$, $y_{pc} = inc(zero)$, $y_{pc} = inc(inc(zero))$, ..., $y_{pc} = inc^k(zero)$. A DF representing all the reachable states would require an infinite number of disjuncts, for k $\rightarrow \infty$.

In some cases non-termination can be avoided by generalizing the set of initial states so as to obtain a larger set of reachable states that is representable by a finite MDG, while still satisfying the condition to be verified. An important case in which this method is applicable is that of simple micropro-cessors and similar circuits that exhibit a cyclic behavior. When comparing two state machines derived from two implementations of a processor, or from an implementation and a specification, the initial state of the product ma-chine can be arbitrary, subject only to two constraints: (i) each machine's control state is the one where the instruction cycle begins, and (ii) the cor-responding visible registers in both machines have the same initial values. Then the set of reachable states usually has a finite representation because, informally speaking, after an instruction has been executed the product ma-chine goes to a state that is a special case of this initial state. In the case discussed above, non-termination would be avoided by letting the value of the program counter be represented by a variable rather than a constant, which would allow the subsumption check to succeed. This method is referred to as *initial state generalization.* We discuss the non-termination problem in more detail in [CZSL94, ZSTC96] and propose several other solutions.

4.2.4 False Negatives. During reachability analysis, it is possible that the invariant holds for the intended interpretation ψ_0 but not for all ψ. The abstract verification will then fail even though the interpreted state machine satisfies the invariant, a false negative result. Yet, when data operations are viewed as *black boxes*, the invariant is expected to hold for every ψ; hence, if the reachability analysis returns "failure", there must be an error in the design. In this sense we say that the verification method is applicable to designs where the data operations are viewed as black boxes.

RTL designs generated by high-level synthesis are usually of this form. This is because high-level synthesis algorithms schedule and allocate data operations without being concerned with the specific nature of the operations.

Another example of well-behaved circuits are processors. A general purpose processor provides data operations for use by the programs running on the processor. It is the programs, not the processor, that make use of the operations. The data operations can be therefore be viewed as black boxes when specifying and verifying the processor. Thus, the class of processor-like circuits is well suited to the above techniques, both from the point of view of termination and from the point of view of false negatives.

We do not know at present whether the problem of verifying that a certain condition holds is decidable when using abstract sorts, completely uninterpreted function symbols and abstract descriptions of state machines.

5. Verification of Benchmark Circuits

In this section we discuss the results of applying abstract implicit state enumeration to three synchronous circuits from the IFIP benchmark suite [Krop94b, Krop94a]. They are the Arbiter, the Greatest Common Divisor (GCD) and the Filter. All the experiments were performed on a SPARC station 20, using our MDG package implemented in Quintus Prolog Version 3.2. The execution times, memory and the number of nodes generated are shown in Table 5.1.

The circuit of the GCD benchmark that we implemented is generic. We used in the datapath abstract signals of type *wordn* to model generic words. The complete circuit is composed of 29 basic components and has a total of 8 state variables. Beside the implementation description, we provided a behavioral specification at the RT level using tabular expressions and abstract state machines. We verified the GCD circuit by checking its equivalence to the behavioral specification. The verification holds for generic words of arbitrary width.

The benchmark FILTER corresponds to the concrete example described in [Krop94a]. It has 5 input values (n = 5) and 3 stages (k = 3). Each stage

Table 5.1. Statistics from benchmark verification

Benchmark	CPU time (in sec)	Memory (in MB)	# MDG Nodes generated
GCD (n-bit)	1.93	1.48	432
FILTER	0.94	0.5	213
Arbiter (4-bit)	1.3	1.4	832
Arbiter (8-bit)	5.0	2.7	2862
Arbiter (16-bit)	4595.0	101.4	202752

is composed of seven components. The circuit has 22 basic components and 12 abstract state variables. Similarly as in the GCD example, we provided a behavioral specification of the FILTER which we checked for equivalence with the the circuit implementation.

The implementation used for the Arbiter appears in [McMi93a]. We constructed 4, 8 and 16 bit versions of this synchronous circuit at the Boolean level (all signals are of a concrete sort). Each cell of the arbiter contains two control registers and a set of logic gates for a total of 10 components per cell. For example, the 16 bit arbiter has 163 basic components and 32 concrete state variables. For each version, we verified two safety properties (properties 1 and 3 in [Krop94a]) which reflect the correct arbitration behavior (Table 5.1 includes the statistics of checking the conjunction of these two properties). Checking liveness properties (property 2 in [Krop94a]) is not currently supported by our MDG tools.

We also experimented with a number of other IFIP benchmark circuits [Krop94a], including the Traffic Light Controller, Adder, Min_Max, Tamarack-3, Multiplier, Divided, and Associative Memory.

For asynchronous circuits, such as Single Pulser and Black-Jack Dealer [Krop94b], we need to verify liveness properties that we cannot do for the moment.

6. Fairisle ATM Switch Fabric: A Case Study in Verification using MDGs

In this section we present a case study of formal verification of the Fairisle 4×4 ATM (Asynchronous Transfer Mode) switch fabric using MDGs. The device is in use for real applications in the Cambridge Fairisle network [Curz94], designed at the Computer Laboratory of the University of Cambridge.

Using a hierarchical approach, we first verified the original gate-level implementation of the switch fabric against an RTL implementation, then we verified the RTL implementation against a behavioral specification given as an abstract state machine (ASM). We thus obtained complete verification from a high-level behavior down to the gate level. We also verified some specific invariants that reflect the behavior of the fabric in its real operating environment.

6.1 The Fairisle ATM Switch Fabric

The 4×4 Fairisle switch consists of three types of components: the input port controllers, the output port controllers and the switch fabric, as shown in Figure 6.1. It switches ATM cells from the input ports to the output ports. A cell consists of a fixed number of bytes.

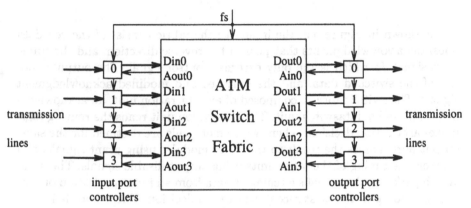

Fig. 6.1. The Fairisle ATM switch

The behavior of the switch is cyclical. In each cycle or frame, the input port controllers synchronize incoming data cells, append control information in the front of the cells in the routing tag (Figure 6.2), and send them to the fabric. The fabric waits for cells to arrive, strips off the tags, arbitrates between cells destined to the same port, sends successful cells to the appropriate output port controllers, and passes acknowledgments from the output port controllers to the input port controllers.

If different port controllers inject cells destined for the same output port controller (as indicated by the route bits in the tag) into the fabric at the same time, then only one will succeed. The others must retry later. The routing tag also includes priority information (priority bit) that is used by the fabric for arbitration which takes place in two stages. High priority cells are given precedence before the remaining cells. The choice within both priorities is made on a round-robin basis. The input controllers are informed of whether their cells were successful using acknowledgment signals. The fabric sends a negative acknowledgment to the unsuccessful input ports, but passes the acknowledgment from the requested output port to the successful input port. The port controllers and the switch fabric all use the same clock, hence bytes are received synchronously on all links. They also use a higher-level cell frame clock – the frame start signal fs. It ensures that the port controllers inject data cells into the fabric synchronously so that the routing tags arrive at the same time. If no input port raises the active bit throughout the frame then the frame is inactive – no cells are processed. Otherwise it is active.

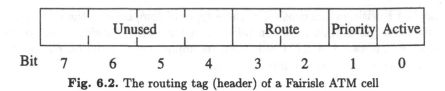

Unused				Route		Priority	Active

Bit   7   6   5   4   3   2   1   0

Fig. 6.2. The routing tag (header) of a Fairisle ATM cell

As shown in Figure 6.3, the inputs to the fabric consist of the cell data lines, the acknowledgments that pass in the reverse direction, and the frame start signal fs which is the only external control signal. The outputs consist of the switched data, and the switched and modified acknowledgment signals. The switch fabric is composed of an arbitration unit, an acknowledgment unit and a dataswitch unit. The arbitration unit reads the routing tags, makes arbitration decisions when two or more cells are destined for the same output port, passes the result to the other modules using grant signals, and controls the timing of the other units using output disable signals. The dataswitch performs the actual switching of data from an input port to an output port according to the most recent arbitration decision. The acknowledgment unit passes appropriate acknowledgment signals to the input ports. Negative acknowledgments are sent until arbitration is completed.

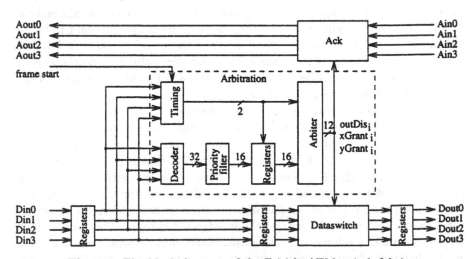

Fig. 6.3. The block diagram of the Fairisle ATM switch fabric

All the design units are repeatedly subdivided until eventually the logic gate level is reached, providing a hierarchy of components. The design has a total of 441 logic gates with two or more inputs and flip-flops.

6.2 MDG Models

6.2.1 Gate and RT Implementations. We described the implementation of the fabric at the gate and the RT levels. In the former case, we directly translated the original Qudos HDL description [Curz94] into our MDG–HDL using the same set of components.

Based on the gate-level description, we produced an RTL implementation by describing the dataswitch using abstract multiplexors instead of logic gates. Here, the data signals $Din_i/Dout_i (i = 0, 1, 2, 3)$ are modeled as n-bit words and are assigned an abstract sort $wordn$. The control fields contained in the cell header, i.e., active, priority and route fields, are extracted from the abstract data signals using cross-operators (Figure 6.4). The ASM model is thus obtained by compiling the abstract description of the RTL implementation.

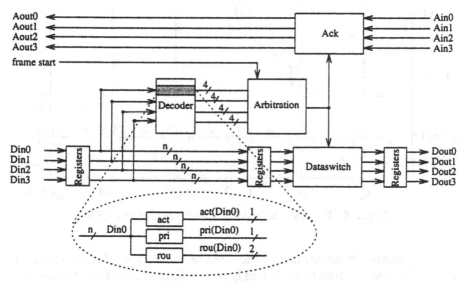

Fig. 6.4. The abstraction model of the switch fabric

6.2.2 Behavioral Specification. The specification of the switch fabric was developed in two forms: a high-level behavioral state machine and a set of invariants reflecting the essential behavior of the switch. The former is described in this section, the latter is discussed in Section 6.3.2.

Starting from a set of timing-diagrams describing the expected behavior of the switch fabric, we derived a complete specification in the form of an abstract state machine. This specification was developed independently of the actual hardware design and includes no restrictions on the frame and cell lengths, and the word width. It reflects the complete behavior of the fabric under the assumption that the environment maintains certain timing constraints on the arrival of the frame start signal and the cell headers.

To verify the RTL implementation against the ASM of the behavioral specification, we make the corresponding input/output signals to be of the same sort and use the same function symbols to extract the control information (active, priority and route fields) from the header.

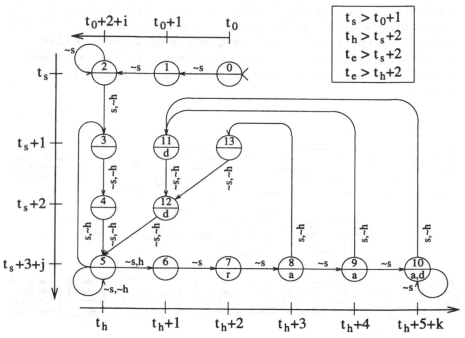

Fig. 6.5. The ASM behavioral specification of the ATM

A schematic representation of the ASM of the 4 by 4 fabric is shown in Figure 6.5. The symbols t_0, t_s, t_h and t_e represent the initial time, the time of arrival of the frame start signal, the time of arrival of the routing bytes and the time of the end of a frame, respectively. There are 14 states: States 0, 1 and 2 along the time axis t_0 describe the initial behavior of the switch fabric. States 2, 3, 4 and 5 along the time axis t_s describe the behavior of the switch on the arrival of the fs signal. States 6 to 13 along the time axis t_h describe the behavior of the switch fabric after the arrival of the headers. The waiting loops in states 2, 5 and 10 are shown by the non-zero natural numbers i, j

and k, respectively. Figure 6.5 also includes many metasymbols used to keep the presentation simple. For instance, the symbols s and h denote the arrival of frame start fs and of the routing tag (header), respectively.

Inside a circle, the symbols r, a and d indicate various operations that take place in the states: round-robin arbitration, the output of acknowledgments and the output of data, respectively. The absence of those symbols means that there is no computation and the default value is output. The operations are defined by separate state machines. We omit their descriptions here since there is nothing special about them and they are quite long to present. What needs to be mentioned, however, is the sort definition of variables. The data signals Din_i and $Dout_i$ ($i=[0..3]$) are defined as an abstract sort *wordn*. The acknowledgement signals Ain_i and $Aout_i$ ($i=[0..3]$) are of sort *bool*. More details can be found in [LTZS96].

6.3 Verification

6.3.1 Equivalence Checking.
For verifying the equivalence of the gate-level implementation and the abstract (RTL) hardware model, the abstract n-bit words were instantiated to 8 bits using *uninterpreted* functions which decode abstract data to Boolean data [TZSC96]. Equivalence checking of the RTL implementation and the behavioral specification was performed for an arbitrary word width n and any frame size and cell length [LTZS96].

By combining the above two verification steps, we hierarchically obtained a complete verification of the switch fabric from the high-level behavior down to the gate-level implementation. The experimental results on a SPARC station 10 are recapitulated in Table 6.1, including the CPU time, memory usage and the number of MDG nodes generated.

Table 6.1. The ATM experimental results for equivalence checking

Verifications	Time (Sec)	Mem (MB)	#Nodes
Gate-Level to RT-Level	183	22	183300
RT-Level to Beh.-Level	2920	150	320556

No errors were discovered in the implementation. For the sake of experimentation, however, we injected several errors into the implementation: (1) We exchanged the inputs to the JK Flip-Flop that produces the output disable signal. This prevented the circuit from resetting. (2) We used the priority information of the input port 0 to control the input port 2. (3) We used an AND gate instead of an OR gate within the acknowledgment unit producing a faulty $Aout_0$ signal. These three errors were detected by verifying the RTL implementation model against the behavioral specification. Table 6.2 shows the experimental results including times for reachability analysis on the specification and counterexample, memory usage and the number of MDG nodes generated.

Table 6.2. Verification of some faulty implementations of the ATM

Experiments	Reachability (Sec)	Counterex. (Sec)	Mem (MB)	#Nodes
Error 1	11	9	1	2462
Error 2	850	450	120	150904
Error 3	600	400	105	147339

6.3.2 Invariant Checking. Although the ASM describes the complete behavior of the switch fabric, we partially validated (in an early stage of the project) the fabric implementation by property checking. This is useful as it gives quick confidence check at low cost. Sample properties are correct circuit reset and correct data routing.

We consider the behavior of the fabric when operating in the intended real Fairisle switch environment. The switch generates frame start signal fs (Figure 6.3) at every 64th clock cycle. Initially, it should wait at least 2 clock cycles to let the fabric reset before it can generate the first fs signal. The header of a cell is generated at the 9th clock cycles after fs is set.

This cyclic behavior can be simulated as an *environment state machine* having 68 states as shown in Figure 6.6. The machine generates the frame

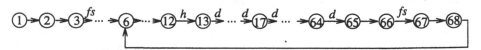

Fig. 6.6. The environment state machine of the ATM

start signal fs, the headers h and the data d in the states as indicated in the figure. Normally, d is a fresh abstract variable representing data in the cell; and h can be instantiated according to the property to be verified. We also assume that the first fs signal is generated at the 3rd clock cycle after power on. States 1 to 5 are related to the initialization of the fabric. States 6 to 68 represent the cyclic behavior of the fabric, where one cycle corresponds to one frame. With this diagram, we can map the time points to states in a similar way as we explained in the preceding section. In this case, $t_s = 3$ or 66; $t_h = 12$; and $t_e = 66$. Then, e.g., $t_h + 5$ to $t_e + 2$ are essentially the states between 17 and 68 when the remaining data of the cell following the header are switched to the output port. It can be checked that this state machine is an instance of the general timing state machine (Figure 6.5) with cell length of 53 and frame size of 64.

Below, we list properties that we verified and give their ITE expressions. The state variable c of the environment state machine is of a concrete sort having the enumeration [1..68].

$P1$: From $t_s + 3$ to $t_h + 4$, the default value is put on the data output port 0.
if $(c \in [6..16])$ then $Dout_0 = $ zero else don't-care.

$P2$: From $t_s + 1$ to $t_h + 2$, the default value is put on the acknowledgment output port 0.
if $(c \in [4..14, 67, 68])$ then $Aout_0 = 0$ else don't-care.

$P3$: From $t_h + 5$ to $t_e + 2$, if input port 0 chooses output port 0 with the priority bit set in the header and no other input port has its priority bit set, then the value on $Dout_0$ will be Din_0' which is the input of Din_0 four clock cycles earlier.
if $(c \in [17..68]) \wedge (priority[0..3] = [1, 0, 0, 0]) \wedge (route[0] = 0)$ then $Dout_0 = Din_0'$ else don't-care. ($priority[0..3]$ are the priority bits from all the input ports and route[0] represents the routing bits for input port 0)

$P4$: From $t_h + 3$ to t_e, if input port 0 chooses output port 0 with the priority bit set in the routing tag, and no other input port has its priority bit set, the value on $Aout_0$ will be the input of Ain_0.
if $(c \in [15..66]) \wedge (priority[0..3] = [1, 0, 0, 0]) \wedge (route[0] = 0)$ then $Aout_0 = Ain_0$ else don't-care.

These invariants can be easily represented using MDGs. To verify them, we compose the fabric with the environment state machine as shown in Figure 6.7. As there is a 4-clock-cycle delay for the cells to reach the output ports, a delay circuit is used to remember the input values that are to be compared with the outputs. Hence, we can state the properties in terms of the equality between Din_0' and $Dout_0$ (e.g. P3). Combining these machines (dashed frame in Figure 6.7), we obtain the required platform for checking the invariants. The above properties easily detected the three introduced design errors. The experimental results are reported in Table 6.3.

Table 6.3. Verification of properties $P1 - P4$

Verifications	Time (Sec)	Mem (MB)	#Nodes
P1	202	15	30295
P2	183	15	30356
P3	143	14	27995
P4	201	15	33001
Error 1 by P1	49	8	16119
Error 2 by P3	77	11	24001
Error 3 by P4	82	11	24274

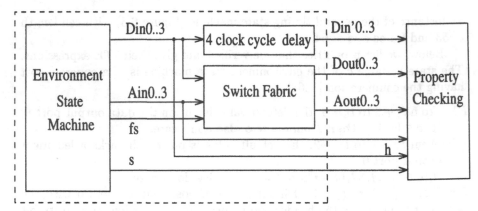

Fig. 6.7. The composite state machine for invariant checking on the ATM switch fabric

7. Conclusions and Future Work

We presented a verification methodology that makes it possible to verify sequential circuits automatically at the RT level, using abstract sorts and uninterpreted function symbols. It is based on a new kind of decision graphs, Multiway Decision Graphs (MDG). This approach allows data signals to be represented by a single variable of abstract sort rather than by 32 or 64 Boolean variables. We also described a set of algorithms for manipulating MDGs, and shown how they can be used for combinational verification, invariant and behavioral equivalence checking of sequential circuits using abstract implicit state enumeration.

Our work has shown that the use of abstract sorts for formal verification can produce interesting results by raising the level of abstraction at which the problem is stated. The contribution of MDGs beyond the use of abstract sorts is that they allow to use at a higher level of abstraction some of the ROBDD techniques that have been successful at the Boolean level.

We provided experimental results for a set of benchmarks obtained using a prototype MDG package implemented in Prolog. We demonstrated that formal verification of a 4×4 ATM switch fabric can be conducted automatically using the MDG tools.

There are many opportunities for further work in formal verification using the MDG representation of first-order formulas:

- We are developing model checking algorithms for an appropriate first-order temporal logic.
- We are exploring the links between theorem proving systems and MDG-based tools. There are two possible approaches to their integration. (1) We can embed the model checker as a specialized decision procedure in a theorem prover. This makes the theorem proving software more efficient and powerful. (2) MDG-based model checking can proceed and complete

successfully when output checking and state set inclusion can be decided by rewriting and syntactic matching. When this is not possible, we could prove the specific subgoal using a theorem prover. For systems containing complex structures, such as loops, we have to combine model checking, inductive proofs, and rewriting to accomplish the verification task effectively.

Acknowledgement. We would like to thank Nancy Boulerice, Ying Xu and Dan Voicu for carrying out the experimental work on benchmarks. The work was partially supported by an NSERC Canada Strategic Grant No. STR0167079 and the experiments were carried out on workstations provided by the Canadian Microelectronics Corporation.

Design Verification Using
SYNCHRONIZED TRANSITIONS

Jørgen Staunstrup

1. Introduction

SYNCHRONIZED TRANSITIONS is a language for modeling the computation of
dedicated electronics, such as an embedded controller or a communication ad-
apter. The realization of such a computation may involve a single (hardware)
device or multiple components in both hardware and software executing in
parallel. A high-level description of a design is an abstraction (model) of its
physical behavior. Abstraction mechanisms enable designers to ignore some
details while working on others. Abstraction does not make the complexity
of large designs vanish, but it allows one to handle it in small manageable
portions.

By creating an abstract, technology independent model, called a **design**,
it becomes possible to analyze decisions early in the design process where
large parts may still not be fixed. It is, however, important to be able to
examine those parts that are known. As an example, consider a bus arbiter
where it is important that at most one device has access to a common bus
(mutual exclusion). This chapter describes techniques for verifying the mutual exclusion property of a partly completed design, for example, the bus
arbiter even if other parts of the design are still missing. This is called **design
verification**.

A commonly used technique for validating a design is to simulate it on a
carefully selected set of test data. Ideally, one would like to do an exhaustive check where *all computations* of the design are exercised. However, this is
seldomly possible in practice, and only a small sample of the actual computations are tested by simulation. Advances in algorithms, data structures, and
design languages have provided verification techniques which are powerful
enough to do an exhaustive check on a significant set of practical examples
[BiMa88, McMi93a]. Even though these techniques are exhaustive and cover
all computations they are not based on an explicit enumeration of all possibilities; but on an implicit representation of the state space. This also opens the
possibility of checking an infinite set of computations, for example, a design
using unbounded integers. In order to use these exhaustive techniques, both
the intended and actual behavior must be expressed in a formal notation, e.g.
as a program in a programming language/hardware description language or
as a logic formula.

The design verification technique described in this chapter is such an exhaustive technique where it is possible to establish consistency between a

precisely described model, the design, and a rigorous formulation of selected key requirements. To illustrate this, consider the bus arbiter example where the requirement is mutually exclusive access. This is expressed as the assertion: $\neg(gr_i \wedge gr_j)$, where gr_i is a boolean signal indicating that device i is allowed to use the bus so the assertion does not allow that any computation has a state where both gr_i and gr_j are true $(i \neq j)$. The design verification technique is supported by various mechanical tools such as translators, a theorem prover, and state space exploration tools.

It is important to realize that formal verification deals with an abstraction and *not* with the physical realization. Formal verification of a requirement does not provide any guarantee against malfunctioning of a physical realization that does not correspond to the model. The very nature of a model is to reduce the amount of detail that needs consideration by providing an abstraction. It is outside the realm of formal methods to ensure that a model adequately reflects physical phenomena such as light or changing voltages; this must be established experimentally. Therefore, formal and experimental techniques supplement each other and both have a role to play in a sound design methodology.

The chapter is structured as follows. In Sect. 3. the notation is introduced using the Black-Jack dealer (one of the common examples used throughout this book, see Appendix for an overview). The Black-Jack dealer is also used as an informal introduction to design verification. Sect. 4. sketches the relationship between a model described with SYNCHRONIZED TRANSITIONS and a hardware realization. This is done by showing how to synthesize circuitry for key constructs. Sect. 5. provides a description of the verification techniques and how it is mechanized. In that section, it is also described how to exploit a modular design description to reduce the verification effort. Sect. 6. describes how to do refinement (implementation verification) using SYNCHRONIZED TRANSITIONS, and in Sect. 7. it is shown how to model a synchronous computation. Finally, Sect. 8. summarizes both qualitative and quantitative results from verifying the four common examples used throughout this book: the Black-Jack, the 1Syst, the Single-Pulser, and the Arbiter.

2. Related Work

The emphasis in this chapter is on design verification showing that a high-level, possibly incomplete, description meets requirements stated by the designer. This is somewhat different from implementation verification [Gord86, Brya85] where one shows that a concrete realization meets a specification, see [Gupt92] for a survey. Although the scope and approach is somewhat different there are also issues that are common to the two approaches.

Specification language: Many approaches are based on manipulating design descriptions formulated in a logic intended for verification [FoHa89,

Gord86, Hunt86]. The approach described in this chapter is somewhat different; by supporting verification in a design language that is not specifically aimed at verification. Hence, *the same textual design* description can be used for a number of purposes, e.g., simulation, synthesis and performance analysis.

Mechanization: In all approaches to formal hardware verification one needs a "proof engine" for carrying out the tedious parts of the verification. Currently, researchers are investigating several different approaches for verification, e.g., model checking [Brya85] and theorem proving [BoMo79, GaGu89]. The tools described in this chapter are based on theorem proving where a proof consists of a number of verification steps. There are a number of mechanical theorem provers available. They differ significantly, for example, in the expressive power of the notation allowed for stating a conjecture. However, the notion of proof used in all the currently available mechanical theorem provers is rather similar.

There are strong conceptual similarities between SYNCHRONIZED TRANSITIONS and UNITY, as developed by Chandy and Misra [ChMi89]. Both describe a computation as a collection of atomic conditional assignments without any explicit flow of control. Chandy and Misra propose this as a general programming paradigm. The main difference is the structuring concepts of SYNCHRONIZED TRANSITIONS: cells, parameters, statics, etc. In addition, there are a number of syntactical differences.

3. Basics

This section introduces the SYNCHRONIZED TRANSITIONS notation and its underlying model that is the basis of the associated formal verification techniques and the tools supporting it. A **design description** in SYNCHRONIZED TRANSITIONS defines a set of computations as a transition system. The transitions operate on a set of state variables. The term **design** is used to cover the description of a computation that can be realized in a wide range of technologies. This term is used to avoid the bias implied by more specific terms such as a hardware description or a program.

In addition to the two fundamental concepts transitions and state variables, SYNCHRONIZED TRANSITIONS has notation for expressing initialization, parameterization, requirements (protocols/invariants), and hierarchy (cells). The Black-Jack dealer is used to introduce these constructs. A more complete description of SYNCHRONIZED TRANSITIONS is given in [Stau94a].

3.1 Notation

The Black-Jack dealer example is used to introduce the notation and key constructs of SYNCHRONIZED TRANSITIONS. The inputs to the Black-Jack dealer are *cardready* (true/false) and *card* (2 of Clubs, ..., Ace of Spades).

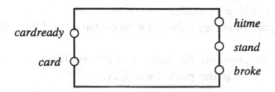

The outputs are *hitme*, *stand*, and *broke* (all boolean). The *cardready/hitme* signals are used for a four-phase handshake with the operator. Cards are valued from 2 to 10 (court cards have the value 10), and aces may be valued as either 1 or 11. The Black-Jack dealer is repeatedly presented with cards. It must assert *stand* when its accumulated score reaches 17; and it must assert *broke* when its score exceeds 21. In either case, the next card starts a new game.

3.1.1 Transitions and State Variables. A design is described as a state transition system consisting of transitions and state variables, for example:

$$\ll idle \wedge \neg (broke \vee stand) \wedge cardready \rightarrow idle := FALSE \gg$$

This is a **transition** describing how the **state variable** *idle* may change (to false) when the **precondition** $idle \wedge \neg (broke \vee stand) \wedge cardready$ holds. All four state variables are boolean. A **precondition** is a boolean expression; in the special case where this expression is just the constant value *TRUE*, it can be omitted.

Sometimes it is necessary to distinguish between the textual **transition description** (such as the ones given above) and the dynamic transition taking place at a certain point of the computation. For example, a transition description such as

$$\ll \neg idle \wedge \neg cardready \rightarrow idle := TRUE \gg$$

may describe a number of transitions in the computation (one for each time *idle* becomes *TRUE*).

A design description has a fixed number of state variables: $s_1, s_2, \ldots, s_n$, and a fixed number of transition descriptions: $t_1, t_2, \ldots, t_m$. Each state variable has a type which defines a fixed domain of values that the state variable can hold. A **state**, s, is a mapping of all state variables to values:

$$(s_1 \mapsto v_1, s_2 \mapsto v_2, \ldots, s_n \mapsto v_n)$$

where v_i $(1 \leq i \leq n)$ is a value in the domain of state variable s_i.

A transition description, t, $\ll c \rightarrow v := e \gg$ defines a binary relation, $t(S_i, S_{i+1})$, where:

- t is **active** if $c \wedge (v \neq e)$ holds,
- S_i, S_{i+1} is a pair of states called the **pre-state**, and S_{i+1} the **post-state** respectively,
- t is active in the pre-state S_i, and the state resulting from doing the state assignment ($v := e$) is the post-state S_{i+1}.

A design defines a set of **computations** as sequences of states: $S_0, S_1, \ldots$, such that S_0 is an initial state, and for each pair: S_i, S_{i+1}, there is a transition t such that the relation $t(S_i, S_{i+1})$ holds.

3.1.2 Initial State. The **initial state** of a computation is specified by giving the initial values of some or all of the state variables, for example:

> *INITIALLY*
> *idle = TRUE hitme = FALSE stand = FALSE broke = FALSE*

It is not required that state variables are given an initial value.

3.1.3 Invariants. Requirements to a design are formalized as predicates, called **invariants**, constraining the state space, for example, that the Black-Jack dealer never allows *broke* and *hitme* to be true simultaneously (similarly *stand* and *hitme* must not be true simultaneously, neither must *idle* and *hitme*, etc.). An invariant is a logical expression possibly prefixed by one or more universal quantitications.

Invariants (and protocols as introduced in Sect. 3.1.4) are formulated by the designer to express requirements on the design. They are part of the design description, for example:

> *INVARIANT*
> $\neg$*(broke* $\wedge$ *hitme)* $\wedge \neg$*(idle* $\wedge$ *hitme)* $\wedge$
> $\neg$*(stand* $\wedge$ *hitme)* $\wedge \neg$*(broke* $\wedge$ *stand)*

Invariants do not influence the behavior of a design, and removing them from a design description does not change the computation. This redundancy is quite similar to the use of declarations in high-level languages that makes it possible to check the consistency between a declaration and the use of a named quantity.

3.1.4 Protocols. Protocols are predicates on pairs of states, *pre, post*, defining a restriction on the allowable transitions between states (to those where the pre- and post-state satisfy the predicate). The following is an example of a protocol, stating that x does not change.

> *PROTOCOL x.pre = x.post*

x.pre denotes the value of x in the pre-state and similarly *x.post* is the value of x in the post-state. In general, a **protocol** is a boolean expression where all variables are suffixed with either *pre* or *post*.

The **four-phase handshake protocol** used in the Black-Jack dealer is specified as follows.

FUNCTION fourphase(a, b: BOOLEAN): BOOLEAN
RETURN ((a.pre≠a.post) ⇒ (a.post≠b.post)) ∧
*   ((b.pre≠b.post) ⇒ (b.post=a.post))*

To meet the protocol *fourphase*, *a* must get the value of ¬*b* (or be unchanged), and when *b* changes it must get the value of *a*. For instance, *idle* and *cardready* follow this protocol:

PROTOCOL fourphase(idle, cardready)

Formal Verification. The requirements expressed as invariants and protocols can be formally verified using the techniques and tools described in Sect. 5. When a designer succeeds in verifying such requirements it can be concluded that no computation will contain states violating the invariant or transitions violating the protocol.

3.1.5 Integers and Subranges. In addition to the boolean state variables used above, a design may contain state variables of type integer or a subrange of integer values, for example:

TYPE cardval = [1..10]

This specifies a type named *cardval* consisting of the integers in the range 1 to 10. Aces are represented by the value 1:

STATIC ace = 1

A type is used to specify the value range of a state variable, for example:

STATE card: cardval

The state variable *card* may hold any of the values indicated by the type *cardval*, in this case the range 1 to 10. A transition description using such a state variable is:

≪ *card* ≠ *ace* → *count:= count+card* ≫

Fig. 3.1 shows all the type declarations needed in the Black-Jack example.

Well-formedness. It is important that the value assigned to a state variable is in the domain defined by the type of the state variable. For example, the state variable *card* must only be assigned values in the range 1 to 10. Usually it is not possible to determine the values of variables statically, and therefore it is not possible to make a static check (during compilation) that the assigned value is within the correct range. For programming language implementations the checking is done at run-time every time a value is assigned to a variable. However, for a design language like SYNCHRONIZED TRANSITIONS where many descriptions are realized as circuits, doing the check at run-time is prohibitive. Instead such checks can be done by formally verifying that all values assigned are in the proper domain. When this is possible, there is no need to do the checking during execution. The specification of a range is similar to an invariant. In fact, the range information can be transformed into an invariant and verified together with other invariants. The tools supporting verification of designs in SYNCHRONIZED TRANSITIONS are capable of automatically generating such range invariants and other similar conditions ensuring against indexing arrays out of bounds, division by zero, etc. These invariants are called **well-formedness invariants** [Mell94]. Verification of the well-formedness invariants for the Black-Jack design are discussed in Sect. 8.1.1.

3.1.6 Records. The types boolean, integer, and subrange are simple types. Structured types like records and arrays are used to group associated values, e.g., a record typically describes a number of different attributes of a data item. In the Black-Jack example, the dealer's hand is described as a record with two components, a flag indicating that the hand has at least one ace and the accumulated score (where aces are counted as 1's):

$handtype = RECORD$
$\qquad anyaces: BOOLEAN$
$\qquad count: scorerange$
$\qquad END$

If *anyaces* is true the dealer might count one of the aces as 11, thereby obtaining a score of $count + 10$ (he will never count 2 aces as 11).

The individual components of a variable, c, of type record are obtained by giving the name of the state variable and the name of the component separated by a period ("."), e.g., $c.anyaces$, as in the transition description

$\ll card = ace \rightarrow c.anyaces := TRUE \gg$

3.1.7 Functions. Functions describe mappings from a list of parameters to a value, for example:

$FUNCTION\ evallow(h: handtype): scorerange$
$\qquad RETURN\ h.count$

$FUNCTION\ evalhigh(h: handtype): extscorerange$
$\qquad RETURN\ h.count + IF\ h.anyaces\ THEN\ 10\ ELSE\ 0$

The body of a **function** is a single expression. Function calls may appear in all expressions both in transitions, invariants, and protocols. This is, for example, a transition for detecting that the dealer is broke

$\ll \neg idle \wedge cardready \wedge evallow(hand) > 21 \rightarrow broke := TRUE \gg$

and this is an invariant using the function $evallow(hand)$

$INVARIANT\ broke \Rightarrow evallow(hand) > 21$

3.1.8 The Asynchronous Combinator. The **asynchronous combinator**, ‖, is used to describe the composition of a number of independent transitions. Consider for example:

− A transition for detecting that the dealer is broke

$$\ll \neg idle \wedge cardready \wedge evallow(hand) > 21 \rightarrow broke := TRUE \gg$$

− A transition for detecting that the dealer must stand

$$\ll \neg idle \wedge cardready \wedge (standrange(evallow(hand)) \vee \\ standrange(evalhigh(hand))) \rightarrow stand := TRUE \gg$$

The function *standrange* is defined in Fig. 3.1.

− A transition for determining that a new card can be dealt

$$\ll \neg idle \wedge cardready \wedge evallow(hand) \leq 16 \wedge \\ \neg standrange(evalhigh(hand)) \rightarrow hitme := TRUE \gg$$

These transition descriptions are composed into an **asynchronous design** as follows:

$$\ll \neg idle \wedge cardready \wedge evallow(hand) > 21 \rightarrow broke := TRUE \gg \;\|$$
$$\ll \neg idle \wedge cardready \wedge (standrange(evallow(hand)) \vee \\ standrange(evalhigh(hand))) \rightarrow stand := TRUE \gg \;\|$$
$$\ll \neg idle \wedge cardready \wedge evallow(hand) \leq 16 \wedge \\ \neg standrange(evalhigh(hand)) \rightarrow hitme := TRUE \gg$$

Each of these transition descriptions (plus another handful) are needed to complete the design.

Such a design, consisting of a number of asynchronously composed transition descriptions, describes the set of computations that can be obtained by repeated nondeterministic selection and execution of active transitions. Note that this implies that transitions are atomic. This does not prevent a realization from performing several transitions simultaneously, see also Sect. 4.

3.1.9 The Product Combinator. The **product combinator** is used to factor a transition description into a number of simpler transitions. Let t_1, t_2 be the following two transition descriptions (where v_1 and v_2 are different state variables):

$$TRANSITION\; t_1 \ll c_1 \rightarrow v_1 := e_1 \gg$$
$$TRANSITION\; t_2 \ll c_2 \rightarrow v_2 := e_2 \gg$$

The product, $t_1 * t_2$, is equivalent to the following transition description:

$$\ll c_1 \wedge c_2 \rightarrow v_1, v_2 := e_1, e_2 \gg$$

From a theoretical viewpoint the product combinator does not add expressive power, but it is a useful construct for structuring a design description.

As an important special case, consider a transition description consisting of a precondition only, for example: $\ll \neg idle \wedge cardready \gg$. The product combinator is used to factor this precondition out of a number of other transition descriptions, for example, the transition descriptions given in Sect. 3.1.8:

$\ll \neg idle \wedge cardready \gg * ($
    $\ll evallow(hand) > 21 \rightarrow broke := TRUE \gg \;||$
    $\ll standrange(evallow(hand)) \vee standrange(evalhigh(hand))$
        $\rightarrow stand := TRUE \gg \;||$
    $\ll evallow(hand) \leq 16 \wedge \neg standrange(evalhigh(hand))$
        $\rightarrow hitme := TRUE \gg)$

There is a third combinator, +, the synchronous combinator. It is used for describing a design where two or more transitions are done simultaneously. Sect. 7.1 describes an example using synchronous composition.

3.2 Design of a Black-Jack Dealer

This section gives a complete description of the Black-Jack dealer using SYN-CHRONIZED TRANSITIONS introduced in sections 3.1.1–3.1.9. A hand is represented by the sum of the card values (where aces are counted as 1's), and by a flag indicating that the hand has at least one ace. A low and a high score of a hand are computed by the two functions *evallow* and *evalhigh*. The low score is computed by counting aces as 1's, i.e., it is just the accumulated score of the hand; the high score is computed by counting one ace (if there is one) as 11. If either the low or the high score is between 17 and 21 (determined by the function *standrange*) the dealer must stand. The following invariant captures the relationship between the state variables *hand*, *hitme*, *stand*, and *broke*:

$INVARIANT$
    $\neg idle \Rightarrow ($
    $(broke \Rightarrow (evallow(hand) > 21)) \wedge$
    $(stand \Rightarrow (standrange(evallow(hand)) \vee standrange(evalhigh(hand)))) \wedge$
    $(hitme \Rightarrow (evallow(hand) \leq 16 \wedge \neg standrange(evalhigh(hand)))))$

Sections 3.1.1–3.1.9 have explained most of the design description of the Black-Jack dealer. In Fig. 3.1 all the pieces have been put together. Most of the constructs have now been introduced and illustrated with the design description for the Black-Jack dealer. However, one important construct is missing, namely the cell concept which enables a designer to break a design into modules (cells). The cell construct and in particular the parameter mechanism has been strongly influenced by the verification technique that is explained in Sect. 5.

```
STATIC ace = 1
TYPE
   cardval = [1..10] scorerange = [0..26] extscorerange = [0..36]
   handtype = RECORD anyaces: BOOLEAN; count: scorerange END
CELL blackjack(cardready, hitme, stand, broke: BOOLEAN;
      card: cardval; hand: handtype);
   FUNCTION evallow(h: handtype): scorerange RETURN h.count
   FUNCTION evalhigh(h: handtype): extscorerange
      RETURN h.count+IF h.anyaces THEN 10 ELSE 0
   FUNCTION standrange(s: extscorerange): BOOLEAN
      RETURN (16 < s) ∧ (s ≤ 21)
   FUNCTION fourphase(a, b: BOOLEAN): BOOLEAN
      RETURN ((a.pre≠a.post) ⇒ (a.post≠b.post)) ∧
      ((b.pre≠b.post) ⇒ (b.post=a.post))
   STATE idle: BOOLEAN
   INVARIANT
      ¬(broke ∧ stand) ∧ ¬(broke ∧ hitme) ∧ ¬(stand ∧ hitme) ∧ ¬(idle ∧ hitme) ∧
      ¬idle ⇒ (
      ( broke ⇒ (evallow(hand) > 21) ) ∧
      ( stand ⇒ (standrange(evallow(hand)) ∨ standrange(evalhigh(hand))) ) ∧
      ( hitme ⇒ (evallow(hand)≤16 ∧ ¬standrange(evalhigh(hand))) ) ) ∧
      idle ⇒ ( ( ¬(stand ∨ broke)) ⇒
         (evallow(hand)≤16 ∧ ¬standrange(evalhigh(hand))) )
   PROTOCOL
      fourphase(idle, cardready) ∧
      (¬(broke.pre ∨ stand.pre) ⇒ fourphase(cardready, hitme)) ∧
      (¬(hitme.pre ∨ stand.pre) ⇒ fourphase(cardready, broke)) ∧
      (¬(hitme.pre ∨ broke.pre) ⇒ fourphase(cardready, stand))
   INITIALLY
      idle = TRUE hitme = FALSE stand = FALSE broke = FALSE
      hand.anyaces = FALSE hand.count = 0
BEGIN
   ≪ idle ∧ (broke ∨ stand) ∧ ¬cardready →
      hand.anyaces, hand.count, broke, stand:= FALSE, 0, FALSE, FALSE ≫ ||
   ≪ idle ∧ ¬(broke ∨ stand) ∧ cardready → idle:= FALSE ≫ *
   ≪ hand.count := hand.count+card ≫ * (
      ≪ card=ace → hand.anyaces:= TRUE ≫ ||
      ≪ card≠ace ≫ ) ||
   ≪ ¬idle ∧ cardready ≫ * (
      ≪ evallow(hand) > 21 → broke:= TRUE ≫ ||
      ≪ standrange(evallow(hand)) ∨ standrange(evalhigh(hand)) →
         stand:= TRUE ≫ ||
      ≪ evallow(hand)≤16 ∧ ¬standrange(evalhigh(hand)) → hitme:= TRUE ≫ ) ||
   ≪ ¬idle ∧ ¬cardready → idle := TRUE ≫ * (
      ≪ hitme → hitme:= FALSE ≫ || ≪ broke ∨ stand ≫)
END blackjack
```

Fig. 3.1. Black-Jack design in SYNCHRONIZED TRANSITIONS.

3.3 Translation

Designs described with SYNCHRONIZED TRANSITIONS can be manipulated and analyzed in a number of ways in order to do:

1. *Design verification:* Proving that transitions maintain properties formulated as invariant assertions and protocols, see Sect. 5.
2. *Implementation verification:* Proving that a concrete program, the realization, is a correct implementation of another program, the abstraction, see Sect. 6.
3. *Simulation:* Experimenting with a design by executing it with various input data. These tools are not discussed in this chapter.
4. *Synthesis:* Transforming a design into a circuit description, e.g., a netlist or a layout, see Sect. 4.1.

A design description in SYNCHRONIZED TRANSITIONS provides a model of a design. Such a model is theoretically equivalent to a single (finite) state machine, however, from a practical point of view this equivalence is not very useful. The finite state machine corresponding to a design description often becomes too large and without the structure needed to make it manageable in practice. Instead the model encourages design structures consisting of a number of (relatively) independent concurrent state machines. This is utilized in the tools supporting SYNCHRONIZED TRANSITIONS which all take advantage of the possibility of breaking a design description into a number of independent parts.

Although the syntactical details of SYNCHRONIZED TRANSITIONS are of less importance, it is important to understand that many constraints are introduced to allow for a simple syntax directed translation in all supporting tools. The next section describes how a design description can be transformed into a hardware circuit. There is a direct correspondence between the parts of the design description and the corresponding parts of the circuit realization. There is a similar correspondence in the formal verification described in Sect. 5.

4. Modeling Hardware

This section shows how to use SYNCHRONIZED TRANSITIONS for modeling hardware with particular emphasis on integrated circuits. First Sect. 4.1 - 4.2 show how to transform a design description in SYNCHRONIZED TRANSITIONS into a circuit realization, and secondly, Sect. 4.3 illustrates how to model and verify existing circuits.

The primary goal of the transformation shown in Sect. 4.1 is to show a transparent relationship between a concrete physical circuit and the abstract models provided by the notation introduced in Sect. 3. The resulting circuit is not particularly efficient; better results can be obtained by using high-level synthesis techniques [Camp85], however, this is outside the scope of this section. There is a one-to-one correspondence between the design description and the circuit; which is therefore called the **direct realization**. It is possible to give a similar transformation for deriving an executable program (software) from a design description in SYNCHRONIZED TRANSITIONS; in fact, there is a compiler called ST2C doing exactly that.

It is important to stress that the direct circuit realizations described in this section are just one possible realization of the abstract models defined by design descriptions in SYNCHRONIZED TRANSITIONS. For example, the abstract computations of a design such as the Black-Jack dealer are sequences of atomic transitions where the order of concurrent transitions in these sequences is determined nondeterministically. This abstraction does not it require a component for making nondeterministic choices during execution nor does it preclude that transitions are executed simultaneously in a concrete realization (explained further in Sect. 4.2 and in [LoSt92]). The direct circuit realizations described in this section use a traditional two-phase clock. It is also possible to realize designs as self-timed circuits, see [Stau94a].

4.1 Direct Realizations

This section describes how to transform (synthesize) a design description into a particular circuit realization called a **direct realization**. In a direct realization, each transition corresponds to an independent sub-circuit that is only operating when the corresponding transition is active. The key idea of the direct realization is to provide separate circuitry realizing the computation of each transition, and to let a wire (bundle of wires for other types than boolean) correspond to each state variable.

4.1.1 Transitions. Consider a generic transition, t

$$TRANSITION\ t \ll c \rightarrow v := e \gg$$

This is realized as a sub-circuit with three parts:

- A combinational network for computing the precondition $\boxed{c}$,
- a combinational network for computing the values of the right-hand sides of the multi-assignment $\boxed{e}$, and
- a driver for storing the new values of the state variables $\boxed{v:=}$.

The three components of a transition, $\boxed{c}$, $\boxed{e}$, and $\boxed{v:=}$ always follow the same pattern, but the internal function is different for different transitions. Fig. 4.1 shows a direct realization of the transition:

$$\ll \neg\ idle \wedge \neg\ cardready \rightarrow idle := TRUE \gg$$

A logic synthesis program [BHMS84] can be used to generate the combinational logic needed to realize the precondition, c, and the right-hand side, e.

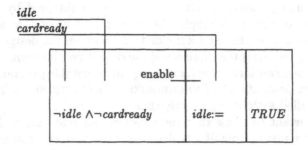

Fig. 4.1. Direct realization of a transition from the Black-Jack dealer

4.1.2 State Variables. State variables are realized as wires carrying the current value of the variable, hence, one can read the value by simply connecting to the wires. Fig. 4.2 illustrates the realization of a single transition, t, connected to the wires carrying the values of the state variables read and written by t.

In addition to the wires, a storage mechanism is required to maintain the current value of the variable. There are a number of ways to realize such a storage mechanism; however, the following two characteristics are important:

- The values of state variables are changed by many different transitions.
- No upper bound can be given on how long the values of state variables must be retained.

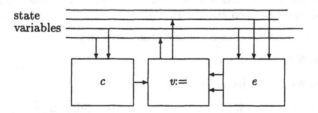

Fig. 4.2. Direct realization of a single transition

The transitions are synthesized as independent sub-circuits; when realizing one particular transition, no knowledge is required of other transitions. This precludes realizations where a state variable is associated with a single transition, because that would require a global analysis of the design description in order to identify all other transitions writing that particular variable. Instead, the value of a state variable is retained on a wire (in case of a boolean variable) or on a number of wires (other types). The capacitance of the wire can hold the value written to the wire for a short period of time; however, some refresh circuitry is needed to keep the value stable between updates.

The exact details of such a **refresh element** depends on which circuit technology is used. Sect. 4.2 shows one possible realization using a traditional two-phase non-overlapping clock.

4.1.3 Combining Transition Realizations. A design with many transitions is realized by synthesizing a sub-circuit like the one shown in Fig. 4.1 for each transition. Each sub-circuit reads some of the state variables and writes new values to one or more state variables. The complete design is realized by connecting all these sub-circuits, i.e., connecting all the wires from all sub-circuits corresponding to a particular state variable. An example is shown in Fig. 4.3.

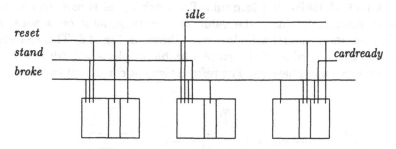

Fig. 4.3. Combining direct realizations

A direct realization usually yields a number of feedbacks caused by the reading and writing of state variables. This may happen directly in a single transition or indirectly in a chain of transitions. Such feedback may lead to hazards in the realization. These potential hazards are avoided in different ways in different technologies. Sect. 4.2 describes how to get a standard clocked realization. There are, however, alternatives such as self-timed realizations [Stau94a].

4.2 Clocked Realizations

Clocked circuitry is based on a periodic signal distributed to all sub-circuits and used for activating them simultaneously. In a direct realization there is an independent sub-circuit corresponding to every transition and the clock is connected to all of these. The execution of transitions consists of two phases: the first is the evaluation of preconditions and right-hand-side expressions, the second is assignment of new values to the state variables appearing on the left-hand side of active transitions. The two phases are implemented by letting every other clock signal, called φ_1, activate the evaluations and letting the other half, called φ_2, activate the assignments (see Fig. 4.4).

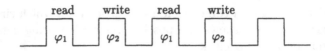

read   write   read   write

φ_1 φ_2 φ_1 φ_2

Fig. 4.4. Two-phase clock

When a design is realized as described above, the preconditions and right-hand sides of all transitions are evaluated in parallel. Strictly speaking, this is only a correct realization of a subclass of designs meeting certain restrictions such as disallowing that two different active transitions write different values to the same state variable. In [LoSt92] this class is defined and it is shown that the realization is sound. The direct realization shown above can also be used to realize synchronous designs such as the ones discussed in Sect. 7.

4.2.1 A Clocked Refresh Element. The clock signal is used to control a refresh element maintaining the value of a state variable on a wire. A schematic diagram of the refresh element is shown in Fig. 4.5. This element reassigns the current value of the state variable in phase φ_1, while a new value can be written in phase φ_2. The refresh element consists of two cascade

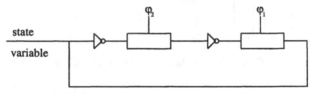

state
variable

φ_2 φ_1

Fig. 4.5. Dynamic refresh element

coupled clocked inverters. Further details are given in [Chri90], which also describes a prototype tool that automatically generates a VLSI layout (using the direct realization) from a design description.

4.3 Modeling Existing Circuits

This section illustrates how to model an existing hardware design (circuit). SYNCHRONIZED TRANSITIONS is not restricted to modeling circuits at a particular level, and it can be used to create both low level models such as a transistor network and high level models at the structural or behavioral level. This is illustrated by the two examples shown below. In [LeGS93] it is, for example, demonstrated how verification of a gate-level design revealed design errors that had not been captured by extensive simulations. In [NiSt95] it is described how a transistor level model of a RAM design is used to verify that the circuit is speed-independent.

4.3.1 A Gate Level Model. It is straightforward to use SYNCHRONIZED TRANSITIONS to describe a netlist of simple gates. This is illustrated by the full adder shown below (assuming that only two-input gates are available):

CELL full_adder(a, b, s, ci, cout: BOOLEAN)
BEGIN
    ≪ *s:= XOR(XOR(a, b), ci)* ≫ ||
    ≪ *cout:= OR(OR(AND(a, b), AND(a, ci)), AND(ci, b))* ≫
END full_adder

Such structural descriptions corresponding to a netlist can be given in all hardware description languages, e.g., VHDL and Verilog. By using SYNCHRONIZED TRANSITIONS it becomes possible to formally verify the design description. This has been done on a number of examples where an existing netlist has been transformed into the required syntax (a very simple transformation), for example [LeGS93]. Another example is the verification [Mell94] of the adder given in [ChMi89] (very similar to the efficient Brent-Kung adder).

4.3.2 A Transistor Level Model. This section illustrates a design on the transistor level that is part of a self-timed RAM [NiSt95]. This design description was used to verify the speed-independence of the transistor network making up the core of the RAM design.

The first transition describes the inactivation of the precharge signal (note that $\overline{Prech}$ means active low), and the second the activation. The third and the fourth transitions describe the activation and inactivation of the select signal, respectively.

Precharge:
    ≪ *Read* ∨ *Write* → $\overline{Prech}$*:= TRUE* ≫
    ≪ ¬*Sel_{ack}* ∧ ¬*(Read* ∨ *Write)* → $\overline{Prech}$*:= FALSE* ≫
Select:
    ≪ *Adr_i* ∧ ¬$\overline{Prech}$ → *Sel_i:= TRUE* ≫
    ≪ ¬*Adr_i* → *Sel_i:= FALSE* ≫

4.4 Evaluation

The direct realization is very simple compared to the sophisticated techniques used in state-of-the-art synthesis tools. In many cases a direct realization is not efficient because a separate copy of the necessary hardware is generated for each transition. In high-level synthesis tools [Camp85], advanced allocation and scheduling methods are capable of multiplexing the same hardware units for many different purposes. These methods could also be used for synthesizing efficient circuits from design descriptions in SYNCHRONIZED TRANSITIONS, although no tools are currently available for doing this.

5. Verification Technique

This section describes a technique for formally verifying the invariants and protocols of a design called **the localized verification technique**. As mentioned in the introduction, the formal verification technique is exhaustive by its very nature; if a certain invariant or protocol has been verified the computation will *never* enter a state violating the invariant or perform a transition violating the protocol. This exhaustiveness should not be misinterpreted as an absolute guarantee against malfunctioning; it is only the requirements captured by the invariants and the protocols that are verified. If these are very weak, one can not make very strong conclusion from a successful verification. This incompleteness also has significance for the verification techniques and tools described in this section. They have been developed with an emphasis on ease of use and quick feedback to a designer (at the expense of completeness).

The verification technique is sketched in Fig. 5.1, it is based on transforming the design description (in SYNCHRONIZED TRANSITIONS) into a number of verification conditions that are subsequently checked. The verification is based on exactly the same design description that is used for other transformations such as synthesis (described in Sect. 4.1) or simulation. The approach is based on doing the verification mechanically, and as indicated in the figure there are several alternative tools, called **proof engines** in the figure. These are mechanical tools capable of proving the verification conditions. Some proof engines are fully automatic using a model checking [McMi93a] approach others are based on theorem proving [BoMo79] which in some cases require manual assistance from the the designer.

Since formal verification is in principle an exhaustive task, one has to be cautious and avoid introducing techniques with an excessive execution time. In order to investigate this aspects and to be able to experiment with alternatives the approach allows one to use several different proof engines. Almost all the results reported in this chapter are based on using a theorem prover (called LP [GaGu89]) as the proof engine. The details and peculiarities of the theorem prover are not important for understanding the approach. Sect. 5.2 gives a brief overview of the theorem prover as well as a few alternatives.

Although the exact details of the theorem are of less significance, the approach is heavily influenced by the decision of using a mechanized tool. A primary goal has been to reduce the **verification effort** needed to formally verify a design description. No attempt is made to formalize the notion of verification effort here, but informally it is a measure of the number and complexity of the verification conditions (implications) needed to verify a particular design. The key contribution of the approach described in this chapter is the localized verification technique utilizing the structure of a modular design [StGG89] where the verification effort is proportional to the (textual) size of the design description. This is described in further detail in Sect. 5.3.

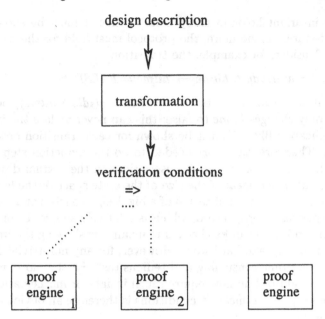

Fig. 5.1. Overview of the verification technique

5.1 Inductive Verification

The localized verification technique for design verification is based on an induction technique proposed by Floyd for verifying loop invariants [Floy67]. Informally, the technique consists of showing that each transition description only leads to states where the invariant holds provided it is started in states where it holds. By also showing that the invariant holds in the initial state one may conclude that the invariant and protocol hold throughout the computation. This is inherently an induction proof (over the length of the computation) where showing that the invariant holds in the initial state corresponds to the base case, and the verification of all transition descriptions corresponds to the inductive step. The localized verification technique consists of verifying each transition description and the initial state. There is, however, no need to do the induction explicitly for each design description.

More formally, assume that I is an invariant, P a protocol, and that t is a transition description of a design. Now $I(pre)$ and $I(post)$ holds if the invariant is satisfied in the pre-state and post-state respectively. Similarly, $t(pre, post)$ holds if the transition t can make a state change from the state pre to the state $post$. Finally, $P(pre, post)$ holds if the protocol P is satisfied by the pair $(pre, post)$.

Definition 5.1. *The transition t meets the invariant and protocol if,*

$$LC : \quad I(pre) \wedge t(pre, post) \Rightarrow I(post) \wedge P(pre, post)$$

i.e., if the invariant holds in the pre-state, then it must be shown to hold in the post-state. Furthermore, the protocol must hold for the pair of states *pre, post*. Consider, for example, the transition

$$\ll \neg idle \wedge \neg cardready \wedge hitme \to hitme := FALSE \gg$$

This transition certainly meets the invariant: $\neg(idle \wedge hitme)$, because the transition only changes *hitme* to false; this can never violate the invariant.

An implication like LC must be shown for each transition description in the design. Therefore, the effort needed to do the induction step is proportional to the number of transition descriptions in the textual description of the design, but *independent* of the size of the state space or the length of the computation. Hence, the verification of a big design consists of a (large) number of independent steps, each of which is relatively simple. In principle the verification conditions can be checked manually by showing the implications using paper, pencil, and hard work. However, for any non-trivial design this would be a very time consuming and tedious task. Furthermore, most of the verification conditions do not require much insight or mathematical ingenuity, and showing the verification conditions is therefore an obvious candidate for automation.

5.2 Mechanization

This section describes how to mechanize the localized verification technique described in Sect. 5.1 (and its generalization to modular designs described in Sect. 5.3) where the verification of a design is done in a number of *independent steps*, one for each transition. A large design is handled by a number of (relatively) simple proofs instead of a few large and complicated ones. This is illustrated in Fig. 5.1; For each transition description t the translator generates a proof obligation like the implication LC.

To mechanize the verification, one needs a proof engine that can handle the verification conditions (implications). In the initial phases of this work we experimented heavily with the exact form of the verification conditions and the general purpose theorem prover LP [GaGu89] was a very convenient proof engine. LP is a theorem prover for a subset of multi-sorted first-order logic. It is designed to work efficiently on large problems and to be used by relatively naive users. LPs design is based on the assumption that initial attempts to state conjectures correctly, and then to prove them, usually fail. As a result, LP is designed to carry out routine (and possibly lengthy) steps in a proof automatically and to provide useful information about *why* proofs fail, if and when they do. Most of our verifications still uses this theorem prover, however, now that the verification technique has stabilized, it would no doubt be more efficient to develop a specialized proof engine optimized towards our application. We have conducted experiments with a BDD based prototype tool that confirms this.

The choice of a first-order logic based proof engine has not been significant for the overall approach. The axiomatization of conjectures and facts might be more elegant in a higher-order logic, however, the emphasis has been on providing efficient machine assistance, and LP has this far satisfied the needs. The translator from SYNCHRONIZED TRANSITIONS to LP, called ST2LP, incorporates several years of experience in exploiting LP's capabilities (and limitations) to obtain efficient proofs with a minimum of manual assistance. Furthermore, ST2LP extracts appropriate declarations for state variables, axioms defining transitions, protocols, invariants, and verification conditions from design descriptions. For example, when formalizing invariance proofs, the translator generates an LP command to initiate a proof by cases on the transition selected for execution. As a result of these idioms, LP usually verifies designs directly from the output of the translator with *little or no manual assistance*. The amount of manual assistance needed depends on the domain (types of state variables). For designs involving only booleans, manual assistance is never required, whereas a design using unbounded integers, arithmetic and inequalities would often require manual help to supplement the standard axioms.

5.2.1 Availability of Tools. All the tools are written in C and runs under Unix. They are available free of charge by anonymous ftp from the Technical University of Denmark: ftp.it.dtu.dk, or by contacting the author: jst@it.dtu.dk. In addition to the tools, one can get a complete up-to-date reference manual and other documentation.

LP is written in CLU and is available free of charge, and without a license, by anonymous ftp from larch.lcs.mit.edu. Prospective users can retrieve an executable version of LP, along with a supporting run-time library containing sample axiom sets and proofs, for DECstations, Sparcstations, MIPS machines, Linux, or Sun workstations, and DEC Alpha. Source code is also available.

5.3 Modular Designs

It is rarely feasible to view a non-trivial design as one flat collection of transitions without any structure. In SYNCHRONIZED TRANSITIONS a design may be modularized by breaking it into parts, each such part is called a **cell**. For example, the Black-Jack dealer is a cell. The Black-Jack dealer takes care of the game rules, etc., and it relies on a deck of cards that supplies (legal) cards at the right time. The deck of cards is another cell that works independently of the Black-Jack cell. The dealer and the deck of cards communicates through shared variables, e.g., the dealer signals that it wants a new card using the state variable *hitme*, and the deck of cards signals that a new card is ready using the state variable *cardready*. The interface consisting of shared variables is illustrated in Fig. 5.2. Breaking down a large design into cells is not only an aid when describing the design; it also makes the design easier

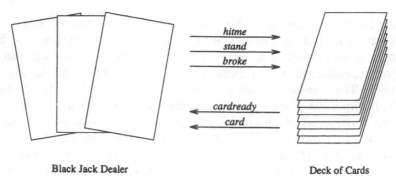

Black Jack Dealer   Deck of Cards

Fig. 5.2. Shared variable interface.

to verify! This section presents a localized technique for verifying modular designs. The technique was originally proposed in [StGG89] and it has been further developed in [StMe95]. Each cell of the design is verified in isolation, showing what is called local correctness. Furthermore, it is verified that each cell does not violate the protocols and invariants of other cells. This is called non-interference, and it is verified solely on the basis of the interface of the cells, without considering their internal details. The localized verification technique is illustrated on the Black-Jack dealer extended with a deck of cards cell.

The deck of cards cell must supply cards as the dealer is requesting them. This is the case if the dealer needs more cards for the hand (signaled with the *hitme* state variable), or when a game has finished and a new game must be started (signaled with the state variables *stand* and *broke*). The deck of cards cell uses the state variable *cardready* to tell when a new card is ready. The deck of cards cell may potentially break the four-phase protocol stated in the Black-Jack dealer cell because it writes the state variable *cardready*. It is now shown how localized verification is used to verify that this is not the case.

5.3.1 Local Correctness. Local correctness is one part of the localized verification technique, where a cell is verified independently of its context. To do this, it is shown that all transitions of a cell maintain the invariant and protocol of the cell. For the Black-Jack dealer cell this corresponds to the verification condition explained in Sect. 5.1. This is called the **local correctness condition.**

The state variables in the two cells are shared; this allows the cells to communicate. But it also means that the transitions of one cell can violate the invariant and protocol of the other cell (interference). For example, the four-phase protocol between *hitme* and *cardready* is verified in the local correctness proof in the Black-Jack cell (where *hitme* is changed), but the deck of cards cell writes *cardready*, and this might violate the four-phase protocol. Hence, local correctness is not sufficient to guarantee global correctness. In addition, it must be shown that cells do not interfere.

5.3.2 Non-interference. Non-interference is the complementary part of the localized verification technique. In this part, it is verified that invariants and protocols involving shared variables (parameters) are preserved. To verify that all invariants and protocols are maintained, it must be shown that the transitions in the deck of cards cell do not interfere with the invariants and protocols in the Black-Jack cell, and it must also be shown that the transitions in the Black-Jack cell do not interfere with the invariants and protocols in the deck of cards cell.

Instead of showing non-interference for *each transition* separately, it is done *once for each cell instantiation*. The key idea is to assume that the internal protocol, the invariant and the protocol of, e.g., the deck of cards cell capture the essential properties of all transitions in a *single* assertion. This is expressed in the following two **non-interference conditions**:

$$I_{deck}(post) \wedge P_{deck}(pre, post) \wedge IP_{deck}(pre, post) \Rightarrow I_{bj}(post) \wedge P_{bj}(pre, post)$$

$$I_{bj}(post) \wedge P_{bj}(pre, post) \wedge IP_{bj}(pre, post) \Rightarrow I_{deck}(post) \wedge P_{deck}(pre, post)$$

(I, P, and IP are the invariant, protocol and internal protocol in the Black-Jack cell (subscript bj) and deck of cards cell (subscript $deck$)). Note that the non-interference conditions avoid considering the individual transitions of the Black-Jack dealer and deck of card cells. For example, it is shown from the internal protocol of the deck of cards cell that this cell does not interfere with the four-phase protocol (P_{bj}). This is shown without considering the transitions of the deck of cards cell; as long as they comply with the internal protocol of the cell (this is verified as local correctness), the deck of cards cell will not violate the four-phase protocol.

The cell instantiations of a design form a tree. It is not necessary to verify the non-interference conditions between any two cells in the design, but only between "neighbor" cells in the instantiation tree. By showing the two non-interference conditions for all the neighbor pairs it is also (implicitly) shown that changes made in accordance with the protocol for one neighboring cell are also in accordance with the protocols for other neighbors. Therefore it circumvents the need to consider separately interferences that might result when parameters are passed through several levels. In [Mell94] the technique is explained in more detail and it is shown to be sound.

5.4 Evaluation

By using the localized verification technique each transition description is only verified once no matter how many times it is instantiated. Each line of a design description gives rise to zero (declarations, headers, etc.), one (local invariance for transitions), or two implications (non-interference of cells). This is the justification for the claim that *the verification effort grows linearly with the size of the textual design description*. For recursively defined cells, the effort needed to show non-interference is independent of the recursion depth, because the recursive instantiation of a cell yields just the two non-interference implications.

However, the efficiency of the localized verification has a price. First of all, the technique is not complete. One can easily construct an example of a correct design where it is not immediately possible to show the required implications. In practice this does not seem to be a unsurmountable problem (because the designer can strengthen the invariant). This is documented by the large number of examples that have been verified [Stau94a, Mell94]. A more significant practical problem is inherent in the inductive approach on which the technique is based. It is based on showing implications such as:

$$I(pre) \wedge t(pre, post) \Rightarrow I(post) \wedge P(pre, post)$$

Note that the invariant, I, appears both as an assumption and as a conclusion. This means that one has to find the right balance when stating an invariant. If it is made very strong ($I(pre)$ very restrictive), it means that $I(post)$ also becomes very strong and hence difficult to prove. On the other hand making the assumptions very weak, can make it difficult to conclude that $I(post)$ holds. Finding the right balance is not always easy. To illustrate this, assume that the designer for some reason only wants to verify the invariant $\neg(broke \wedge stand)$, i.e. that *broke* and *stand* are not simultaneously true. In order to verify this using localized verification, it is necessary for the designer to formulate a stronger invariant, e.g., the invariant stated in Fig. 3.1. Identifying and formulating this is often a significant part of the verification effort. Hence, the linear growth of the verification effort has its price, namely an added effort by the designer.

5.5 Liveness Properties

Both invariants and protocols are examples of safety properties, i.e. requirements stating that something "bad" will *not happen*. There is also considerable interest in liveness properties stating that some desirable thing *will eventually happen*. An example of the latter would be the obvious requirement to the Black-Jack dealer that when presented with a new card, it will sooner or later indicate whether it is broke, wants a new card (*hitme*) or stands.

It is currently not possible to state such liveness properties using the notation and tools described in this chapter. There is nothing inherent in the approach preventing this and in UNITY, that in many respects is very similar, it is possible to state and verify liveness properties. However, the syntax currently supported by the SYNCHRONIZED TRANSITIONS tools does not include constructs for stating and verifying liveness.

The most important reason why the tools do not handle liveness is to simplify the verification procedure. Formal and mechanized verification of safety properties is already a challenge in particular for larger examples than those shown in this book.

One may well ask: "what is the point of verifying some safety properties, if it cannot be ensured that the final realization will perform even a single step of the specified computation?" Two answers are offered: First, there are many important and desirable properties that are not verified formally, for example, power consumption, timing, and noise sensitivity. Presently, liveness of designs in SYNCHRONIZED TRANSITIONS is among these. Secondly, a safety proof ensures that if a realization ever delivers an output, then it meets the verified properties. This is a significant guarantee. In practice, many design errors manifest themselves as circuits giving the wrong answer rather than as not giving any answer at all.

6. Refinement

As stated in the introduction SYNCHRONIZED TRANSITIONS provides a notation for describing an abstract model of a computation. In order to construct an efficient realization, it is usually necessary to refine the abstraction into a concrete realization. It is, for example, helpful to use data types such as sets, list, tables, reals, and integers in an abstract design, but these must be refined into a concrete data representation in the realization. This refinement typically includes a number of restrictions enforced by physical limitations and other practical constraints. It is, for example, necessary to restrict integer values to a certain range in order to represent the values in a fixed number of bits. This section describes a formal approach to refinement enabling a designer to rigorously verify that a concrete realization correctly implements a given abstract model. The approach requires that both the abstraction and the realization are described in SYNCHRONIZED TRANSITIONS and they are called the **abstract design** and the **concrete design**, respectively. The Single-Pulser (see Appendix) is used as an illustrative example. Its observable behavior is specified as follows [JoMC94]:

> The Single-Pulser has one (boolean) input, i, and one output, o (boolean). For each pulse received on i, a single unit-time pulse is emitted on o.

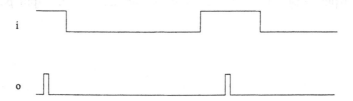

An abstract design describes the computation with as few constraints as possible. This gives the implementor maximal freedom in choosing an efficient realization. In the Single-Pulser example, the abstract design allows for the widest possible choice of when to emit the output pulse. In this particular example, it turns out that the abstract design is a bit more complex than the concrete; usually it is the other way around.

6.1 Design of the Single-Pulser

In the design description of the Single-Pulser behavior given below the input i and output o are state variables. The design allows the output pulse to appear anywhere in the neighborhood of the input pulse. Neighborhood is the interval from just after[1] i has changed to high and until just after i has changed to low. Another possible interpretation of neighborhood would be to allow the output pulse to occur also after i becomes low (but before it turns high the next time). Both choices are possible and an important aspect of any abstraction is to document such choices in order to analyze their consequences. So the abstraction makes an important design decision surface. If one had gone directly to the realization, the designer might not have realized that he had a choice. In this case the first interpretation is used. This is a somewhat arbitrary decision, it would be equally easy to describe the other behavior.

The abstract design leaves a range of possible realizations open. In the Single-Pulser there are, for example, a range of correct choices for when to emit the output pulse. In the concrete design one of the possible realizations is chosen. Therefore, a concrete design may not exhibit all the behaviors of the abstract design. Informally, the **behavior** of a design is the set of computations that are externally visible, therefore, changes to local (internal) state variables are not directly reflected in the behavior of a cell. A more rigorous definition of behavior is given in [Mare95].

6.1.1 Abstract Design of the Single-Pulser. This section describes an abstract design of the Single-Pulser. This design allows the output pulse to appear anywhere in the neighborhood of the input pulse (the interpretation of neighborhood is discussed above). The design has two internal state variables x and p. Informally, x keeps track of the last value of i and p indicates whether an output pulse has been generated for the most recent input pulse.

---

[1] A precise definition of "just after" would require a model of time. For example, in a traditional clocked model, it could be within one clock tick.

CELL abstract_pulser(i, o: BOOLEAN)
  STATE x, p: BOOLEAN
  INITIALLY i = FALSE x = FALSE o = FALSE p = FALSE
  INVARIANT p $\Rightarrow$ x
BEGIN
  $\ll$ *x:= i* $\gg$ * (
  ($\ll$ *o $\rightarrow$ o:= FALSE* $\gg$ *
    ($\ll$ *$\neg i \wedge p \rightarrow p:= FALSE$* $\gg$ || $\ll$ *$\neg(\neg i \wedge x)$* $\gg$)) ||
  ($\ll$ *$\neg o$* $\gg$ *
    (($\ll$ *$\neg i \wedge x \rightarrow p:= FALSE$* $\gg$ *
    ($\ll$ *$\neg p \rightarrow o:= TRUE$* $\gg$ || $\ll$ *p* $\gg$)) ||
    ($\ll$ *$\neg(\neg i \wedge x)$* $\gg$ *
    ($\ll i \wedge \neg p \rightarrow o, p:= TRUE, TRUE$ $\gg$ || $\ll$ *TRUE* $\gg$)))))
  END abstract_pulser

Note that the transition $\ll$ *x:= i* $\gg$ is a common factor "multiplied" to all other transitions. Therefore, *i* is copied to *x* in every step of the computation. The design description consists of two cases, one where *o* is true, and one where it is false. When *o* is true, it is always set to false immediately and depending on whether $\neg i \wedge p$ holds, *p* is set to false. The different cases for *o* being false can be traced in a similar way.

6.1.2 Realization of the Single-Pulser. This section describes two of the many possible realizations of the Single-Pulser, called *concrete_pulser1* and *concrete_pulser2* [JoMC94]. The difference is:

1. *concrete_pulser1*: emits a single pulse at the earliest possible time (after *i* goes high), and
2. *concrete_pulser2*: emits a single pulse just after *i* goes low.

CELL concrete_pulser1(i, o: BOOLEAN)
  STATE x: BOOLEAN
  INITIALLY i = FALSE x = FALSE o = FALSE
  INVARIANT o $\Rightarrow$ x
BEGIN
  $\ll$ *o, x:= i $\wedge \neg x$, i* $\gg$
END concrete_pulser1

CELL concrete_pulser2(i, o: BOOLEAN)
  STATE x: BOOLEAN
  INITIALLY i = FALSE x = FALSE o = FALSE
  INVARIANT o $\Rightarrow \neg x$
BEGIN
  $\ll$ *o, x:= $\neg i \wedge x$, i* $\gg$
END concrete_pulser2

Both of these are correct realizations in the sense that they only exhibit behaviors that could also be exhibited by the abstract design. The next section describes a technique for verifying (formally) that this is the case.

6.2 Formal Verification of Refinement

This section present a formal notion of refinement and a mechanized technique for verifying that one design is a refinement of another.

Informally, the definition of refinement requires that any behavior exhibited by the concrete design could also be exhibited by the abstract. Note that it is not necessary that it can exhibit *all* behaviors.

Definition 6.1. *A concrete design is a **refinement** of an abstract design if the set of behaviors of the concrete design is a subset of the behaviors of the abstract design.*

Using this definition implies that the concrete design can be substituted for the abstract in any environment where the abstract is used, see also [Stau94a, AbLa88]. However, the definition does not directly suggest a useful way of verifying refinement because it requires that all computations of both the abstract and concrete designs are described in a form that makes it possible to show that one is a subset of the other. Below a more practical verification technique for showing refinement is presented; as is the case with the other verification techniques described in this chapter, it represent a compromise between completeness and practical feasibility.

6.2.1 Refinement Mappings. To be a refinement, the concrete design must resemble the abstract one. Typically, the representation of data is different, the abstract design could for example be expressed using integers, whereas the concrete design uses bit-vectors. This relationship is captured by an **abstraction function**.

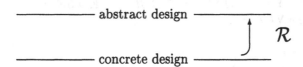

Formally, an abstraction function is a mapping from the concrete state space, S_c, to the abstract state space, S_a.

In the Single-Pulser example, the refinement mappings for both realizations maps the concrete instances of i, o, and x to the corresponding abstract variables. The state variable p is mapped as follows:

abstract_pulse - concrete_pulser1: $p \leftarrow x$

abstract_pulse - concrete_pulser2: $p \leftarrow FALSE$

6.2.2 The Refinement Condition. To show that a concrete design is a refinement, it must be shown that any behavior exhibited by the concrete design is also a possible behavior of the abstract design. This is ensured if the initial state of the concrete design is mapped to an initial state of the abstract, and if every execution of a concrete transition, t_c, either is invisible or there is a corresponding abstract transition, t_a, with a similar effect. This is formalized by the **refinement condition**.

Definition 6.2. *Two designs, D_a and D_c, meet the refinement condition, if the following implication holds for all transitions, t_c, of D_c*

$$t_c(pre_c, post_c) \Rightarrow \mathcal{R}(pre_c) = \mathcal{R}(post_c) \vee \exists t_a \in D_a : t_a(\mathcal{R}(pre_c), \mathcal{R}(post_c))$$

and if

$$\mathcal{R}(init_c) \cong init_a$$

The first requirement is met if each transition in the concrete design, t_c, has a corresponding transition, t_a, in the abstract design, i.e., if t_c can change the state of the concrete design from pre_c to $post_c$, then t_a must change the abstract state from $\mathcal{R}(pre_c)$ to $\mathcal{R}(post_c)$ or if there is no visible change of the abstract state $\mathcal{R}(pre_c) = \mathcal{R}(post_c)$. The second requirement forces the concrete variables to be initialized as prescribed by the abstract design. More precisely, the symbol $\cong$ means that all state variables, v, that are given an initial value in the abstract design must be given a corresponding initial value in the concrete design.

Typically, a concrete design includes details that are not in the abstract design, for example, extra internal state variables such as the carry bits in an adder or the internal state of a finite state machine. Changes in these are not reflected at the abstract level, which means that $pre_c \neq post_c$, but $\mathcal{R}(pre_c) = \mathcal{R}(post_c)$.

6.2.3 Mechanization. As is the case with verifying invariants and protocols, most refinement proofs are quite tedious and they are therefore candidates for mechanization. The refinement condition suggests a verification technique which is very similar to the one used for design verification, i.e., breaking the verification into a number of independent steps, one for each transition (of the concrete design). Each step consists of showing an implication (defined in the refinement condition). The mechanization is very similar to the one used for design verification; a translator STREF [Mare94] generates the necessary verification conditions for showing the refinement condition given two designs and an abstraction function. These tools have been used to verify a number of examples [Mare95] including the Single-Pulser, see Sect. 8.2.

6.3 Evaluation

The refinement condition is only a sufficient condition. There are examples of designs where one is a refinement of the other but the refinement cannot be verified with the technique described here. It is formulated this way to allow for a verification by many (small) steps; one for each transition in the concrete design. It is possible to formulate weaker conditions, however, for practical verification, the strength of the condition is only one aspect; the ease of use and possibilities for mechanical support are equally important aspects. This has clearly influenced the formulation of the refinement condition given in Sect. 6.2.2.

By defining refinement as "subset of behaviors" one gets the nice property that in any environment it is possible to replace an abstract design by one of its refinements. If it is a refinement, the concrete design only exhibits behaviors that the abstract design also has, hence the environment cannot tell any difference between the two. However, in many cases it seems too restrictive to capture the notion of a correct implementation.

7. Synchronous Designs

This section describes constructs for designing synchronous computations. The designs presented so far have all been asynchronous computations and the nondeterminism inherent in these often leads to simple and succinct descriptions. However, there are also examples where the asynchronous approach is awkward, e.g., when describing computations with components that operate in lock step as for example systolic algorithms. Therefore, SYN-CHRONIZED TRANSITIONS also has constructs for describing **synchronous designs**. Syntactically, these are similar to asynchronous designs and all the constructs introduced in sections 3. and 5.3 can also be used for describing synchronous designs, for example, transitions, functions, and cells.

The 1Syst from the common benchmarks provides a nice illustration of a synchronous design.

7.1 The 1Syst

The 1Syst example (FIR filter) is a simple, but typical, example of a synchronous design. It produces a sequence of output data, $Y = y_1, y_2, \ldots, y_i, \ldots$ from a sequence of input data, $X = x_1, x_2, \ldots, x_i, \ldots$ by computing an inner product of the form:

$$y_i = \sum_{j=1}^{j=k} a_j \cdot x_{i+j-1}$$

The order k is fixed. The filter computation is described as a linear array

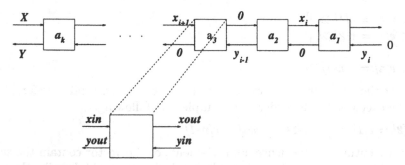

```
CELL multacc(xin, xout, yin, yout: element)
  STATE a: element
BEGIN
  ≪ yout:= yin + a*xin ≫
  ≪ xout:= xin ≫
END multacc
```

Fig. 7.1. Structure of 1Syst

of simple processing elements (see Fig. 7.1). In every other iteration, a new value of X is inserted in the array (at the point indicated with X in Fig. 7.1). In the remaining iterations the value 0 is inserted. Similarly, a new element of the output sequence, Y, is computed in every other iteration (the output appears at the point indicated with Y in Fig. 7.1). At the point indicated with the rightmost 0 a zero value is input in every iteration and all cells compute a multiplication and accumulate the result. Furthermore, the x-value is propagated which makes the sequence of inputs, X, move from left to right. Each cell receives an element of X and a partially computed element from Y. It propagates the x-value to the right and the updated y-value to the left.

It is awkward to describe the 1Syst as an asynchronous design; instead Sect. 7.3 shows how to do it using synchronous composition which is introduced in the next section.

7.2 The Synchronous Combinator

This section introduces the synchronous combinator, $+$. It plays a role similar to the asynchronous combinator, $\|$. To motivate the use of the synchronous combinator, consider a simple shift register for shifting n-bit vectors.

```
TYPE vector = ARRAY [1..n] OF BOOLEAN
STATE v: vector
```

The bit vector, v, must act like a shift register, i.e., $v[2]$ gets the value of $v[1]$, $v[3]$ the value of $v[2]$, etc. How is such a shift register described in SYNCHRONIZED TRANSITIONS? One may suggest the following design, but it is not correct:

$\ll v[2]:= v[1] \gg \|$
$\ll v[3]:= v[2] \gg \|$
...
$\ll v[n]:= v[n-1] \gg$

because the $n-2$ transitions are composed asynchronously and therefore, they can be executed in any order, for example, the following:

$v[2]:= v[1] \; v[3]:= v[2] \ldots v[n]:= v[n-1]$

After executing this sequence, all n elements of the vector contain the same value which was not the intention. Alternatively, one could describe the shift register as follows:

$\ll v[2], v[3], \ldots, v[n]:= v[1], v[2], \ldots, v[n-1] \gg$

This describes the correct computation, because the value of $v[1]$ is assigned to $v[2]$, $v[2]$ to $v[3]$, etc; but syntactically it is clumsy and not easy to generalize to an arbitrary value of n. A better solution is obtained by using the the **synchronous combinator**, $+$, that describes computations where a number of transitions are executed simultaneously, for example:

$\ll v[2]:= v[1] \gg +$
$\ll v[3]:= v[2] \gg +$
...
$\ll v[n]:= v[n-1] \gg$

This describes a computation where the $n-1$ values, contained in $v[1]$, $v[2]$, ..., $v[n-1]$ are simultaneously transferred to $v[2]$, $v[3]$, ..., $v[n]$. In general, all the expressions on the right-hand-side expressions of *all* transitions are evaluated, and these values are assigned to the state variables appearing on the left-hand side of the assignments.

7.2.1 Exclusive Write. In order to avoid conflicting assignments, synchronous composition is only defined if the variables on the left-hand sides of all assignments are distinct. This is expressed more precisely by the exclusive write condition given below.

To compose a set of transitions, $t_1, t_2, \ldots, t_n$, synchronously, the set must meet the **exclusive write condition, EW**.

Definition 7.1. *The transitions $t_1, t_2, \ldots, t_n$ meet the Exclusive Write Condition if*

$$\forall i, j \in [1..n] : W^{t_i} \cap W^{t_j} \neq \emptyset \Rightarrow \neg(c_i \wedge c_j)$$

W^{t_i} is the write set of transition t_i, i.e., the set of variables appearing on the left-hand side of its multi-assignment and c_i is the precondition of t_i. The exclusive write condition requires that transitions with intersecting write sets are not enabled simultaneously.

The shift register meets the exclusive write condition because all the variables appearing on the left-hand sides are distinct, which means that all write sets are disjoint ($W^{t_i} \cap W^{t_j} = \emptyset$).

7.2.2 Computational Model. The synchronous composition of the n transitions: $t_1 + t_2 + \ldots + t_n$ describes an iterative computation where each iteration consists of two phases. During the first phase, all the n preconditions, c_i, and right-hand side expressions, e_i, are evaluated (in the pre-state). In the second phase, all the variables, v_i, on the left-hand side of the enabled transitions are updated with the new values specified by the corresponding expressions e_i. The preconditions and the right-hand side expressions are evaluated *before* any of the state variables are changed.

To verify a synchronous design, it is necessary to modify the verification technique sketched in Sect. 5.1. Further details are given in [Mell94].

7.2.3 Relating Synchronous and Asynchronous Composition. From a theoretical point of view, synchronous composition does not add any additional expressive power. It is possible to translate any design description using synchronous composition into an equivalent form using asynchronous composition only. Consider, for example, the synchronous composition of the three transitions: $t_1 + t_2 + t_3$. The following asynchronous design describes the same computation:

$$
\begin{aligned}
&\ll c_1 \,\wedge\, c_2 \,\wedge\, c_3 \rightarrow v_1, v_2, v_3 := e_1, e_2, e_3 \gg \| \\
&\ll c_1 \,\wedge\, c_2 \,\wedge\neg c_3 \rightarrow v_1, v_2 := e_1, e_2 \qquad\quad \gg \| \\
&\ll c_1 \,\wedge\neg c_2 \,\wedge\, c_3 \rightarrow v_1, v_3 := e_1, e_3 \qquad\quad \gg \| \\
&\ll c_1 \,\wedge\neg c_2 \,\wedge\neg c_3 \rightarrow v_1 := e_1 \qquad\qquad\qquad \gg \| \\
&\ll \neg c_1 \wedge c_2 \,\wedge\, c_3 \rightarrow v_2, v_3 := e_2, e_3 \qquad\quad \gg \| \\
&\qquad\qquad\qquad \ldots \\
&\ll \neg c_1 \wedge \neg c_2 \,\wedge\neg c_3 \gg
\end{aligned}
$$

This way of describing a computation is clumsy and impractical for larger designs. So although the asynchronous combinator is sufficient in theory, the synchronous combinator is needed to make practical designs tractable.

The iterative nature of the synchronous model corresponds to repeated selection in the model of asynchronous designs. However, in the synchronous model there is no nondeterministic selection of the transition to execute, instead *all* enabled transitions are executed. In particular, transitions that are always enabled, e.g., because their preconditions are *TRUE*, will be executed in every iteration. Synchronous designs are used to model synchronous circuits controlled by a global clock and each iteration corresponds to a clock cycle.

7.3 Quantification

Fig. 7.2 shows the complete design of the filter parameterized by the value of k. This design description uses a new construct, called **quantified instantiation**

$$\{ + \ i: \ [2..k\text{-}1] \ | \ multacc(x[i], \ x[i\text{-}1], \ y[i\text{-}1], \ y[i]) \ \}$$

This describes $k-2$ instantiation of the cell *multacc*, one for value in the range *[2..k-1]*, i.e,

multacc(x[2], x[1], y[1], y[2])
multacc(x[3], x[2], y[2], y[3])
...
multacc(x[k-1], x[k-2], y[k-2], y[k-1])

The filter illustrates how synchronous computations are described using the synchronous combinator. It captures the essence of a synchronous computation where the functional behavior relies on exact timing. Generally, it is quite complicated to design such computations because timing and functionality are mixed.

```
CELL filter(X, Y: element; STATIC k: INTEGER)
TYPE range = [1..k-1]
STATE
  x, y: ARRAY range OF element
  zero, dummy: element
INITIALLY zero = 0

CELL multacc(xin, xout, yin, yout: element)
  STATE a: element
BEGIN
  ≪ yout:= yin + a*xin ≫ +≪ xout:= xin ≫
END multacc
BEGIN
  multacc(X, x[k-1], y[k-1], Y) +
  {+ i: [2..k-1] | multacc(x[i], x[i-1], y[i-1], y[i]) } +
  multacc(x[1], dummy, zero, y[1])
END filter
```

Fig. 7.2. Description of the 1Syst design

8. Experimental Results

This section summarizes the experience gained from using SYNCHRONIZED TRANSITIONS and the associated tools on a number of examples, including the four common examples used throughout this book.

8.1 Verifying the Black-Jack Design

Sect. 3.2 contains a design description of the Black-Jack Dealer. The requirements to the design, e.g., the interface between the design and its environment are formalized as invariants and protocols. These have been mechanically verified using the translator ST2LP and the theorem prover LP. As explained in Sect. 5. the tools verify that for a given invariant I and protocol P:

- I holds in the initial state and
- for each transition, T: $I(pre) \land t(pre, post) \Rightarrow I(post) \land P(pre, post)$

The translator generates these verification conditions: one for the initialization and one for each of the transitions in the design. All of the verification conditions are verified independently and they are handled without any user interaction between the theorem prover and the user. The total verification time on an Alpha workstation is less than two minutes.

8.1.1 Well-formedness. The Black-Jack dealer illustrates an interesting use of formal verification to ensure that the design description is well formed, e.g., that no out of range values are assigned to variables of a subrange type. This is partly done by the translator that performs as many static checks as possible. However, some checks cannot be done by ordinary static analysis and instead they are done at run-time. The tools supporting verification of designs in SYNCHRONIZED TRANSITIONS automatically extract these dynamic properties and formulate them as (well-formedness) invariants (I_{wf}). For instance, $hand.count$ should keep a value in the integer range specified by its type, $scorerange$:

$$I_{wf}: 0 \leq hand.count \leq 26$$

The invariants for well-formedness are generated by the translator and yield a set of proof obligations similar to those for the ordinary invariants and protocols. These are verified at the same time as the ordinary invariants by defining the invariant as the conjunction of the ordinary and well-formedness invariants. It is our experience that it is more practical to do these proofs independently. In this way, it is possible to concentrate the initial efforts in the verification process on the ordinary invariants which often capture the more interesting and essential properties of the design.

8.2 Single Pulser

Sect. 6.1.2 describes a realization of the Single Pulser. It has been shown to be a refinement of the abstract Single Pulser (Sect. 6.1.1) using the translator STREF [Mare94] and LP.

All the refinements and the design verification proofs (of invariants) are done without any manual interaction; the verification time on an Alpha workstation is less than one minute. It took approximately one day to describe and debug the two designs (concrete and abstract).

8.3 1Syst

The 1Syst example (FIR filter) is described in Sect. 7.1. At this point the verification technique for synchronous designs is not as well developed as for asynchronous. Therefore, no attempt has been made to invent requirements that could be verified using design verification and no refinements have been verified.

8.4 FIFO Queue

This section describes the results from verifying a rather simple FIFO queue. The simplicity of the queue makes it possible to focus on the quantitative verification effort needed to verify queues of different sizes.

The values stored in the FIFO design are called **elements**. In this simple FIFO an element has one of three values: E (for empty), T (for true), and F (for false). Elements are inserted into the queue as sequences of E, T, and F values such that any T and F are separated by at least one E, for example, $ETTTTEEEFFFETETTE$ representing the sequence of values T, F, T, T. Given a state variable, s, the predicate $e(s)$ is true if s has the value E and false otherwise.

The FIFO queue is realized as a number of state variables (each of which can hold an element) and a number of transitions for moving elements from one state variable to the next. When an element is inserted, it moves down the queue. Meanwhile, further elements can be inserted and several elements can move in parallel. The following transition describes how elements move:

$$\ll e(i) \neq e(s) \rightarrow o := i \gg$$

The three state variables, i, o, and s hold three adjacent elements.

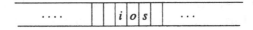

The elements move in a worm-like fashion, where a particular value can stretch over several state variables, or it can be compressed into a single state variable surrounded by E values. This worm-like behavior makes the FIFO a key component in asynchronous circuits [SpSt93].

The FIFO consists of a number of state variables linked by transitions such as the one shown above, and the size of the queue is easily varied by changing the number of state variables (and hence the number of transitions). To experiment with the quantitative verification effort the queue is constructed from a number, m, of identical cells, *fifo*, where each cell contains a number, n, of state variables/transitions. In the description shown below $n = 6$ and $m = 4$.

To verify local correctness, one must show the local correctness condition LC for each transition in the cell *fifo*. Hence, it must be expected that the

verification effort needed to show local correctness grows linearly with the
number of transitions in the design description of the cell.

TYPE link = (E, F, T)
FUNCTION w(i, o: link): BOOLEAN RETURN (i=E)=(o=E)⇒ i=o
TRANSITION e(i, o, s: link) ≪ ((i=E)≠(s=E)) → o:= i ≫
CELL fifo(i, s1, o, s: link)
  STATE s2, s3, s4, s5: link
  INVARIANT
    w(i, s1) ∧ w(s1, s2) ∧ w(s2, s3) ∧ w(s3, s4) ∧
    w(s4, s5) ∧ w(s5, o) ∧ w(o, s)
BEGIN
  e(i, s1, s2) ‖ e(s1, s2, s3) ‖ e(s2, s3, s4) ‖
  e(s3, s4, s5) ‖ e(s4, s5, o) ‖ e(s5, o, s)
END fifo

The complete FIFO consists of four (*m*) instances of the cell *fifo*, e.g.,

STATE i11, i12, i21, i22, i31, i32, i41, i42, i51, o1, o2, o3, o4: link
INVARIANT
  w(i11, i12) ∧ w(o1, i21) ∧ w(i21, i22) ∧ w(o2, i31) ∧
  w(i31, i32) ∧ w(o3, i41) ∧ w(i41, i42) ∧ w(o4, i51)
BEGIN
  fifo(i11, i12, o1, i21) ‖ e(o1, i21, i22) ‖
  fifo(i21, i22, o2, i31) ‖ e(o2, i31, i32) ‖
  fifo(i31, i32, o3, i41) ‖ e(o3, i41, i42) ‖
  fifo(i41, i42, o4, i51)
END

To show non-interference, one must show the two non-interference condition
for each instantiation of the cell *fifo*. Hence, it must be expected that the
verification effort needed to show non-interference grows linearly with the
number of instantiations in the design description.

    Fig. 8.1 shows the execution time needed to verify the simple queue for
various combinations of n and m. The two curves show two set of experiments,
the lower curve gives the execution times for various values of m with $n = 6$,
and the upper curve shows the times for $n = 12$. When $m = 0$ there is no
non-interference so the only verification is local invariance. The largest queue
($n = 12, m = 4$) holds more than 50 elements. It is verified in a few minutes
on an Alpha workstation.

    The design description of the simple FIFO makes it easy to experiment
with the quantitative effort needed for verification. As illustrated in Fig. 8.1
this effort grows linearly with the size of the design description. This claim is
supported by all other experiments using SYNCHRONIZED TRANSITIONS and
the tools described here. In this particular case there is, however, a much
more efficient way of verifying a large queue. The simple FIFO queue can be
described recursively as shown below:

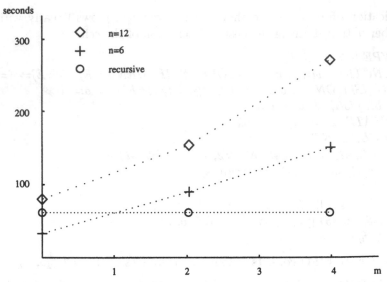

Fig. 8.1. Execution time needed for verifying the simple FIFO

CELL fifo(i, i1, o, s: link; STATIC n: INTEGER)
  STATE i2: link
  INITIALLY i1=E
  INVARIANT w(i, i1) ∧ w(i1, i2) ∧ w(o, s)
BEGIN
  e(i, i1, i2) ||
  { n > 0 | fifo(i1, i2, o, s, n-1) } ||
  { n = 0 | e(i2, o, s) }
END fifo

STATE i, i1, o, s: link
INITIALLY i=E o=E s=E
INVARIANT w(i, i1) ∧ w(o, s)
BEGIN
  fifo(i, i1, o, s, n)
END

In this recursive version of the FIFO queue it is not necessary to bind the
size *n* to a particular value in order to do the verification. The tools do
the verification in less than a minute (on an Alpha). The running time is
independent of the size (*n* and *m*), and it is depicted as a line parallel with
the x-axis on Fig. 8.1. This confirms that the verification effort depends on
the size of the design description and not on the size of the state space or
length of the computation.

8.5 Arbiter

In this section two different arbiters are described and verified. The first is a tree structured arbiter that illustrates the localized verification technique. The second is the arbiter from the common examples (originally posed by McMillan [McMi93a]).

8.5.1 Tree-arbiter. The tree-arbiter presented in this section illustrates the localized verification technique.

Structure. The tree-arbiter is structured as a binary tree in which all nodes (including the root and the leaves) are identical. An external client using the arbiter is connected to a leaf of the tree, see Fig. 8.2. The arbitration algorithm is based on propagating a privilege from the root to a requesting leaf.

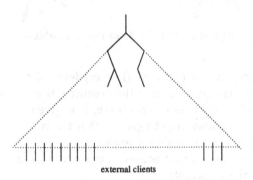

external clients

Fig. 8.2. Structure of the tree-arbiter

Operation. Each node of the tree is connected to three other nodes: the parent and two children. Each connection consists of a pair of state variables, *req* and *gr*, standing for request and grant. Such a pair is used according to the four-phase protocol (see also Sect. 3.1.4): A node requests the privilege by setting *req* to true. When *gr* becomes true, the node has the privilege and may propagate it down the tree. The privilege is passed back by setting *req* to false. When *gr* becomes false, a new request can be made. To operate correctly, the arbiter may assume that all request/grant pairs follow the four-phase protocol that is illustrated below.

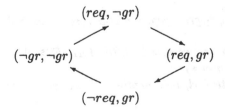

The formalization of this protocol is given in Sect. 3.1.4.

Fig. 8.3 shows a few nodes of the arbiter tree and their interconnections. The state variables, *reqp* and *grp*, of a node are connected to *reql* and *grl*

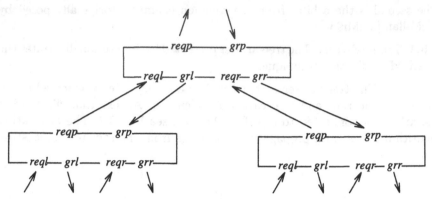

Fig. 8.3. Two levels of the tree-arbiter

(or to *reqr* and *grr*) at the previous (higher) level of the tree. When both children of a particular node request the resource, it is given first to the left child; when that child releases the resource, it is given to the right child. A transition at the root copies *reqp* to *grp*, so that the root is able to propagate a grant immediately after it makes a request.

The arbiter must never grant access to two nodes simultaneously and this is expressed with the invariant:

INVARIANT $\neg(grr \wedge grl)$

The complete design description is shown below.

```
CELL arbiter(grp, reqp: BOOLEAN; STATIC depth: INTEGER)
   STATE grl, reql, grr, reqr: BOOLEAN
BEGIN
   (* requestparent *)
   ≪ ¬ grp ∧ (reql ∨ reqr) → reqp:= TRUE ≫ ||
   (* grantleft *)
   ≪ grp ∧ reqp ∧ reql ∧ ¬ grr → grl:= TRUE ≫ ||
   (* grantright *)
   ≪grp∧reqp∧reqr∧¬grl∧¬reql→ grr:= TRUE≫||
   (* doneleft *)
   ≪ ¬ reql ∧ grl→ grl, grr, reqp:= FALSE, reqr, reqr≫||
   (* doneright *)
   ≪ ¬ reqr ∧ grr → grr, reqp:= FALSE, FALSE ≫ ||
   (* instantiate children *)
   { depth > 0 | arbiter(grl, reql, depth-1) || arbiter(grr, reqr, depth-1) }
END arbiter
```

Verification. The verification of the arbiter consists of showing the invariant and protocols mentioned above. This was among the first examples verified using mechanized tools for SYNCHRONIZED TRANSITIONS [GaGS88]. The example illustrates several typical aspects of our approach to design verification.

The recursive design description is generic and the same textual description is used for all sizes of the arbiter. To vary the size of the arbiter, the only change needed is to change the value of the actual parameter given in the instantiation of the arbiter tree, for example:

arbiter(gr, req, 8)

instantiates an arbiter tree of depth 8 with 2^8 leaves. Because the size of the design description is independent of the size of the arbiter tree, the number of verification conditions needed to show that the invariants and protocols hold is also independent of the size. A similar observation was made for the FIFO queue in Sect. 8.4.

To use the localized verification technique, the designer must supply the invariants and protocols formalizing the requirements that are to be verified, for example, the mutual exclusion of a node in the arbiter tree: $\neg(grl \wedge grr)$. However, as noted in Sect. 5.4 it is often necessary to come up with a stronger assertion to use the localized technique. In the case of the tree-arbiter, it is for example necessary to strengthen the invariant as follows:

$$\neg \ (grl \wedge grr) \wedge (\ (grl \vee grr) \Rightarrow (grp \wedge reqp) \ )$$

This is the invariant that has been mechanically verified.

8.5.2 McMillan Arbiter. The McMillan Arbiter is one the common examples (see Appendix), it consists of a number of identical elements each of which has two parts:

The synchronous part administers the flow of the token, in each step the token can move from one element to the next.

The asynchronous part is a combinational computation (function) determining the value of the acknowledge signal.

The combination of the synchronous parts of all elements works like a cyclic shift register making the token rotate. The design descriptions of the two parts of an element are shown in Fig. 8.4.

It is straigtforward to formalize the mutual exclusion requirement. However, it is also required that every persistent request is eventually acknowledged. This is an example of a "liveness property" which cannot be expressed in the notation for expressing invariants and protocols, see also Sect. 5.5.

8.6 Evaluation

It should be stressed that the localized verification technique is just one way of verifying that an assertion holds throughout a computation. As just illustrated, one consequence of using this technique is that it often requires *the*

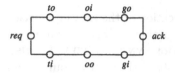

(* synchronous part *)   (* asynchronous part *)
$\ll t := ti \gg +$ $\ll oo := oi \lor (w \land t) \gg \|$
$\ll to := t \gg +$ $\ll go := gi \land \neg\ req \gg \|$
$\ll w := (ti \land req) \lor (w \land req) \gg$ $\ll ack := req \land (gi \lor (w \land t)) \gg$

Fig. 8.4. A single element of the McMillan Arbiter

designer to come up with an assertion that is somewhat stronger than the requirement to be verified. There are techniques that overcome the difficulty of finding invariants by automatically computing the strongest of them all: *A predicate characterizing exactly the set of reachable states*. Having computed this set, the verification task is reduced to checking that the predicate characterizing the set implies the property of interest. The set of *reachable states* is the subset of the state space that can be reached by a sequence of transitions from any of the initial states. This set is often computed as an increasing sequence of approximations starting with the initial states and in each step adding what can be reached by making one further transition [ClEm81]. If the system is finite, this sequence of approximations will always converge to the full set of reachable states in a finite number of steps. This computation requires choosing a representation for sets of states. Using an explicit representation very quickly leads to a combinatorial explosion of the number of states generated, resulting in poor performance. However, *implicit* representations of state sets with clever datastructures such as Reduced Ordered Binary Decision Diagrams [Brya86], ROBDDs. ROBDDs are not guaranteed to avoid the combinatorial explosion but they do on very many examples, providing one of the most successful heuristics currently known.

9. Conclusion

SYNCHRONIZED TRANSITIONS and the associated tools have been used for specifying and verifying a number of examples ranging from small tricky designs, such as arbiters, to larger realistic examples, such as an Ethernet Bridge and parts of an ATM switch. The tools are also used in a semester course on formal techniques for hardware design. This chapter contains a first introduction to the notation and the associated tools illustrated by the common examples used throughout this book.

A design description in SYNCHRONIZED TRANSITIONS encourages design structures consisting of a number of (relatively) independent concurrent state machines. This is utilized in the tools supporting SYNCHRONIZED TRANS-ITIONS which all take advantage of the possibility of breaking a design description into a number of independent parts. Two sets of tools have been illustrated, one for transforming a design description into a circuit realization, and one for transforming it into a set of verification conditions to be verified by a proof engine. By using this verification technique *the verification effort grows linearly with the size of the textual design description.* For recursively defined cells, the effort needed is independent of the recursion depth (and hence independent of the size of the state space or length of the computation). This efficiency has its price, namely an added effort is required by the designer to state invariants and protocols used for showing non-interference between distinct cells.

Acknowledgement. SYNCHRONIZED TRANSITIONS has been developed over a period of more than 10 years. Anders P. Ravn played a key role in shaping the fundamental concepts and a number of people have assisted in its further development. Christian D. Nielsen, Niels Maretti, Niels Mellergaard, Henrik Hulgaard, Per Henrik Christensen, and Jacob Christensen have all contributed with improvements of the notation, design methods, and tools. The work on mechanical verification has been done in cooperation with John Guttag, Steve Garland, Niels Maretti, and Niels Mellergaard.

Niels Mellergaard and Niels Maretti have written the translator that forms the basis of all currently available tools supporting SYNCHRONIZED TRANSITIONS. They have also verified a large number of examples providing invaluable experience.

Last but not least, Mark Greenstreet has contributed to all aspects of the development of SYNCHRONIZED TRANSITIONS. His enthusiasm and creativity have been a constant source of inspiration. I am very grateful to Mark for many stimulating discussions over a period of almost ten years during which SYNCHRONIZED TRANSITIONS has evolved.

Sect. 5.3 is a minor modification of a similar section in the paper [StMe95]. Henrik Hulgaard and Thomas Kropf made numerous useful suggestions for improvement of earlier versions of this chapter.

Financial support for developing SYNCHRONIZED TRANSITIONS has been given by Århus University, The Danish Technical Research Council, Digital External Research, Mogens Balslev's fond, and The Danish Natural Science Research Council.

Hardware Verification Using PVS *

Mandayam Srivas, Harald Rueß, and David Cyrluk

1. Introduction

The past decade has seen tremendous progress in the application of formal methods for hardware design and verification. Much of the early work was on applying proof checking and theorem proving tools to the modeling and verification of hardware designs [Gord83b, Hunt89]. Though these approaches were quite general, the verification process required a significant amount of human input. More recently, there has been a large body of work devoted to the use of model checking, language containment, and reachability analysis to finite-state machine models of hardware [ClGr87a, BCMD92, BCLM94]. The latter class of systems work automatically but they do not yet scale up efficiently to realistic hardware designs. The challenge then is to combine the generality of theorem proving with an acceptable level of effective and efficient automation.

Our main thesis is that in order to achieve a balance between generality, automation, and efficiency, a verification system must provide powerful and efficient primitive inference procedures that can be combined by means of user-defined, general-purpose, high-level proof strategies. This design philosophy has formed the guiding principle for the implementation of the PVS system [OwRS92, ORRS96, ORSH95]. It combines an expressive specification language with an interactive proof checker that has a reasonable amount of theorem proving capabilities. PVS is designed to automate the tedious and obvious low-level inferences while allowing the user to control the proof construction at a meaningful level. Exploratory proofs are usually carried out at a level close to the primitive inference steps, but greater automation is achieved by defining high-level proof strategies. When compared to other proof checkers, the primitive inference steps of PVS are very powerful as they are implemented using a set of powerful decision procedures.

The domain of problems that have been investigated with PVS involves verification of industrial-strength microprocessors [MiSr95, SrMi95a, Cyrl96], protocol verification [Hoom95, HaSh96, PaDi96a, Shan92], arithmetic circuits [RuSS96, Rues96, MiLe96], real-time properties [Shan93, Hoom94], fault tolerance [ViBu92, Mine93, Rush93], and clock synchronization [Shan92, Rush94, MiJo96].

---

* The development of PVS was funded by SRI International through IR&D funds. Various applications and customizations have been funded by NSF Grant CCR-930044, NASA, ARPA contract A721, and NRL contract N00015-92-C-2177.

Although PVS is a general purpose theorem prover, it supports the specific needs of hardware verification through the use of an expressive specification language, a bit-vector library, decision procedures for equality, linear arithmetic, and arrays, propositional simplification based on binary decision diagrams, and integration of symbolic model checking.

While PVS is capable of verifying a wide variety of hardware circuit designs, most of the large verifications we have performed are in the area of pipelined microprocessors and complex arithmetic circuits at register transfer level. The reason we have concentrated our effort on datapath-intensive circuits at register transfer level is because theorem-proving techniques are most effective in these domains. Also, the inadequacy of conventional simulation-based CAD tools is most pronounced at register transfer levels and higher for complicated designs involving, for example, pipelining. So, we will devote most of this chapter to describing the approaches to verification in these domains of applications.

This chapter is organized as follows. Sect. 2. contains a comparison of PVS with related theorem proving systems, and in Sect. 3. we describe the basic features of the PVS specification language and the PVS prover. Predicative and functional styles of hardware descriptions in PVS are discussed in Sect. 4. In that section we also demonstrate the capabilities of the PVS specification language to model generic hardware components. The next two sections are devoted to specifications, methodologies, and proofs for verifying microprocessors and arithmetic circuits. Sect. 5. includes a description of the basic methodology of processor verification together with the verification of toy processors including the Tamarack processor and a discussion of how these techniques scale up for verifications of industrial-strength processors. Sect. 6. provides a description of a hierarchical verification of a combinational multiplier. In that section, we also outline the verification of an SRT division circuit that is similar to the one in the Pentium microprocessor. Finally, Sect. 7. summarizes the experiments we have performed on verifying the circuits (single pulser, arbiter, Black Jack, FIR filter) used throughout this book.

2. Related Work

The PVS system is engineered by combining a number of theorem proving techniques some of which were pioneered and proven effective in other systems. For example, NUPRL [Cons86] and VERITAS [HaDa92a] provide predicate subtypes and dependent types, and the theorem proving techniques draw on LCF [GoMW79], the Boyer-Moore prover [BoMo79, BoMo88], and on earlier work at SRI [Shos84]. Historically, theorem proving systems have made a trade-off between expressiveness of the logic/specification language supported and the degree of effective automation provided. PVS differs from others in its aggressive use of decision procedures and in tightly integrating capabilities that usually occur separately; this design philosophy has allowed PVS to provide a very expressive specification language and powerful and effective mechanization within a classical framework.

The Boyer-Moore theorem prover, NQTHM, is the best known of the batch-oriented theorem proving systems used in hardware verification [BoMo79, BoMo88]. Many of its deductive components are quite similar to those in PVS. The system uses a fast propositional simplifier, and also includes a rewriter and a linear arithmetic package. On the other hand, a fully automatic batch-oriented theorem prover has the drawback of being a toolkit with only a single tool. Doing exploratory proof development with such a theorem prover is tedious because of the low bandwidth of interaction. It is difficult to reconcile efficiency with generality in a fully automated theorem prover since a single proof strategy is being applied to all theorems.

The primitive inference steps in PVS are a lot more powerful than in HOL [GoMe93]. So, it is not necessary to build complex tactics to handle tedious lower level proofs in PVS. A user knowledgeable in the ways of PVS can typically get hardware proofs to go through mostly automatically by making a few critical decisions at the start of the proof. However, PVS does provide the user with the equivalent of HOL's tacticals, called *strategies*, and other features to control the desired level of automation in a proof.

3. Basics

This section introduces the notations and the characteristic features of the PVS specification language and the prover. For more gentle and more complete introductions see the tutorials [CORS95] and [RuSt95].

3.1 The Specification Language

The PVS specification language builds on classical typed higher-order logic with the usual base types bool, nat, rational, real among others and the function type constructor [A -> B]. The type system of PVS is augmented with *dependent types* and *abstract data types*.

A distinctive feature of the PVS specification language is *predicate subtyping*. A subtype {x:A | P(x)} consists of exactly those elements a of type A satisfying predicate P(a). For example, the predicate subtypes below[n] and subrange[-m, n] respectively define the integer ranges $[0, \ldots, n)$ and $[-m, \ldots, n]$. Predicate subtypes are used to explicitly constrain the domain and ranges of operations in a specification and to define partial functions. Thus, predicate subtypes bring great richness of expression to a logic of total functions but require theorem proving to ensure type correctness. In general, type-checking with predicate subtypes is undecidable because the predicate characterizing the subtype can be arbitrarily complex. The type-checker generates *type correctness conditions* (TCCs) by analyzing the usage of the predicate subtypes in expressions. A large number of TCCs are discharged by specialized proof strategies that can be automatically invoked by the typechecker. A PVS expression is not considered to be fully type-checked unless all generated TCCs have been proven correct.

PVS specifications consist of a number of theories. A theory is a collection of declarations: types, constants (including functions), axioms that express properties about the constants, and theorems and lemmas to be proved. Theories may import other theories and may be parametric in types and constants. Every entity declared in a parameterized theory is implicitly parameterized with respect to the parameters of the theory. A built-in *prelude* and loadable *libraries* provide standard specifications and proved facts for a large number of theories. Since bit-vectors are used heavily in hardware verification and throughout this text, we describe the structure of the PVS bit-vector library in more detail.

Bits are represented as elements of the predicate subtype upto[1] which consists of the numbers 0 and 1, and an N-bit bit-vector is represented as an array bvec[N], i.e. a function from the type below[N] of natural numbers less than N to bit. The parameter N is constrained to be a positive natural number posnat since bit-vectors of length 0 are not permitted.

Many basic operations and facts about bit-vectors are predefined in the bit-vector library of PVS. Concatenation of bit-vectors xv and yv, for example, is denoted by xv o yv, function application xv(i) denotes selection of the ith bit, and extraction of (i-j+1) many bits i through j from bit-vector xv of type bvec[N] is denoted by $xv^\wedge$(i,j). Obviously, extraction is restricted, using predicate subtyping, to the cases (i < N) and (i >= j).

The function bv2nat defines the unsigned interpretation of bit-vectors of length N as natural numbers.

```
bv_nat[N: posnat]: THEORY
BEGIN
  IMPORTING bitvectors@bv[N]

  bv2nat_rec(n: upto[N], bv:bvec[N]): RECURSIVE nat =
    IF n = 0 THEN 0
    ELSE exp2(n-1) * bv(n-1) + bv2nat_rec(n - 1, bv) ENDIF
  MEASURE n

  bv2nat(bv:bvec[N]): below[exp2(N)] = bv2nat_rec(N, bv)

  CONVERSION bv2nat
END bv_nat
```

It is defined in the theory bv_nat that is parameterized with respect to the length N of bit-vectors. bv2nat_rec is an auxiliary function used only to define bv2nat; it interprets the lower portion of a bit vector as a natural number. Recursive function definitions in PVS must have an associated MEASURE function to ensure termination. The type-checker automatically generates type correctness proof obligations to show that the argument to every recursive call is decreasing according to the given measure. Moreover, the CONVERSION directive causes the type-checker to implicitly inject the conversion bv2nat whenever it finds a bit-vector instead of the required number. In this way it is possible to *overload* bit-vectors xv with corresponding unsigned interpretations bv2nat(xv). This kind of overloading is commonly used in computer arithmetic textbooks to express concise and readable circuit descriptions.

3.2 The Proof Checker

The PVS proof checker is intended to serve as a productive medium for debugging specifications and constructing readable proofs. The human verifier constructs proofs in PVS by repeatedly simplifying a conjecture into subgoals using prover *commands*, which can be primitive inference steps or pre-defined strategies, until no further subgoals remain. A proof goal in PVS is represented by a sequent. PVS differs from most proof checkers in providing primitive inference rules that are quite powerful, including decision procedures for ground linear arithmetic inequalities with real coefficients, propositional logic and equalities over expressions with uninterpreted functions. The primitive rules also perform steps such as quantifier instantiation, rewriting, beta-reduction, and boolean simplification. PVS has a simple strategy language for combining inference steps into more powerful proof strategies. In interactive use, when prompted with a subgoal, the user types in a proof command that either invokes a primitive inference rule or a compound proof strategy. For example, the skolem command introduces constants for universal-strength quantifiers while the inst command instantiates an existential-strength quantifier with its witness. The lift-if command invokes a primitive inference step that moves a block of conditionals nested within one or more sequent formulas to the top level of the formula. The prop command invokes a compound propositional simplification strategy (or tactic). Since the prop command was too inefficient for hardware verification, a new primitive command bddsimp was implemented which invokes an efficient *off-the-shelf* BDD-based simplifier. Various other commands are discussed below.

Proofs and partial proofs can be saved, edited, and rerun. It is possible to extend and modify specifications during a proof; the finished proof has to be rerun to ensure that such changes are benign.

3.3 Ground Decision Procedures and Rewriting

The ground decision procedures of PVS are used to simplify quantifier-free Boolean combinations of formulas involving a combination of arithmetic inequalities and equality over expressions involving uninterpreted functions, and to propagate type information. Consider a formula of the form

```
f(x) = f(f(x)) IMPLIES f(f(f(f(x)))) = f(x)
```

where the variable x is implicitly universally quantified. This is really a *ground* (i.e. variable-free) formula since the universally quantified variable x can be replaced by a newly chosen constant, say c. We can then negate this formula and express this negation as the conjunction of literals: f(c) = f(f(c)) AND NOT f(f(f(f(c)))) = f(c). We can then prove the original formula by refuting its negation. To refute the negation we can assert the information in each literal into a data structure until a contradiction is found. In this case, we can use a *congruence closure* data structure to rapidly propagate equality information [Shos84, CyLS96].

' The PVS decision procedures combine congruence closure over interpreted and uninterpreted functions and relations with refutation procedures for linear arithmetic, arrays, and tuples. Ground arrays are important in hardware examples, where memory can be represented as a function from addresses to data. For example, the decision procedure can deduce the application of an updated function (func WITH [(j) := val])(i) to be equal to func(i) under the assumption i /= j.[1]

However, PVS does not merely make use of decision procedures to prove theorems but also to record type constraints and to simplify subterms in a formula using any assumptions that *govern* the occurrence of the subterm. The prover uses an internal data structure shared by all the decision procedures to record the current state of the assumptions and inferred facts. These governing assumptions can either be the test parts of surrounding conditional (IF - THEN - ELSE) expressions or type constraints on governing bound variables. Such simplifications typically ensure that formulas do not become too large in the course of a proof.

PVS supports automatic conditional rewriting. Any definition, axiom, or lemma that is in the form of a *conditional rewrite rule* (*cond* => *lhs* → *rhs*) can be entered by the user by means of a prover command (auto-rewrite, auto-rewrite-theory, etc.) into the internal rewrite rule data base of the prover for future rewriting. The prover commands that perform rewriting apply the current set of rewrite rules in the data base to simplify the expressions in the current goal. Also important is the fact that automatic rewriting is tightly coupled with the use of decision procedures, since many of the conditions and type correctness conditions that must be discharged in applying a rewrite rule succumb rather easily to the decision procedures. The primitive prover command assert, which is one of the most commonly used workhorses in the proofs illustrated throughout this chapter, combines all the decision procedures with rewriting to discharge or simplify the current goal. The decision procedures can be invoked without rewriting using the command (simplify).

---

[1] The WITH expression is an abbreviation for the result of updating a function at a given point in the domain value with a new value.

3.4 The Power of Interaction

The following example illustrates the power that a close interaction between rewriting and the decision procedures can provide for the user in PVS. Such a close interaction is not as easily accomplished if the decision procedures and rewriting were implemented as separate tactics or strategies.

```
t: NAT
s: VAR state

MAR_t0: AXIOM
  t /= 0 IMPLIES dest(IR(s)) = MAR(s)

MDR5(s): data =
  IF t <= 2 THEN
    IF p(t) THEN
      MDR(s)
    ELSE
      rf(s) WITH [(MAR(s)) := MDR(s)](dest(IR(s)))
    ENDIF
  ELSE
    somedate
  ENDIF

  property: THEOREM t < 3 & p(0) IMPLIES MDR(s) = MDR5(s)
```

In the PVS specification shown above, the goal is to prove theorem property from the axioms MAR_t0 and the definition of MDR5, where the constants p, MDR, MAR, etc. are declared elsewhere. The constant rf is a function that maps addresses to data. Assuming that the definition of MDR and MAR_t0 have been entered as rewrite rules (via the command auto-rewrite), property can be proved (after skolemizing the variable s and flattening the implication) by simply using the command assert on the resulting sequent twice. The first invocation of assert is able to rewrite MDR5(s) to the IF-THEN-ELSE expression beginning at p(t) since, when t is a nat, t <= 2 can be deduced from t < 3 by the decision procedure. The second invocation of assert attempts to rewrite each branch of the resulting IF-THEN-ELSE expression. The p(t) case is trivially true. In the NOT(p(t)) case, the decision procedures deduce that t = 0 is false from p(0). This triggers the rewrite rule MAR_t0 so that the goal becomes MDR(s) = rf(s) WITH [(MAR(s)) := MDR(s)](MAR(s)), which the equality procedures simplify to true.

3.5 Integration of Model Checking into PVS

In the theorem proving approach to program verification, one verifies a property P of a program M by proving $M \supset P$. The model checking approach verifies the same program by showing that the state machine for M is a satisfying model of P, namely $M \models P$. For control-intensive approaches over small finite states, model checking is very effective since a more traditional

Hoare logic style proof involves discovering a sufficiently strong invariant. These two approaches have traditionally been seen as incompatible ways of viewing the verification problem. In recent work [RaSS95], we were able to unify the two views and incorporate a model checker as decision procedure for a well-defined fragment of PVS.

This integration uses the μ-calculus as a medium for communicating between PVS and a model checker for the propositional μ-calculus. Model checking in PVS relies on an external BDD-based decision procedure for Park's relational μ-calculus. CTL formulas—with and without fairness—are encoded in the μ-calculus (see, for example, [EmLe86a, BCMD92]), and the prover command **model-check** decides these formulas, giving PVS similar capabilities as SMV [McMi93a]. The idea here is that a Kripke model is captured by defining a state type, typically given by a record type construction, and a next-state relation N on this state type which defines the state transitions or the edges in the Kripke model. Temporal properties such as those expressible in the branching-time temporal logic CTL can be characterized by fixed-points computed over such a Kripke model using the least and greatest fixed-point operators mu and nu over monotonic predicate transformers. Such fixed-point operators can easily be defined using the higher-order logic of PVS.

```
N: VAR [state, state -> bool]

EX(N, f)(u): bool = (EXISTS v: f(v) AND N(u, v))

EG(N, f)    : pred[state] = nu!(Q): f AND EX(N, Q)

EU(N, f, g): pred[state] = mu!(Q): g OR (f AND EX(N, Q))
```

A CTL formula characterizes a predicate over states in this Kripke model. CTL formulas are built from the propositional atoms using the propositional connectives and the modalities EX, EG, and EU. With respect to a given Kripke model whose transition relation is captured by the binary predicate N, the CTL formula EX(N, f) holds on those states that have a single edge leading to a state where the formula f holds. The CTL formula EG(N, f) holds on those states that have an infinite outgoing path of edges along which the formula f always holds. The CTL formula EU(N, f, g) holds on those states that have a path of zero or more edges leading to a state where g holds so that f holds along each intermediate state on the path. The other CTL modalities can be defined in terms of these modalities.

Model checking CTL properties of a given finite transition system is then achieved by first rewriting the CTL operators into the μ-calculus definitions. Then, the finite state type and its transition relation are binary encoded so that the result is a Boolean μ-calculus expression. The validity of the translated μ-calculus expression is decided by invoking an external BDD-based boolean μ-calculus checker. PVS employs a BDD-based μ-calculus validity checker due to Janssen [Jans93]. This entire procedure is encapsulated by the

PVS prover command `model-check`, which also checks that the state type of the transition system is constructed from finite types.

CTL cannot express fairness but one can once again define fair versions of the CTL operations in the μ-calculus. For example, the operator `fairEG(N, f)(Ff)` is the predicate that holds of a state u such that there is a *fair path*, i.e. one where the predicate `Ff` holds infinitely often, such that the predicate `f` holds on every state on the path. This is expressed in the μ-calculus using the greatest fixed-point nu of the predicate transformer below.

```
fairEG(N, f)(Ff): pred[state] =
  nu(LAMBDA P: EU(N, f, f AND Ff AND EX(N, P)))
```

Simple examples of symbolic model checking in PVS are given in Sect. 7. More advanced uses of the combination of theorem proving and model checking in PVS are described by Havelund and Shankar [HaSh96] to verify safety properties for a bounded retransmission protocol using abstraction and model checking. Shankar [Shan96] combines model checking, abstraction, induction, and compositionality to simplify the verification of unbounded-state systems.

3.6 High-Level Strategies

As described above, the PVS proof checker provides powerful primitive inference steps that make heavy use of decision procedures, but proof construction solely in terms of these inference steps can be quite tedious. PVS therefore provides a language for defining high-level strategies. This language includes recursion, a `let` binding construct, a backtracking `try` strategy construction, and a conditional `if` strategy construction. Typical strategies include those for heuristic instantiation of quantifiers (`inst?`), repeated skolemization, rewriting, and induction followed by simplification and rewriting. The strategy `ground` combines `assert` and `bddsimp` to discharge ground formulas that have disjunctions. The high-level strategy `grind`, for example, repeatedly applies skolemization (`skolem`), instantiation (`inst?`), lifting of conditions (`lift-if`), propositional simplification using BDDs (`bddsimp`), and a combination of rewriting with decision procedures (`assert`) until nothing works. The high-level induction strategy `induct-and-simplify` inducts on a given variable by selecting an appropriate induction scheme and the resulting cases are simplified using the set of enabled rewrites.

4. Modeling Hardware

Formal description styles of hardware components may be broadly classified according to two distinct approaches; namely, *predicative* and *functional* styles of specifications. In this section we illustrate these styles of specification in

PVS. Moreover, we describe the use of theory parameterization to model *generic hardware components.*

The specification techniques described in this section are suitable for modeling hardware designs at a register-transfer level where the clock is implicit. The time period of the clock is assumed to long enough that every combinational component can be considered to have zero delay, and every register (latch) to have a unit delay. At this level, the behavior of a hardware circuit can be modeled as a state transition machine by choosing an appropriate type to represent the state of the circuit and defining a "next state function or relation" corresponding to a single clock cycle. This state-machine style of description, which is popular among model-checkers, such as SMV [McMi93a], can also be used with theorem provers. We employ the state-machine style for illustrating the Tamarack verification in Sec. 5. A disadvantage of the state-machine style is that the hardware structure of the circuit is lost in the specification. The specification approaches described below, however, are designed to reflect as closely as possible the structure of the circuit in the specification.

4.1 Predicative Style Specifications

The behavior of hardware components can be specified in higher-order logic by defining *predicates* that state which combinations of values can appear on their external ports. The behavior of devices built by wiring together smaller devices is represented by conjoining the predicates that specify the behaviors of their components with logical conjunction and using existential quantification to hide internal signals. This style of specifying hardware components in higher-order logic has been popularized by Gordon [Gord86].

Consider, for example, the predicative specifications below. The description INV(i, j) of a two-port inverter[2] may always have the Boolean values on its ports as inverses of each other, and the relations AND2 and OR2 model two-input combinational conjunction and disjunction, respectively.

```
components: THEORY
BEGIN
   i, j, i0, j0: VAR bool

   INV(i, j)      : bool = (j = NOT i)
   AND2(i0, j0, i): bool = (i = (i0 AND j0))
   OR2(i0, j0, i) : bool = (i = (i0 OR j0))
   ...
END components
```

Of particular importance to modeling hardware behavior is the use of predicates whose arguments are functions over discrete time, so-called **signals**. Discrete time is simply modeled by the natural numbers, and the type

---

[2] The specification of the inverter is equivalent to (j IFF NOT i).

```
time: THEORY                    signal[A: TYPE]: THEORY
BEGIN                           BEGIN
                                  IMPORTING time
  time: TYPE = nat
                                  signal: TYPE = [time -> A]
END time                        END signal
```

Fig. 4.1. Signal Specification

signal declared in the parameterized theory signal is a parametric type denoting a function that maps time (a synonym for nat) to the type parameter A. Predicates are then used to define the relationship which must exist between the current and post values on the ports of a device. The relation DELAY(x, y), for example, specifies the behavior of a unit-delay register.

```
DELAY(x, y: signal[bool]): bool =
  FORALL (t: time): NOT(y(0)) & (y(t + 1) = x(t))
```

Given predicates that specify the behaviors of the components of a design, the composite behavior of connected components is specified by forming a conjunction of the individual behaviors and by identifying variables that represent connected ports. Moreover, existential quantification is used to hide internal lines. In this way, the predicative specification detect110_imp formally represents the 110 detector circuit in Fig 4.2. The formula describing the circuit is guaranteed to be satisfiable if every feed-back loop in the circuit has at least one sequential component, i.e., the circuit has no combinational loops. The absence of such loops is not automatically guaranteed by PVS. It can be postulated as a well-formedness condition and be proved separately.

```
detect110: THEORY
BEGIN
  IMPORTING components, signal

  t    : VAR time
  e, a: VAR signal[bool]

  detect110_imp(e, a): bool =
    (EXISTS (11, 12, 13, q4, 15, 16, q7: signal[bool]):
      (FORALL t: INV(e(t), 11(t))
               & INV(15(t), 16(t))
               & AND2(e(t), 15(t), 12(t))
               & AND2(e(t), 16(t), 13(t))
               & DELAY(12, q4)
               & DELAY(13, q7)
               & AND2(11(t), q4(t), a(t))
               & OR2(q4(t), q7(t), 15(t))))
```

The abstract specification of the device's intended behavior is given by the formula detect110_spec below, and the detector circuit defined by

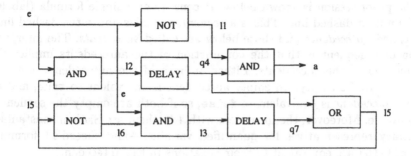

Fig. 4.2. 110 Detector.

detect110_imp can be proved correct with respect to this specification by proving the refinement theorem detect110_correct. This theorem is universally quantified with respect to e and a. Using a set-theoretic interpretation, it expresses the fact that the input/output relation detect110_imp is contained in the detect110_spec relation. Therefore detect110_imp is considered to be an *implementation* or a *refinement* of the specification detect110_spec.

```
% -- Specification
  detect110_spec(e, a): bool =
    FORALL t: a(t + 2) = (NOT e(t + 2) & e(t + 1) & e(t))

  IMPORTING quant_rules

  detect110_correct: THEOREM
    detect110_imp(e, a) IMPLIES detect110_spec(e, a)
END detect110
```

The proof of detect110_correct proceeds by introducing constants for the universally quantified variables e and a, followed by unfolding the definitions from the theories detect110 and components, and by rewriting with some basic facts about quantifiers from theory quant_rules in order to simplify the resulting subgoal.[3] The final skosimp* in the proof script below introduces constants for the internal lines of the 110 detector circuit.

```
(then (skosimp)
  (auto-rewrite-theories "detect110" "components" "quant_rules")
  (assert)
  (skosimp*))
```

then is a strategy constructor that composes a sequence of commands and successively applies commands to generated subgoals. The resulting subgoal

---

[3] Actually, only the rules (FORALL t: A) = A and (FORALL t: A(t) & B(t)) = ((FORALL t: A(t)) & (FORALL t: B(t))) are used.

in the proof session is shown below. It consists of a single formula (labeled {1}) under a dashed line. This is a *sequent*; formulas above the dashed lines are called *antecedents* and those below are called *succedents*. The interpretation of a sequent is that the conjunction of the antecedents implies the disjunction of the succedents. Either or both of the antecedents and succedents may be empty; an empty antecedent is equivalent to true, and an empty succedent is equivalent to false, so if both are empty the sequent is unprovable. Moreover, the identifiers with ! in them are Skolem constants—arbitrary representatives for quantified variables—and succedent formulas like q4!1(0) are equivalent to their negations in the antecedent.

```
{-1}    FORALL t: 11!1(t) = NOT e!1(t)
{-2}    FORALL t: 16!1(t) = NOT 15!1(t)
{-3}    FORALL t: 12!1(t) = (e!1(t) & 15!1(t))
{-4}    FORALL t: 13!1(t) = (e!1(t) & 16!1(t))
{-5}    FORALL t: q4!1(1 + t) = 12!1(t)
{-6}    FORALL t: q7!1(1 + t) = 13!1(t)
{-7}    FORALL t: a!1(t) = (11!1(t) & q4!1(t))
{-8}    FORALL t: 15!1(t) = (q4!1(t) OR q7!1(t))
  |-------
{1}     q4!1(0)
{2}     q7!1(0)
{3}     a!1(2 + t!1) = (NOT e!1(2 + t!1) & e!1(1 + t!1) & e!1(t!1))
```

Notice that the execution of the initial sequence of proof commands yields a *functional style* description of the 110 detection circuit, since every signal that is an output of a component is now specified as a function of the signals appearing at the inputs to the component. In our experience, a proof of correctness for a predicative style specification usually involves executing a few additional steps at the start of the proof to essentially transform this specification to an equivalent functional style. After that, the proof proceeds similar to that of a proof in a functional specification.

Returning to our example, it is a simple matter to finish the proof by rewriting with the antecedent formulas followed by propositional simplification using BDDs. In the following, `auto-rewrite-antecedents*` denotes a strategy that enters every rewrite rule in the antecedent into the rewrite rule data base for future rewriting.

```
(then (auto-rewrite-antecedents*)
      (assert)
      (bddsimp))
```

4.2 Functional Style Specifications

Now, we use a *functional* style of specification to model register-transfer level digital hardware in logic. In this style, the inputs to the design and the outputs of every component in the design are modeled as signals, and every signal that is an output of a component is specified as a function of the signals appearing at the inputs to the component.

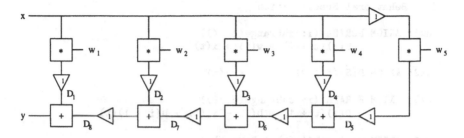

Fig. 4.3. FIR Filter ($n = 5$)

The example to demonstrate functional style specifications in PVS involves the verification of the design of the FIR (Finite Impulse Response) filter in Fig. 4.3, defined at the arithmetic level. This filter produces a sequence of output data from a sequence of input data and weights w_i by computing the inner product:

$$y(t) \; = \; w_1 \cdot x(t-1) + w_2 \cdot x(t-2) + \ldots + w_5 \cdot x(t-5)$$

The PVS specification of the FIR filter in Fig. 4.3 is packaged in the theory fir_filter below, and the formal parameter of this theory specifies the weights w(i) of the computation.

```
fir_filter5[w: [subrange(1, 5) -> real]]: THEORY
BEGIN
   IMPORTING time, signal[real]

   t: VAR time
```

Again, discrete time is simply modeled by the natural numbers, and signals over reals are defined to be functions with domain time and codomain real.

The filter computation is modeled as a collection of signals, one per wire, so that at time tick t for (t >= 0), the value of signal D(i) for i in subrange(1, 8) is D(i)(t).

```
% -- Signal Declarations

   x: signal
   y: signal

   D(i: subrange(1, 8)): signal
```

The second part describes the (synchronous) state transitions of the machine and consists of a set of AXIOMs that specifies the values of the signals over time. For example, the signal value at the internal register D(5) at time t + 2 is defined to be that of the value of input signal x at time t multiplied by the weight w(5).

```
% -- Behavioral Specification

ax1: AXIOM FORALL (i: subrange(1, 4)):
            D(i)(t + 1) = w(i) * x(t)

ax2: AXIOM D(5)(t + 2) = w(5) * x(t)

ax3: AXIOM FORALL (i: subrange(6, 8)):
            D(i)(t + 1) = D(10 - i)(t) + D(i - 1)(t)

ax4: AXIOM y(t) = D(1)(t) + D(8)(t)
```

In PVS, we can use a descriptive style of definition, as illustrated in this example, by selectively introducing properties of the constants declared in a theory as AXIOMs. An advantage of the descriptive style is that it gives control over the degree to which an entity is defined (for example, the value of signal D(5) at times t = 0 and t = 1 is undefined). Alternatively, we can use the definitional forms provided by the language to define some constants. An advantage of using definitions is that a specification is guaranteed to be consistent. So, it is possible to guard completely against creating combinational loops or, at least, isolate their presence to well-defined regions of the specification, if one used a definitional style. On the other hand, the resulting specification may be overly specific (i.e. over-specified), since function definitions in PVS are required to be total. In the descriptive style, the burden of checking the absence of inconsistencies in the specification falls on the specifier.

In the present example, the main correctness property (fir_filter_char) is stated as asserting that for all t >= 5, the value y(t) of the output signal at time t equals the inner product of the weight vector and the vector of the 5 most recent inputs.[4]

```
% -- Invariant

IMPORTING sum

fir_filter_char: THEOREM
  FORALL (t: upfrom(5)):
    y(t) = sum!(i: subrange[1, 5]): w(i) * x(t - i)

END fir_filter5
```

This correctness property (fir_filter_char) is proved by eliminating the universal quantifier, installing the behavioral equations as rewrites and calling the decision procedures.

```
(then (skosimp)
      (auto-rewrite-theories "sum[1, 5]" "fir_filter5")
      (assert))
```

---

[4] The inductive definition of the sum function is obvious and not described here. Moreover, (sum!(x): ...) is an abbreviation for (sum(LAMBDA x: ...)).

4.3 Functional Description of Combinational Circuits

We exemplify the use of functional style specifications in PVS for describing combinational hardware circuits by means of the standard examples of specifying and verifying a 1-bit full-adder and a carry-propagate adder. These developments are also used in Sect. 6.1 to verify an iterative array multiplier.

A full adder is specified in the theory full_adder as a function FA with input bits x, y and a carry-in bit cin. This function is described at the gate-level and computes a pair of bits consisting of the carry and the sum bit.[5]

```
full_adder: THEORY
BEGIN
  IMPORTING bitvectors@bit

  x, y, cin: VAR bit

  FA(x,y,cin): [# carry, sum: bit #] =
    (# carry := (x AND y) OR ((x XOR y) AND cin),
       sum   := (x XOR y) XOR cin #)

  FA_corr: LEMMA
    sum(FA(x, y, cin)) = x + y + cin - 2 * carry(FA(x, y, cin))

END full_adder
```

Correctness of the 1-bit full-adder is stated in terms of a simple arithmetic relationship FA_corr which holds between its inputs and outputs. This simple relationship states that the sum of the three input values x, y, and the carry-in cin equals the value of the carry-out multiplied by two plus the value of the sum output. To prove that this relationship holds between the inputs and outputs of the 1-bit adder, we start by unfolding definitions followed by propositional simplification (based on BDDs). Thus, the strategy grind proves lemma FA_corr automatically.

As usual, a carry propagate adder is specified as a row of full adders. The theory cpa shown below describes the implementation and correctness statement of a carry-propagate adder. It is parameterized with respect to the length of source and target bit-vectors; such circuit descriptions are called *parametric*.

The carry bit that ripples through the full adders is specified recursively by means of the function nth_cin. The function cpa defines the carry output to be the N-th carry, and the n-th sum output of the carry propagate adder is defined to be the sum output of the n-th full adder applied to the n-th bits of the input bit-vectors and the n-th carry.

```
cpa[N: posnat]: THEORY
```

---

[5] [# carry, sum: bit #] denotes a record type with named accessors carry and sum; expressions of this type are of the form (# carry := ..., sum := ... #).

```
BEGIN
  IMPORTING full_adder, bitvectors@bv[N], bitvectors@bv_nat

  xv, yv: VAR bvec[N]; n: VAR below[N]; j: VAR upto[N]

  nth_cin(j, xv, yv): RECURSIVE bit =
  IF j = 0 THEN 0
  ELSE carry(FA(xv(j - 1), yv(j -1),
                    nth_cin(j - 1, xv, yv))) ENDIF
  MEASURE j

  cpa(xv, yv): [# carry: bit, sum: bvec[N] #] =
    (# carry:= nth_cin(N, xv, yv),
       sum  := LAMBDA n: sum(FA(xv(n), yv(n), nth_cin(n,xv,yv)))
    #)
```

Correctness of the N-bit adder is expressed in terms of an arithmetic equation. In this case, the N-bit input and output buses are interpreted as the unsigned binary representation bv2nat of a number (see Sect. 3.1). For the declaration of bv2nat as an implicit coercion, the type-checker implicitly injects the conversion bv2nat whenever it finds a bit-vector instead of the required number expression.

```
cpa_inv: LEMMA
  bv2nat_rec(j, sum(cpa(xv, yv)))
  = bv2nat_rec(j, xv) + bv2nat_rec(j, yv)
    - exp2(j) * nth_cin(j, xv, yv)

cpa_char: THEOREM
    sum(cpa(xv, yv)) = xv + yv - exp2(N) * carry(cpa(xv, yv))
END cpa
```

Theorem cpa_char expresses the conventional correctness statement of an adder row on the *word-level*, and relates the gate-level implementation of cpa with the functional level. This characteristic theorem of cpa is an immediate consequence of the invariant cpa_inv. The proof of this invariant proceeds by induction on the variable j using an induction scheme for the type upto[N] (of j) from the library.

```
(induct-and-simplify "j" :exclude "FA"
                         :theories "full_adder")
```

In this proof, the use of the parameters exclude and theories yields a certain economy of proof construction, since they direct the induction strategy to apply the arithmetic characterization of the full adder as specified in the theory full_adder above instead of blindly unfolding the definition of the function FA.

The proof of the carry-propagate adder shows that the behavioral specification of the full adder is sufficient to specify and reason about higher-level building blocks, since implementation details are irrelevant.

4.4 Generic Hardware Components

Developments on higher levels should not only abstract from the width of datapaths as demonstrated in Sect. 4.3 but also from specific gate-level implementations of the basic building blocks, since, in many cases, behavioral specifications of many hardware components are sufficient to reason about higher-level building blocks. We call components that are parameterized with respect to behavioral specifications of other components *generic hardware components*. The main advantage of generic components is that they allow verification of circuits to be done in a hierarchical fashion.

Parameterized theories in PVS support the specification of generic hardware components, and behavioral constraints associated with the generic hardware components are specified in the ASSUMING part. Any instantiation of the formal parameters incurs the obligation to prove all of the ASSUMPTIONs.

The theory for the carry-propagate adder in Sect. 4.3, for example, could be parameterized with respect to correct implementations of 1-bit full adders FA as described below in order to make it independent of specific implementations.

```
cpa[N: posnat, FA: [bit,bit,bit -> [# carry, sum: bit #]]]: THEORY
BEGIN
  ASSUMING
   FA_char: ASSUMPTION
     sum(FA(x, y, cin)) = x + y + cin - 2 * carry(FA(x, y, cin))
  ENDASSUMING

  ...
 END cpa
```

An alternative to using explicit ASSUMPTION constraints for the generic hardware components is to parameterize the theory with respect to predicate subtypes. The type full_adder, for example, characterizes the set of valid full adders.

```
full_adder_char: THEORY
BEGIN

  full_adder: TYPE =
    { FA: [bit, bit, bit -> [# carry, sum: bit #]] |
        sum(FA(x, y, cin)) = x + y + cin - 2*carry(FA(x, y, cin)) }

END full_adder_char
```

Using this type definition, one may construct the header of a generic hardware component cpa by importing the theory full_adder_char and parameterizing with respect to functions of type full_adder.

```
cpa[N : posnat,      (IMPORTING full_adder_char)
    FA: full_adder
]: THEORY ...
```

5. Microprocessor Verification

PVS has been used to verify a number of significant microprocessor designs. All of these verifications use a common correctness methodology that can be used for relating two state machines. In this section, we first describe this common methodology used for microprocessor verification and illustrate this methodology using the benchmark Tamarack processor ([Joyc88b], see also Appendix). We also describe extensions of the methodology for pipeline verification and briefly describe our experience in using our methodology for the verification of industrial-strength microprocessors.

5.1 A Historical Perspective of Microprocessor Verification

Most microprocessor and microprogram verification exercises relate the programmer's view of a processor to its hardware implementation. A number of microprocessor designs have been formally verified [BeBr94, Hunt94, CaJB78, Cook86, SGGH94, SrBi90, Wind90]. Formal verification of microcode for microprocessors was pioneered by Bill Carter [LeCB74] at IBM in the 1970's and applied to elements of NASA's Standard Spaceborne Computer [LeCB74]; in the 1980's a group at the Aerospace Corporation verified microcode for an implementation of the C/30 switching computer using a verification system called SDVS [Cook86]; and a group at Inmos in the UK established correctness across two levels of description (in Occam) of the microcode for the T800 floating-point unit using mechanized transformations. Similarly, several groups have performed automated verifications of non-microcoded processors, of which Warren Hunt's FM8501 [Hunt94] and subsequent FM9000 [HuBr92] are among the most substantial. The problems of pipeline correctness were also studied previously by Srivas and Bickford [SrBi90], by Saxe and Garland [SGGH94], Burch and Dill [BuDi94], and Windley and Coe [WiCo94]. The Tamarack processor in the benchmark suite was considered quite a challenge not so long ago [Joyc88a]. PVS is able to verify the microcode of Tamarack and Saxe's pipelined processor completely automatically in a matter of minutes [CRSS94].

The AAMP5 processor that PVS was used to partially verify is significantly more complex, at both the macro and micro-architecture levels, than most other processors for which formal verification has been attempted; it has a large, complex instruction set, multiple data types and addressing modes, and a microcoded, pipelined implementation. Of these, the pipeline and autonomous instruction and data fetching present special challenges. One measure of the complexity of a processor is the size of its implementation. In the case of the AAMP5, this is some 500,000 transistors, compared with some tens of thousands in previous formally verified designs and 3.1 million in an Intel Pentium [INTE93].

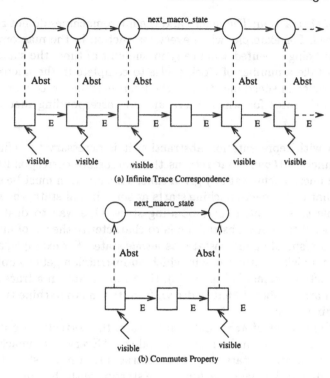

Fig. 5.1. General Microprocessor Correctness

5.2 General Microprocessor Correctness

In most mechanical verifications of microprocessors, the general correctness criterion used is based on establishing a correspondence, as shown in Fig. 5.1(a), between the execution traces of two state machines that the specifications at the macro and micro levels denote. In this figure, E and next_macro_state denote the micromachine and macromachine state transition functions, respectively, and the circles and squares denote the macro and micromachine states, respectively. The objective of the correctness criterion is to ensure that the micromachine does not introduce any behaviors not allowed by the macromachine. The macromachine uses two kinds of abstraction to hide details of the micromachine:

– *Representation* abstraction: Not every component of the micromachine state is visible at the macro level. Even the visible part of the micromachine state can take on quite a different form at the macro level. For example, in a stack-oriented processor, the process stack, which is viewed as residing entirely in the data memory in the macromachine, can actually be split between the memory and internal registers in the micromachine.

— *Temporal* abstraction: The macro and the micromachines do not run at the same speed. For example, while every instruction in the macromachine is viewed as being executed atomically in one unit of time, the same instruction may take a number of clock cycles to complete in the micromachine. The difference in speed means that the micromachine trace may have "intermediate" states for which there are no corresponding macromachine states.

To deal with representation abstraction it is necessary to define an *abstraction* function (**Abst**) that returns the macrostate corresponding to the state of the micromachine at any given time. This function must be surjective to ensure that if the macromachine starts at some initial state, the micromachine is able to start at a corresponding state. One way to deal with the consequences of temporal abstraction is to characterize the set of microstates that have corresponding macrostates as *visible* states. For example, a possible definition of a visible state is one in which an instruction gets completed and a new instruction begins. In Fig. 5.1(a), the visible states in a trace are those in which an arrow labeled **Abst** is drawn from the micromachine trace to the macromachine trace.

Given the notions of **Abst** and visible state, the correctness criterion described by Fig. 5.1(a) can be stated as follows: "Every micromachine trace starting with an initial state (S0) that abstracts to a macrostate (s0) must be abstractable to the macromachine trace starting with the initial state s0." To prove such a correspondence between two infinite traces, it is sufficient to prove the *commuting* property formally stated below and depicted graphically in Fig. 5.1(b). The commuting property captures the correspondence between the traces for one step of the macromachine, i.e. between two successive visible states.

5.3 A General PVS Framework for Verifying Microprocessors

We now provide a general methodology for specifying state machine systems in PVS and verifying that one state machine correctly implements another. The third author has developed a general framework [Cyrl93] for proving correspondence between state machines in PVS and has shown its application for pipelined microprocessors. The characterization of microprocessor correctness described here is similar to the visible state approach given in [Cyrl93]. Windley [Wind90] also develops a general framework for microprocessor correctness, although it is not applicable to pipelined processors. The methodology consists of three steps:

1. Provide the *specification* state machine.
2. Provide the *implementation* state machine.
3. Prove that the implementation machine satisfies the specification machine.

The state machines are specified by providing the following information:

- The type of the *machine state*.
- The *next-state* function.

The state type is typically a record consisting of all the state variables. In a hardware circuit the state variables would be the registers, memory, and other clocked components of the circuit.

The machine's behavior is specified by providing a next-state function. This function takes the current state, and returns the next state. The specification of this function may be quite complicated. It typically requires the specification of other constants, functions, and theories. It is usually defined in terms of auxiliary functions that return the values of the state variables in the next state, given the current state.

5.4 Formalizing the Correctness Condition

Once the specification and implementation state machines have been provided, the user must prove that the implementation state machine *satisfies* the specification state machine.

Before defining this correctness condition we provide some background. The execution history is an infinite sequence of states of the machine state type. In PVS, infinite sequences are functions from natural numbers to some type. Now we define the execution trace of a state machine as the sequence of machine states that results from iterating the next-state function for that machine. The execution trace is constructed by providing a function that for any natural number, n, returns the state that the machine will be in after iterating the next-state function n times starting at some initial state, and given an input sequence. The function `trace`, when given a next-state function and an initial state generates the execution trace for the machine defined by the next-state function. The PVS theory that defines these functions is presented in Fig. 5.2.
Now, given two state machines we would like to say that they are equivalent just in case, for any initial state, the traces of the two machines are identical. However, this is not quite enough. The problem is that we have not specified a way of relating the implementation state to the specification state. For example, in the above definition, what does it mean for the implementation machine and specification machine to start from the same initial state?

Thus, the user must provide an *abstraction* function that when given an implementation state returns a specification state. We denote this function as `Abst`. This function must be surjective to ensure that if a specification trace starts at some initial state, an implementation trace is able to start at a corresponding state. Given the `Abst` function, we can now state formally the condition that taking the abstraction of the implementation trace starting in state s is identical to the specification trace starting in state `Abst`(s). This is stated formally in equation 5.1.

```
traces[state: TYPE]: THEORY
BEGIN
  i, n: VAR nat

  next_type: TYPE = [state -> state]

  next: VAR next_type
  s    : VAR state

  next_to_the_n (next, s, n): RECURSIVE state =
    IF n = 0 THEN s
    ELSE next_to_the_n(next, next(s), n - 1) ENDIF
  MEASURE n

  trace(next, s): sequence[state] =
    LAMBDA i: next_to_the_n(next, s, i)
END traces
```

Fig. 5.2. Traces

(5.1) $\forall s : \mathrm{map}(\mathrm{Abst}, \mathrm{Itrace}(s)) = \mathrm{Atrace}(\mathrm{Abst}(s))$

Here, s is an implementation state, I the implementation next-state function, A the specification next-state function, and `Itrace`, `Atrace` are the corresponding traces generated from the `trace` function in Fig. 5.2. Furthermore, `map` is a function that given a function and a sequence returns a new sequence by applying the given function to every element in the given sequence.

In order to prove the higher-order formula 5.1 it is sufficient to prove the first-order formula `commutes`.

(5.2) $\forall s, input : \mathrm{Abst}(\mathrm{I}(s)) = \mathrm{A}(\mathrm{Abst}(s))$

We have captured the notion of correctness stated in equation 5.1, as well as the sufficient first-order condition `commutes` that implies this equation in a PVS theory `trace_equiv`. The theory is parameterized by the specification state machine, the implementation state machine, and the abstraction function, which is required to be surjective. It contains equation 5.1 as a theorem and its ASSUMING section consists of the `commutes` condition sufficient to prove the correctness condition. This theory is presented in Fig. 5.3.
Whenever this theory is instantiated, it will generate as *typecheck correctness conditions (TCCs)* two verification conditions:

1. The surjectiveness of `Abst`.
2. An instantiation of the `commutes` lemma.

There is no need for the user to prove the general higher-order verification condition, as this is proved once and for all in the generic `trace_equiv` theory.

```
trace_equiv [
  Astate: TYPE,
  A     : [Astate -> Astate],
  Istate: TYPE,
  I     : [Istate -> Istate],
  Abst  : (surjective?[Istate, Astate])
]: THEORY
BEGIN

  ASSUMING
    commutes: ASSUMPTION
      FORALL (s: Istate): Abst(I(s)) = A(Abst(s))
  ENDASSUMING

  IMPORTING traces[Astate], traces[Istate]

  r : VAR Astate
  s : VAR Istate

  Atrace(r): sequence[Astate] = traces[Astate].trace(A, r)
  Itrace(s): sequence[Istate] = traces[Istate].trace(I, s)

  trace_equiv: THEOREM
    map(Abst, Itrace(s)) = Atrace(Abst(s))
END trace_equiv
```

Fig. 5.3. Trace Equivalence

5.5 Verifying the Tamarack Processor

We illustrate our method on verifying the Tamarack microprocessor since it is part of the TPCD benchmark circuits [Krop94b] and also because it has features found in real designs such as microcoded control.

The externally visible state of the microprocessor consists of the program counter PC, accumulator ACC and external memory. Unseen by the user, the internal architecture involves several more registers used by the microcoded control to implement instruction level operations. In addition to registers, the datapath includes a 3-function ALU implementing addition, subtraction, and incrementation. Datapath components are interfaced to a single bus. The control unit is implemented by combinational logic, a latch for the 5-bit address of the current microinstruction and the microcode ROM.

We model and verify an implementation of the Tamarack at a register transfer level. The microcode is represented symbolically as a record with a finite number of fields. The microcode ROM is modeled as an array with an index ranging over natural numbers less than 15. The microprocessor has eight instructions consisting of an opcode field (modeled in our specification as an enumerated type), and an address field. In Gordon's original specification, the address field was thirteen bits wide but we consider a more general specification which is parameterized with respect to the types of the address field and the data word stored in the memory.

5.5.1 Tamarack Specification. We use a functional style to specify the behavior of the processor. The instruction, i.e. the programmer's, level behavior of the processor is specified by defining a function softstep that describes the effect of executing the *current instruction*, i.e. the one pointed to by the PC, on the externally visible state of the processor. Softstep can be considered as the *next-state* function for the abstract level state transition machine of the processor. Fig. 5.4 shows parts of the PVS formalization of Tamarack's macromachine. The style used for the Tamarack specification is slightly more abstract than the styles illustrated in Sect. 4. in that the *next-state* function is specified directly as a function on the state type. To extend the verification described here to another style of specification, all one needs to do is to show that the *next-state* function defined by the other specification is equivalent to the *next-state* function used in our verification.

```
soft[wordt, addrt: TYPE]: THEORY
BEGIN
 IMPORTING wordth[wordt, addrt]

 current_instr((pc: wordt), (mem: memt)): wordt =
   mem(addr_part(pc))

 current_instr_type((pc: wordt), (mem: memt)):
   opcodet = instr_part(current_instr(pc, mem))

 current_instr_addr((pc: wordt), (mem: memt)):
   addrt = addr_part(current_instr(pc, mem))


   .....

  softstep((soft_state: soft_statet)): soft_statet =
    LET
       omem = mempart(soft_state),
        opc = pcpart(soft_state),
       oacc = accpart(soft_state)
    IN
      (# mempart := soft_update_mem(omem, opc, oacc),
         pcpart := soft_update_pc(omem, opc, oacc),
        accpart := soft_update_acc(omem, opc, oacc) #)
END soft
```

Fig. 5.4. Tamarack Specification Machine

The microarchitecture level behavior of the processor is specified by defining a function hardstep (see Fig. 5.5) that describes the effect of executing a single microinstruction in the microcode ROM addressed by the microprogram counter mpc. **Hardstep** is itself defined hierarchically in terms of a set of *update* functions; one for every state sensitive component—such as a register, memory, bus, etc.—used in the register transfer level design. Thus the hardware structure of the register transfer level design of the processor is implicit in the functional dependency among the update functions. **Hardstep** is the *next-state* function for the implementation level state transition machine of the processor.

```
hard2[wordt, addrt: TYPE]: THEORY
BEGIN
  IMPORTING wordth[wordt, addrt]

  srcNdstn:
    TYPE = {pc, mem, ir, acc,       % source and destn for bus
       buf,                         % source only
       mar, arg, alu_add, alu_sub, alu_inc,% destn only
       none}

  alu((aluop: srcNdstn), (arg1: wordt), (arg2: wordt)): wordt =
    CASES aluop OF
       alu_inc: word_add1(arg2),
       alu_add: word_plus(arg1, arg2),
       alu_sub: word_diff(arg1, arg2)
       ELSE anything
    ENDCASES

  hard_statet: TYPE =
    [# memp: memt, pcp, accp, irp, marp, argp, bufp: wordt,
                                      mpcp: microaddrt #]

  hardstep((hard_state: hard_statet)): hard_statet
    = LET omem = memp(hard_state),
          opc = pcp(hard_state),
          oacc = accp(hard_state),
          .....
      IN (# memp := update_mem(omem, minstrn, omar, busvalue),
           pcp := update_pc(opc, busvalue, minstrn),
           accp := update_acc(oacc, busvalue,  minstrn),
           .....
           mpcp := update_mpc(ompc, oacc, oir, microrom(ompc)) #)
    ...
END hard2
```

Fig. 5.5. Tamarack Implementation Machine

5.5.2 Formalizing Tamarack Correctness. We formalize the correctness of the Tamarack processor by instantiating the general theory for relating two state machines developed in section 5.3. The framework presented in the trace_equiv theory assumes that the state machines at the two levels run at

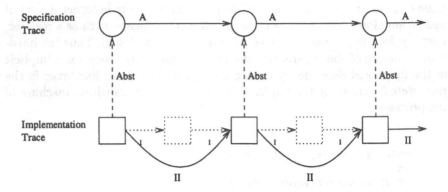

Fig. 5.6. Speeding up the Implementation Machine

the same rate, i.e. it assumes there is an A-step for every I-step. But in most microprocessors, including Tamarack, the specification and implementation state machines run at different *rates*. For example, for Tamarack, a `softstep` transition is implemented by one or more transitions of `hardstep` where the number of transitions depends on the type of the current instruction. One way to handle this temporal abstraction between the two levels is to conceptually "speed up" the implementation machine as shown in Fig. 5.6.

We call the approach in which we "speed up" the implementation machine, the *visible-state* approach. Another approach is to "slow down" or "stutter" the specification machine. For more details about the stuttering approach see [Cyrl93]. We call those states that have a corresponding specification state *visible* states. Those states that do not have a corresponding specification state we call *non-visible* states. Now, the main idea in the *visible-state* approach is to define a *visible* next-state function in the implementation machine. The visible next-state function *combines* the transitions from a *visible* state through zero or more *non-visible* states to the next *visible* state into one transition. (In Fig. 5.6 this function is II.)

One way to define the *visible* next-state function is for the user to provide the following items:

1. A predicate (`start_condn`) on the implementation state that characterizes the subset of visible states. For Tamarack, a visible state is one in where the processor is at the beginning of executing a new instruction. This is characterized by the condition that the microprogram counter `mpc` should be pointing to the beginning of the microcode ROM.
2. A function which, when given the current visible state will return the number of non-visible states the machine will cycle through before reaching the next visible state. We call this function `oracle`. The `oracle`

function for Tamarack is constructed by counting the number of micro instructions in the microcode for each Tamarack instruction.

Then the new next-state function can be defined as follows, where the function next_to_the_n(I,s,n) returns the state resulting from applying I to s n times

```
visible_state: TYPE = {is: Istate | start_condn(is)}

visible_I(vhs: visible_state): visible_state =
    next_to_the_n(I, s, oracle(s)).
```

An alternative to using an oracle function is to define the visible_I function in terms of a predicate that characterizes the visible states. Very often the proof of correctness needs to make use of such a predicate anyway. Nevertheless, using an oracle predicate is useful in that it provides guidance to the theorem prover on how to direct its search. The rest of the *visible-state* approach follows the basic methodology except that visible_I is used in place of I. The theory verification in Fig. 5.7 shows the instantiation of trace_equiv theory for modeling the correctness of Tamarack.

The main verification conditions to be proved for verifying the correctness of Tamarack are automatically generated by PVS's typechecker as part of the type correctness conditions for the theory verification shown in Fig. 5.7. The generated verification conditions are shown in Fig. 5.8. The condition I_TCC1 requires that the start_condn is preserved by every visible state; verif_TCC1 ensures that Abst is surjective; verif_TCC2 ensures the commutes condition. The second condition is fairly trivial to prove. The first and second conditions can be proved automatically by PVS using a version of the *core* strategy that we have devised for microprocessor verification. This strategy is described in section 5.7.

5.6 Verifying Pipelined Microprocessors

The basic idea behind pipelining is to increase the instruction execution rate of a processor by executing different stages of more than one instruction within the same clock cycle. Thus, although execution of an instruction is still spread over multiple clock cycles, it is possible to complete up to one instruction every cycle.

The general correctness criterion shown in Fig. 5.1 is still applicable for pipelined processors with some adjustment as long as instructions get completed in the same order as they enter the pipeline. An adjustment is needed to handle the fact that when a new instruction enters the pipeline, the results of the previous uncompleted instructions may not yet be at their destinations. The definition of the abstraction function Abst may have to peek to a future microstate, possibly nonvisible, following the current visible state (as shown in Fig. 5.9) to get correct values for some of the macrostate components.

```
verification [wordt, addrt: TYPE]: THEORY
BEGIN
  IMPORTING trace_equiv, wordth[wordt, addrt], soft[wordt, addrt],
      hard2[wordt, addrt], verification_rewrites[wordt, addrt],
      microrom_rewrite[wordt, addrt]

  next_to_the_n(I: [hard_statet -> hard_statet],
               hs: hard_statet, cycle: nat)
  : RECURSIVE hard_statet =
      IF cycle = 0 THEN hs
      ELSE next_to_the_n(I, I(hs), cycle-1) ENDIF
  MEASURE cycle

  start_condn ((hs: hard_statet)): bool = (mpcp(hs) = 0)

  visible_hard_state: TYPE = {hs: hard_statet | start_condn(hs)}

  oracle(vhs: visible_hard_state): nat =
    LET opcode = instr_part((memp(vhs))(addr_part(pcp(vhs))))
     IN IF jump_op?(opcode) THEN 4
        ELSIF jump_zero_op?(opcode)
          THEN IF is_zero(accp(vhs)) THEN 5 ELSE 6 ENDIF
        ELSIF load_op?(opcode) THEN 6
        ELSIF store_op?(opcode) THEN 6
        ELSIF add_op?(opcode) THEN 8
        ELSIF nop?(opcode) THEN 5
        ELSE 8 ENDIF

  visible_I(vhs: visible_hard_state): visible_hard_statet =
    next_to_the_n(hardstep, vhs, oracle(vhs))

  Abst (vhs: visible_hard_state): soft_statet
    = (# mempart := memp(vhs), pcpart   := pcp(vhs),
         accpart := accp(vhs)                          #)

  verif: THEORY = trace_equiv[soft_statet, softstep,
                              visible_hard_state, visible_I, Abst]
END verification
```

Fig. 5.7. Implementation Verification of Tamarack

Typically, the values for all but the program counter must be obtained from the future state. Hence, in Fig. 5.9, the function Abst is split into Abs_PC for the program counter and Abs_REG for the register file as those two state components have different latencies. We refer to the distance into the future from the current visible state where the information is to be obtained as the *latency* for the abstraction function.

In principle, it should be possible to define the abstraction function as a function of the current visible state alone, because the information necessary to compute the result of the to-be-completed instructions is stored in

```
% Subtype TCC generated (line 41) for
% next_to_the_n(hardstep, vhs, oracle(vhs))
I_TCC1: OBLIGATION (FORALL (vhs):
  start_condn(next_to_the_n(hardstep, vhs, oracle(vhs)))

% Subtype TCC generated (line 54) for Abst
verif_TCC1: OBLIGATION
  surjective?[visible_hard_state, soft_statet[wordt, addrt]](Abst);

% Assuming TCC generated (line 52) for
% trace_equiv[soft_statet, softstep,
%                   visible_hard_state, visible_I, Abst]
verif_TCC2: OBLIGATION FORALL (s: visible_hard_state):
  Abst(visible_I(s)) = softstep(Abst(s));
```

Fig. 5.8. Generated TCCs for Tamarack Verification

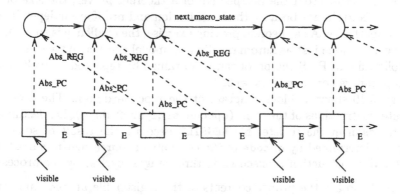

Fig. 5.9. Impact of Pipelining on Abstraction

various hidden registers in the micromachine. In the MiniCayuga verification [SrBi90], which is one of the earliest efforts in the mechanical verification of a pipelined processor, the abstraction function was defined in this fashion. However, it is easier and conceptually clearer to define the abstraction in a "distributed fashion" using the future visible or nonvisible states, which is the approach taken for the AAMP5. The pipelined verifications described in [Cyrl93], [SGGH94], [TaKu93b], and most recently [WiCo94], also define the abstraction function in a distributed fashion. Burch and Dill use a slightly different approach in [BuDi94] to handle the time skew between the two levels. They "run" the micromachine longer by an appropriate number of cycles by streaming in NOP instructions before relating the states of the macro and micromachines.

In Fig. 5.9, the distance between two consecutive visible states can vary for a number of different reasons, although a pipelined design strives to keep the distance to one cycle as often as possible. For example, an instruction requiring a memory access may take longer than a register-to-register instruction, or an instruction that signals an exception may take a few extra cycles for completion. The distance can usually be expressed as a function of the class of the new instruction entering the pipeline in the visible state.

5.6.1 A Simple Pipelined Processor. In this section we develop a complete proof of a correctness property of the controller logic of a simple pipelined processor design described at register-transfer level. The design and the property verified are both based on the processor example given in [BCMD92]. The example has been used as a benchmark for evaluating how well finite state-enumeration based tools, such as model checkers, can handle datapath-oriented circuits with a large number of states by varying the size of the datapath. From the perspective of a theorem prover, the size of the datapath is irrelevant because the specification and proof are independent of the datapath size. As a theorem proving exercise, the challenge is to see if the proof can be done just as automatically as a model checker. As we will see in the following, in PVS the proof can be obtained by repeatedly invoking one of its primitive commands `assert`.

Fig. 5.10 shows a block diagram of the pipeline design. The processor executes instructions of the form (`opcode src1 src2 dstn`), i.e. destination register `dstn` in the register file `REGFILE` becomes the result of some ALU function determined by `opcode` of the contents of source registers `src1` and `src2`. Every instruction is executed in three stages (cycles) by the processor:

1. *Read*: Obtain the proper contents of the register file at `src1` and `src2` and clock them into `opreg1` and `opreg2`, respectively.
2. *Compute*: Perform the ALU operation corresponding to the opcode (remembered in `opcoded`) of the instruction and clock the result into `wbreg`.
3. *Write*: Update the register file at the destination register (remembered in `dstndd`) of the instruction with the value in `wbreg`.

The processor uses a three-stage pipeline to simultaneously execute distinct stages of three successive instructions. That is, the read stage of the current instruction is executed along with the compute stage of the previous instruction and the write stage of the previous-to-previous instruction. Since the `REGFILE` is not updated with the results of the previous and previous-to-previous instructions while a read is being performed for the current instruction, the controller "bypasses" `REGFILE`, if necessary, to get the correct values for the read. The processor can abort, i.e. treat as `NOP`, the instruction in the read stage by asserting the `stall` signal true. An instruction is aborted by inhibiting its write stage by remembering the `stall` signal until the write stage via the registers `stalld` and `stalldd`. We verify that an instruction entering the pipeline at any time gets completed correctly, i.e. will write the

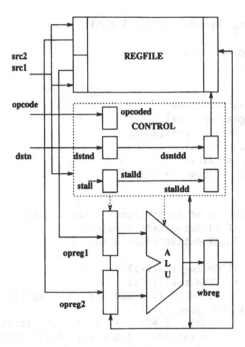

Fig. 5.10. A Pipelined Microprocessor

correct result into the register file, three cycles later, provided the instruction is not aborted.

5.6.2 Formal Specification. The microprocessor specification is organized into three theories, selected parts of which are shown in Fig. 5.11. (The complete specification can be found in [CORS95].) Theory `pipe` contains a specification of the design and a statement of the correctness property to be proved. The theories `signal` and `time` (see Sect 4.1) imported by `pipe` declare the types `signal` and `time` used in `pipe`.

The theory `pipe` is parameterized with respect to the types of the register address, data, and the opcode field of the instructions. By importing the theory `signal` uninstantiated in `pipe`, we have the freedom to create any desired instances of the type `signal`.

The microprocessor specification in `pipe` consists of two parts. The first part declares all the signals used in the design—the inputs to the design and the internal wires that denote the outputs of components. The composite state of REGFILE, which is represented as a function from `addr` to `data`, is modeled by the signal `regfile`. The signals are declared as uninterpreted constants of appropriate types. The second part consists of a set of AXIOMs that specify the values of the signals over time. (To conserve space, we have

```
pipe[addr, data, opcode: TYPE]: THEORY
BEGIN
 IMPORTING signal, time

 t: VAR time

 % -- Signal declarations
  opcode : signal[opcodes]
  src1, src2, dstn: signal[addr]
  stall  : signal[bool]
  aluout : signal[data]
  regfile: signal[[addr -> data]]
      ...

 % -- Specification of constraints on the signals
  dstnd_ax  : AXIOM dstnd(t+1) = dstn(t)
  dstndd_ax : AXIOM dstndd(t+1)= dstnd(t)
      ...
 regfile_ax: AXIOM regfile(t+1) =
  IF stalldd(t) THEN regfile(t)
    ELSE regfile(t) WITH [(dstndd(t)) := wbreg(t)] ENDIF
 opreg1_ax: AXIOM opreg1(t+1) =
  IF src1(t) = dstnd(t) & NOT stalld(t) THEN aluout(t)
    ELSIF src1(t) = dstndd(t) & NOT stalldd(t) THEN wbreg(t)
    ELSE regfile(t)(src1(t)) ENDIF
 opreg2_ax: AXIOM ...

 aluop: [opcodes, data, data -> data]

 ALU_ax: AXIOM
   aluout(t) = aluop(opcoded(t), opreg1(t), opreg2(t))


   ...
END pipe
```

Fig. 5.11. Formalization of Pipelined Microprocessor

only shown the specification of a subset of the signals in the design.) For example, the signal value at the output of the register dstnd at time t+1 is defined to be that of its input a cycle earlier. The output of the ALU, which is a combinational component, is defined in terms of the inputs at the same time instant.

In the present example, the specifications of the signals opreg1 and opreg2 are the most interesting of all. They have to check for any register collisions that might exist between the instruction in the read stage and the instructions in the later stages and bypass reading from the register file in case of collisions. The regfile signal specification is recursive since the register file state remains the same as its previous state except, possibly, at a single register location. Note that the function aluop that denotes the operation ALU performs for a given opcode is left completely unspecified since it is irrelevant to the controller logic.

The theorem correctness to be proved states a correctness property about the execution of the instruction that enters the pipeline at t, provided the instruction is not aborted, i.e. stall(t) is not true.

```
correctness: THEOREM
  FORALL (t: time): NOT(stall(t)) IMPLIES
    regfile(t + 3)(dstn(t))
  = aluop(opcode(t), regfile(t + 2)(src1(t)),
                     regfile(t + 2)(src2(t)))
```

The equation in the conclusion of the implication compares the actual value (left hand side) in the destination register three cycles later, when the result of the instruction would be in place, with the expected value. The expected value is the result of applying the aluop corresponding to the opcode of the instruction to the values at the source field registers in the register file at t+2. We use the state of the register file at t+2 rather than t to allow for the results of the two previous instructions in the pipeline to be completed.

5.7 Mechanization of Proofs of Verification Conditions

Given our style of hardware specification, proof of a formula relating two states of a state machine that are a fixed distance apart usually follows a standard pattern consisting of a sequence of proof tasks shown below:

Quantifier elimination: Eliminate the universally quantified variable (t) by *skolemization* and simplify the preconditions. Skolemization consists of replacing the universally quantified variable by a new constant symbol denoting an arbitrary value for that variable. This technique is a simple and general way to prove a property for all values in a set that a variable ranges over.

Unfolding definitions: Simplify the selected expressions and defined function symbols in the goal by rewriting using definitions, axioms or previously established lemmas in the micromachine specification.

Case analysis and simplification: At the end of the unfolding step, the original goal will have been simplified to an equation on two nested IF-THEN-ELSE expressions, not necessarily identical, involving user-defined as well as primitive function symbols. The IF-THEN-ELSEs are introduced by the unfolding of the defined function symbols. To prove such an equation, it may be necessary to split the proof, based on selected boolean expressions in the current goal, and further simplify the resulting goals.

The following short PVS proof strategy, called the *core* strategy, can accomplish the above proof tasks:

```
1:  (then
2:      (skosimp*)
3:      (auto-rewrite-theories ''list of specification theories'')
4:      (repeat (assert))
5:      (apply (then* (repeat (lift-if))
6:                    (ground)
7:                    (assert))))
```

The proof command on Line 2 performs the skolemization task. The command on Line 3 makes rewrite rules out of all the definitions and axioms in our specification, after which, the command on Line 4 rewrites all the expressions in the goal until no further simplification is possible. Assert performs rewriting as well as simplifications using arithmetic and equality decision procedures. In the case of our verification conditions, this rewriting step has the effect of reducing all expressions into ones that involve only values of signals in the initial state. The compound proof step on Lines 5 through 7 performs the case-analysis and further simplification task. The case analysis is performed by lifting all the IF-THEN-ELSE structures to the top and then simplifying the resulting expression propositionally (bddsimp). (Apply applies a compound proof step as an atomic step.)

We have used the core strategy (or slight variants of it) to prove with a high degree of automation the correctness of several microprocessor designs that have served as informal hardware verification benchmarks for theorem provers. See Cyrluk et al. [CRSS94] for more details about application of this strategy for hardware verification. In general, user's interaction might be necessary at primarily two points. First, in Line 3, where auto-rewrite rules are installed, the user might have to be selective in tuning the rewrite rules especially to keep the simplification step from blowing up. Second, the implementation (Lines 5 through 7) of the "Case analysis and simplification" part of the core strategy is guaranteed to succeed in constructing a proof only if the "Unfolding" step (Line 4) simplifies the original goal into a relation on expressions involving either operations on primitive PVS types that the decision procedures are capable of handling or user-defined functions that can be left uninterpreted. If not, the user's involvement is necessary to complete the proof.

The core strategy is adequate to automatically prove the verification condition of the pipeline example shown in Sect. 5.6.1 and all but the surjective verification condition of the Tamarack microprocessor in Sect. 5.5. The surjective verification condition, which involves an existential quantifier, requires the user to provide a simple instance (record) of the implementation state.

5.8 Verification of Industrial-Strength Microprocessors

The basic approach described in the previous sections was applied to model a commercial avionics processor AAMP5 and to verify a subset of its instructions [SrMi96]. AAMP5 is a microcoded pipelined processor built at the Collins Avionics Division of Rockwell International for Avionics applications. It is a complex CISC processor containing more than half a million transistors and is designed to execute a stack-oriented machine. The verification revealed several errors: some unknown to Rockwell and some planted by Rockwell engineers as a challenge. Compared with other microprocessor verifications [Hunt85, SrBi90, WiCo94, SGGH94], the verification of the AAMP5 is significantly more complex than any other processor for which formal verification has been attempted; it has a large, complex instruction set, multiple data types and addressing modes, and a microcoded pipelined implementation.

Besides the sheer size of the design, the main technical challenge in mechanizing the verification of AAMP5 was the fact that the core strategy was not adequate to automatically prove the correctness of AAMP5. The main reasons that the core strategy could not automatically prove AAMP5 correctness are described below.

The core strategy in Sect. 5.7, when unsuccessful in proving a verification condition, simplifies the original goal into equations on bit-vector expressions. These expressions can be shown to be equivalent only by using properties about the bit-vector operations as lemmas. To automate the bit-vector equivalence checking step of the proof, we formulated a number of bit-vector lemmas in the form of rewrite rules. Most of these lemmas have been proved as theorems by Rick Butler using the bit-vector library. A subset of these rules can be used to first attempt to simplify the bit-vector expressions into a common normal form. If the rules are unsuccessful in normalizing the expressions into a common form, then another set of rules can be used to convert the bit-vector equivalence into an equivalence on natural numbers that can most likely be handled by the decision procedures.

A significant portion of our verification effort was devoted to formulating the bit-vector simplification rules. The rules formulated are parameterized with respect to the size of the bit-vectors involved, but are not complete enough to decide equivalence for all possible bit-vector expressions. They were able to successfully decide equivalence of expressions in most of our proofs. Even when they were unsuccessful, we had to perform only a few standard case-splits manually to complete the proof.

1. The number of cycles between successive visible states for AAMP5 can be indeterminate (although finite) due to pipeline stalling caused by memory wait states.
2. The AAMP5 specification, both at the micro and macro levels, uses a number of complex bit-vector operations, such as concatenation (o), subsetting (^), shifting, etc. All bit-vector operations are specified as parameterized functions that are defined recursively in the bit-vector library. PVS's decision procedures cannot automatically decide bit-vector equalities.

3. The abstraction function relating the implementation state to the specification state is much more complicated than the simple projection function used for Tamarack.

One way to facilitate automation, is to transform the verification problem into one of proving a set of properties each of which relates a pair of states of the micromachine that are a fixed and finite distance apart in time. To accomplish such a transformation of the verification problem, we had to decompose the proof of the commuting lemma into three parts.

1. A part that reasons exclusively about the stalling behavior of the pipeline at the micromachine level.
2. A part that reasons about the instruction correctness in the absence of stalling at the micromachine level.
3. A part that combines the first two parts along with the task of relating the micromachine to the macromachine by applying the abstraction function.

This decomposition was natural since pipeline stalling is implemented by a set of special-purpose microroutines that are invoked as a subroutine during an instruction execution. The correctness of pipeline stalling was characterized by a set of formulas called *general verification conditions* since they are common to all instructions. The correctness of instructions in the absence of stalling were formalized by a separate set called *instruction-specific* verification conditions. Proofs in the first and the third parts were done using the core strategy with a few augmentations as described below. These proofs are automatic enough for a nonexpert to perform. Proofs in the general verification conditions involved combining the core strategy with induction.

6. Verification of Arithmetic Circuits

The well-publicized FDIV error of initial releases of the Pentium [INTE93] floating-point divider sparked renewed interest in the verification of arithmetic circuits [BrCh94, Brya95, LeLe95, AaSe95, KaSu96, ClGZ96, RuSS96, MiLe96]. Verifying functional correctness results about arithmetic circuits, however, poses some serious challenges to current hardware verification techniques, since BDD representations of important arithmetic functions like multiplication grow exponentially with the word size. Consequently, it is not feasible to directly apply BDD-based methods to prove the overall correctness of many interesting arithmetic circuits with wide datapaths. Theorem-proving based approaches, on the other hand, deal with the state explosion problem by exploiting the regular structure of arithmetic circuits and using inductive reasoning to establish correctness results for arbitrary word sizes.

Here, we desribe an exemplary verification, for arbitrary word sizes, of gate-level implementations of iterative array multipliers. The layout of these

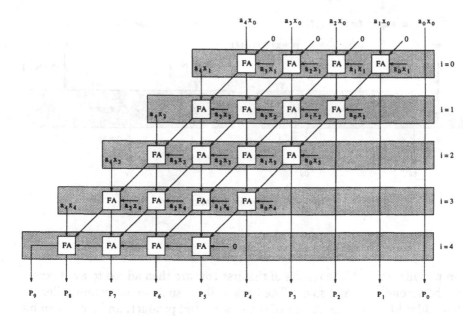

Fig. 6.1. Iterative Array Multipliers

two-dimensional combinational multipliers are described by means of a hierarchy of recursive functions (see also Sect. 4.3), and the hierarchical structure of this description leads to a decomposition of the proof into manageable pieces. Similar decomposition techniques are also applicable for other arithmetic circuits like Booth multipliers or non-restoring division circuits.

The second part of this section includes an outline of the verification of a radix-4 SRT division circuit that is similar to the fixed-point core of the floating-point division algorithm used in the Pentium microprocessor [INTE93]. A more elaborate description of this verification, including a verification for arbitrary radix SRT division algorithms, can be found in [RuSS96].

6.1 Verification of Iterative Array Multipliers

Iterative array multipliers [Kore93] are based on the grade school principle of computing partial products and adding partial products to obtain the required results. To illustrate the operation of an iterative array multiplier for unsigned numbers, examine the 4×5 parallelogram shown in Fig. 6.1. This circuit multiplies two $N = 5$ wide operands, say av and xv. It adds the first two partial products (i.e. av $*$ xv(0) and av $*$ xv(1)) in row one after

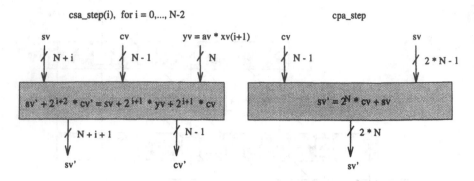

Fig. 6.2. Vertical Refinement

proper alignment.[6] The results of the first row are then added to av * xv(2) in the second row, and so on. The basic cell for such an array multiplier is a full adder FA accepting one bit of the new partial product, and a carry-in bit. In the first four rows there is no horizontal carry propagation. In other words, a *carry-save* type addition is performed in these rows, and the accumulated partial product consists of intermediate sum and carry bits. A horizontal carry propagation is allowed only in the last row. The last row of cells in this figure is a *ripple-carry adder*. Consequently, an iterative array multiplier circuit with operands of size N is abstracted in terms of a series of (N - 1) carry-save steps followed by a carry-propagate step. Block diagrams of these stages of the *vertical decomposition* or vertical refinement are depicted in Fig. 6.2 together with word-level equations between appropriately shifted input and output vectors. The word-level condition sv' = 2^N * cv + sv of the carry-propagate step, for example, states that the (unsigned) interpretation of the output sum vector sv' equals the sum of the input sum vector sv and the carry-in vector cv shifted by N bits.

Moreover, looking at the circuit in Fig. 6.1, every vertical stage can naturally be decomposed into different functional units. The *horizontal decompositions* of csa_step(i) and cpa_step are depicted in Fig. 6.3. For every vertical stage, the decomposition consists of three functional units. The left-most unit simply computes the up-most bit of the next partial sum av(N - 1) * xv(i + 1), the right-most unit shifts through the lower (i + 1) bits of the current partial sum sv, and the central unit consists of a carry-save adder row of length (N - 1). The final step of the multiplier is decomposed in a similar way (see Fig. 6.3) with the central unit being a (N - 1)-bit carry-propagate adder row. These observations lead to the fol-

---

[6] In Fig. 6.1, we write x_i (resp. a_i) instead of xv(i) (resp. av(i)) in order to save space.

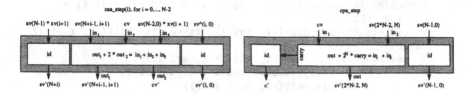

Fig. 6.3. Horizontal Refinement

lowing decomposition of the correctness proof. First, assuming correct adder rows, we prove the correctness of the horizontal decompositions. Second, assuming the correctness of the horizontal decomposition we prove the correctness of the vertical decomposition, and, consequently, the overall correctness of the circuit.

Formalization of Horizontal and Vertical Composition. Formalization of the carry-propagate step as depicted in Fig. 6.3 is straightforward. The function cpa_step defined below calls the carry-propagate adder on the carry vector cv and the upper half of the sum vector sv, and the resulting output sum vector is created by concatenating the carry output with the sum output and the lower half of the input sum.[7]

```
cpa_step((cv: bvec[N - 1]), (sv: bvec[2 * N - 1])): bvec[2 * N] =
  LET cpa = cpa(sv^(2 * N - 2, N), cv) IN
    carry(cpa) o sum(cpa) o sv^(N - 1, 0)
```

The encoding of carry-save steps csa_step(i) is slightly more complicated and requires the concept of *dependent types*, since the types of the input parameters sv and cv and the resulting record expression depend on the value of the index i;[8] this formalization is a straightforward transliteration of the block diagram of the carry-save step in Fig. 6.3.

```
csa_step(i:below[N-1])((sv:bvec[N+i]),(cv:bvec[N-1]),(yv:bvec[N]))
    : [# sum: bvec[N + i + 1], carry: bvec[N - 1] #] =
    LET csa = csa(sv^(N + i - 1, i + 1), yv^(N - 2, 0), cv) IN
      (# sum   := yv(N - 1) o sum(csa) o sv^(i, 0),
         carry := carry(csa)                           #)
```

Finally, the carry-save and the carry-propagate stages are combined to form an iterative array multiplier.[9]

```
connect_csa((i: upto[N - 2]), (av, xv: bvec[N])):
```

---

[7] Here, a conversion from bit to bvec[1] is assumed.
[8] The definition and correctness proof of carry-save adder row csa is omitted, since it is similar to the encoding of carry-propagate adder rows in Sect. 4.3.
[9] bvec0 denotes a constant 0 bit-vector of appropriate length.

```
    RECURSIVE [# carry: bvec[N - 1], sum: bvec[N + i + 1] #] =
    (IF i=0 THEN
        csa_step(0)(bit_mult(av, xv(0)), bvec0, bit_mult(av, xv(1)))
    ELSE
        csa_step(i)(sum(connect_csa(i - 1, av, xv)),
                    carry(connect_csa(i - 1, av, xv)),
                    bit_mult(av, xv(i + 1)))
    ENDIF)
    MEASURE i

iam(av, xv: bvec[N]): bvec[2 * N] =
  cpa_step(carry(connect_csa(N - 2)(av, xv)),
           sum(connect_csa(N - 2)(av, xv)))
```

The iterative array multiplier iam consists of $(N - 2)$ stages of appropriately wired carry-save steps csa_step(i) followed by a final carry-propagate step. This finishes the formalization of this multiplier.

Correctness. The correctness proofs of the horizontal refinements need the following decomposition and composition lemmas. Let n, m be positive natural numbers, and xv, yv be bit-vectors of sizes n and m, respectively, and i of type subrange[1,n-1]; then:

```
    decompose: LEMMA xv o yv = xv * exp2(m) + yv
    compose  : LEMMA xv^(i - 1, 0) + xv^(n - 1, i) * exp2(i) = xv
```

First, the correctness of the horizontal refinement of the carry-save step is expressed by the following lemma:

```
    csa_step_char: LEMMA
      FORALL ((i:below[N - 1]), (sv: bvec[N + i]),
              (cv: bvec[N - 1]),(xv: bvec[N])):
        sum(csa_step(i, sv, cv, xv))
        = sv + exp2(i + 1) * xv + exp2(i + 1) * cv
          - exp2(i + 2) * carry(csa_step(i, sv, cv, xv))
```

The PVS prover needs some guidance to apply the compose lemma above in an appropriate way, since the rewrite engine of PVS does not apply *commutative-associative* matching.

Second, the formula cpa_step_char (compare with Fig. 6.2) for the carry-propagate step is simply proved by quantifier elimination followed by unfolding the definition cpa_step and repeated calls to the decision procedures.

```
    cpa_step_char: LEMMA
      FORALL (cv: bvec[N - 1], sv: bvec[2 * N - 1]):
        cpa_step(cv, sv) = sv + cv * exp2(N)
```

Third and finally, the lemmas csa_step_char and cpa_step_char permit an easy induction proof of the behavioral correctness iam_char of the multiplier.

```
    av, xv: VAR bvec[N]; i: VAR upto[N - 2]
```

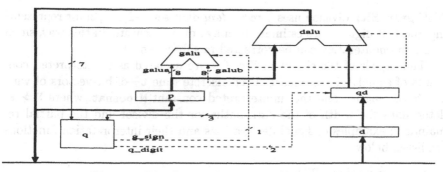

Fig. 6.4. The data path for the division circuit.

```
iam_inv: LEMMA
  sum(connect_csa(i, av, xv))
  = av * bv2nat_rec(i + 2, xv)
    - exp2(i + 2) * carry(connect_csa(i, av, xv))

iam_char: THEOREM
  iam(av, xv) = av * xv
```

The invariant iam_inv, and consequently the functional correctness iam_char of iterative array multipliers, is proved by induction on i. Both the base case and the induction step are easily proved with an appropriate combination of quantifier reasoning, rewriting, and calls to the decision procedures.

Altogether, the combination of the correctness proofs of the full adder and adder rows in Sect. 4.3, and the correctness of the high-level design of the iterative array multiplier yields verified implementations of iterative array multipliers on the gate-level.

6.2 Verification of SRT Division

SRT division algorithms [Toch58, Robe58, Sorl61] are among the most popular methods for implementing floating-point division and related operations in high-performance arithmetic units. Even though the theory of SRT division has been extensively studied [Atki68], the design of dividers still remains a serious challenge [ObFl94], and it is easy to make mistakes in its implementation—as was highlighted by the much publicized FDIV error in the Intel Pentium chip.

The data path for the radix-4 SRT division circuit in Fig. 6.4 has been developed by Taylor [Tayl81], and in [RuSS96] we describe a mechanized verification of this circuit. Here, we outline some of the results of this verification.

Given a dividend d and a partial remainder p—initialized with the divisor—the dalu in Fig. 6.4 computes a quotient digit q and a new partial remainder p in each iteration. Since the computation of the new partial remainder depends on the computation of the quotient digit, SRT circuits precompute an "estimation" P of the next partial remainder using a guess

ALU galu. SRT division uses a *redundant* digit set $[-2, \ldots, 2]$ for representing quotient digits, since estimations may not be accurate. In this way, small errors in one iteration can be corrected in subsequent iterations.

The signals of the circuit in Fig. 6.4 are declared as uninterpreted constants of signals—i.e. sequences over discrete time t—of bit-vectors of various fixed lengths, and the uninterpreted constant N:posnat, where N > 8, determines the width of the data paths for the divisor and the partial remainders; examples of signal declarations and their interpretation functions are listed below.

```
d    : signal[bvec[N]]
d(t): rational = fp[1, N - 1].val(d(t))

P    : signal[bvec[7]
P(t): rational = fp2c[4, 3].val(P(t))
```

The divisor signal d has a fixed-point interpretation with 1 leading and (N - 1) residual bits, and the estimation P of the next partial remainder has a 2's-complement fixed-point interpretation with 4 leading bits and 3 residual bits.

The hard part in each iteration is to select an appropriate quotient digit from the computed estimation of the next partial remainder and a truncation of the divisor d. SRT division algorithms usually use lookup tables for this task. Triangular-shaped regions at top and bottom of these tables are never referenced by the algorithm; the Pentium error was that certain entries believed to be in this inaccessible region, and containing arbitrary data, were, in fact, sometimes referenced during execution [Prat95].

The quotient lookup table in Fig. 6.5 is a particularly compact one [Tayl81]. It computes the next quotient digit from the truncation D of type bvec[3] of the divisor to the three leading bits and the estimation P of type bvec[7] of the next partial remainder. Bits 6 down to 2 of P are used as a table index and the remaining bits are used in some cases to compute the resulting value.

The formalization of the resulting lookup table in Fig. 6.5 uses the TABLE construct of the PVS specification language [OwRS95]. This construct was added to the PVS specification language in order to provide visually appealing two-dimensional tabular specifications in the manner advocated by Parnas and others [Parn95]. It proved adequate to express the lookup table of this SRT circuit in a concise and perspicuous way. In particular, blank entries in the lookup table in Fig. 6.5 model partiality and cause the type-checker to generate TCCs which ensure that viable arguments D, P never point to such a blank entry. Furthermore, the table construct requires that the lookup is functional and ensures this by generating *disjointness* and *coverage* TCCs. The TCCs generated from the definition of this lookup table in Fig. 6.5 can be proved largely automatically in PVS.

Injection of an error similar to that in the Pentium leads to a failed TCC proof whose sequent is a counterexample that highlights the error. Miner

```
LET a= -(2 - P(1) * P(0)), b= -(2 - P(1)), c= 1 + P(1),
    d= -(1 - P(1)), e= P(1)
 IN TABLE  P^(6,2), D
```

	000	001	010	011	100	101	110	111
01010								2
01001						2	2	2
01000					2	2	2	2
00111			2	2	2	2	2	2
00110		2	2	2	2	2	2	2
00101	2	2	2	2	2	2	2	1
00100	2	2	2	2	c	1	1	1
00011	2	c	1	1	1	1	1	1
00010	1	1	1	1	1	1	1	1
00001	1	1	1	1	e	0	0	0
00000	0	0	0	0	0	0	0	0
11111	0	0	0	0	0	0	0	0
11110	-1	-1	d	d	0	0	0	0
11101	-1	-1	-1	-1	-1	-1	-1	-1
11100	a	b	-1	-1	-1	-1	-1	-1
11011	-2	-2	-2	b	-1	-1	-1	-1
11010	-2	-2	-2	-2	-2	-2	b	-1
11001	-2	-2	-2	-2	-2	-2	-2	-2
11000			-2	-2	-2	-2	-2	-2
10111				-2	-2	-2	-2	-2
10110						-2	-2	-2
10101							-2	-2

Fig. 6.5. Quotient Lookup Table

and Leathrum have used this capability in PVS to develop several new SRT tables [MiLe96], each in less than three hours.

We prove in [RuSS96], basically using the grind strategy, the following lemmas about the circuit in Fig. 6.4.

```
lemma1: LEMMA p(t + 1) =  4 * (p(t) - q(t) * d(t))
lemma2: LEMMA P(t) <= p(t + 1) & p(t + 1) < P(t) + 3/16
```

lemma1 is the basic recurrence relation of SRT algorithms, and lemma2 states that P is an underestimator of the next partial remainder where the maximal error is 3/16. These conditions are sufficient, as we prove in [RuSS96], to show convergence of this circuit.

```
taylor_convergence: THEOREM
  LET residue = p(0) / d(0) - val(t + 1, q) IN
   - 2 / (3 * 4^t) <= residue & residue <= 2 / (3 * 4^t)
```

Since val(t + 1, q) denotes the (radix-4) value of the first t + 1 quotient digits, the value of the accumulated quotient digits converges to the actual quotient p(0) / d(0).

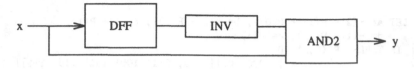

Fig. 7.1. Abstract Model of Single Pulser

7. Experimental Results

This section describes the use of PVS for verifying some additional examples from the TPCD benchmark not covered in the earlier sections. We hope these examples provide a comparison of the effectiveness of PVS with respect to other verification systems including model checkers. In some of the examples, we have specified the designs at a higher level than at which they are described in the benchmark for the sake of keeping the level of modeling roughly uniform, namely close to register transfer level, for all the examples in this chapter.

7.1 Verifying the Single Pulser

The *single pulser* in Fig. 7.1 is a clocked sequential device with one boolean input x and a boolean output y. For each pulse received on x, a single pulse is emitted on y, and the simple design in Fig. 7.1 emits a single pulse at the earliest possible time (after x goes high). For a discussion on various formal models of the single pulser, including one in the PVS language, see [JoMC94].

Here, we describe a formalization of the *single pulser* (see Appendix) by means of a finite transition relation. The required properties have been encoded into CTL with fairness constraints and proved automatically using symbolic model checking.

```
singlepulser: THEORY
BEGIN
  state: TYPE = [# x, dff: bool #]

  s, curr, next: VAR state

  x(s)  : bool = x(s)     % Accessors are not functions in PVS
  dff(s): bool = dff(s)

  IMPORTING MU@connectives[state]
```

A state of the single pulser model consists of a boolean input x and the value of the D-flipflop dff. Furthermore, functions corresponding to the accessors of the state record are defined, and the imported theory connectives from the PVS library defines liftings of the boolean connectives to the type pred[state].

```
init: pred[state] = NOT dff

N(curr, next): bool = (next = (curr WITH [dff := x(curr)]))

y: pred[state] = (x AND NOT dff)
```

The init predicate and the binary relation N characterize the set of initial states and the transitions from the current state to the next state, respectively. Informally, the WITH construct states that state next is reachable in one step from state curr if they coincide in every component except for the dff component, which must have the value of the current input x(curr).

```
IMPORTING MU@ctlops[state], MU@fairctlops[state]

rising_edge: pred[state] = (x AND NOT dff)

is: VAR { s: state | init(s) }

char1: LEMMA AG(N, rising_edge IMPLIES AF(N, y))(is)

char2: LEMMA
  AG(N, y IMPLIES AX(N, fairAU(N, NOT(y), rising_edge)
                        (rising_edge)))
    (is)

char3: LEMMA
  AG(N, rising_edge IMPLIES
          (NOT(y) IMPLIES AX(N, AU(N, NOT(rising_edge), y))))
    (is)

END singlepulser
```

Lemma char1 specifies the—for the given implementation rather trivial— liveness property that rising edges eventually lead to an output pulse, char2 states that there is at most one output pulse for each rising edge, and char3 specifies the fact that there is at most one rising edge for each output pulse. Thus, Lemmas char2 and char3 together establish the correspondence between input and output pulse. Notice that the specification of lemma char2 requires a *fair*-CTL operator fairAU with the fairness predicate rising_edge (see also [JoMC94]). Lemmas char1 and char3 are proved using the model-check command of the PVS prover (see Sect. 3.5). In addition, for the presence of fair-CTL operators, the proof of char2 requires unfolding of the definitions in fairctlops[state].

```
(then (auto-rewrite-theory "fairctlops[state]")
      (model-check))
```

7.2 Verifying the Arbiter

The arbiter described below (see also [McMi93a]) consists of 4 identical cells each of which consists of a local state transition function for the registers token and persistent, and combinational functions determining override, grant, and acknowledge values ack. Furthermore, definitions like persistent(i) introduce convenient accessors for the value of the persistent register of the ith cell.

```
arbiter: THEORY
BEGIN

  i, j: VAR below[4]

% -- State

  state: TYPE =
    [below[4] -> [# request, persistent, token: bool #]]

  s, s0, s1: VAR state

  request(i)   : pred[state] = LAMBDA s: request(s(i))
  persistent(i): pred[state] = LAMBDA s: persistent(s(i))
  token(i)     : pred[state] = LAMBDA s: token(s(i))

  IMPORTING MU@connectives[state]
```

The combinational structure of values is described by means of the recursive functions override and grant. Notice that every grant value depends only on the grant values of lower index, while override values depends on override values of larger index; moreover, every grant value depends on every override value.

```
% -- Definitions

  override(k: upto[4]): RECURSIVE pred[state] =
    IF k = 4 THEN FALSE
    ELSE override(k + 1) OR (token(k) AND persistent(k)) ENDIF
  MEASURE (LAMBDA (k: upto[4]): 4 - k)

  grant(i): RECURSIVE pred[state] =
    IF i = 0 THEN NOT override(0)
    ELSE grant(i - 1) AND NOT request(i - 1) ENDIF
  MEASURE i

  ack(i): pred[state] =
    request(i) AND ((persistent(i) AND token(i)) OR grant(i))
```

Finally, the ith acknowledge value is computed from the local state of the ith cell and the ith grant value. This concludes the encoding of the combinational functions of the arbiter.

The state transition system is characterized, as usual, by a state predicate init constraining the possible initial values and a relation next between current and successor states.

```
% -- Init
```

```
init(s): bool =
   (NOT persistent(0)(s)) & ... & (NOT persistent(3)(s))
 & token(0)(s) & (NOT token(1)(s)) & ...

% -- Next state

next(s0, s1): bool =
      token(0)(s1) = token(3)(s0) & token(1)(s1) = token(0)(s0)
    & token(2)(s1) = token(1)(s0) & token(3)(s1) = token(2)(s0)
    & persistent(0)(s1)
      = (request(0)(s0) AND (persistent(0)(s0) OR token(0)(s0)))
    & ...
    & persistent(3)(s1)
      = (request(3)(s0) AND (persistent(3)(s0) OR token(3)(s0)))
```

From the definition of next it becomes obvious that the arbiter state machine works like a cyclic shift register making the token rotate. Shorter and more concise specifications of both the initial predicate and the next-state relation are possible by capturing the regularity of the arbiter specification and by using universal quantification.

The arbiter must never grant access to two nodes simultaneously and this is expressed by the safety property th1 . The liveness property th2 states that every persistent request is eventually acknowledged, and th3 states that acknowledge is not asserted without request. Notice that theorems th1 through th3 are universally quantified with respect to free occurrences of i and j.

```
% -- Specs

IMPORTING MU@ctlops[state]

th1: THEOREM
   init(s) IMPLIES
      i /= j IMPLIES AG(next, NOT(ack(i) AND ack(j)))(s)

th2: THEOREM
   init(s) IMPLIES
      AG(next, AF(next, request(i) IMPLIES ack(i)))(s)

th3: THEOREM
   AG(next, ack(i) IMPLIES request(i))(s)

END arbiter
```

We only consider the proof script for theorems th2 and th3. The proof of th1 is identical except for the case split. First, skosimp introduces the constant i!1 for the universally quantified variable i, case* case splits over the possible values of i!1, and assert simplifies the resulting subgoals.

```
(then (skosimp*)
      (case* "i!1 = 0" "i!1 = 1" "i!1 = 2" "i!1 = 3")
      (assert)
```

```
(replace*)
(model-check))
```

In a next step `replace*` replaces, for each resulting case, the constant `i!1` with the numeric value of `i!1`, and, finally, the `model-check` command discharges each subgoal by unfolding the CTL-Operators into their μ-calculus encodings, decoding elements of the finite state type, followed by μ-calculus model checking.

This proof demonstrates a rather crude interaction of symbolic model checking techniques with theorem proving capabilities in order to prove a class of theorems in one step. A more advanced combination of theorem proving and model checking in PVS is employed, for example, by Shankar [Shan96] in order to use symbolic model checking in the induction step of an N-process mutual exclusion algorithm problem.

7.3 Verifying the Black-Jack Design

Our PVS formalization of the design of the Black Jack dealer benchmark circuit (see page 334) closely follows the description in [MeST94]. Given the record type `state` for the Black Jack dealer machine, the set of legitimate initial states is characterized by a state predicate `pred[state]` and transitions are modeled as elements of `pred[[state,state]]` as in the formalizations of the single pulser and the arbiter above. In the case of the Black Jack dealer, however, single transitions are combined asynchronously by means of logical disjunction. For the resulting logical description of the state machine we were able to prove the invariants stated in [MeST94] by applying the core strategy from Sect. 5.7 to the subgoals generated from an initial case split. This case split permits proving the invariant for each transition separately. The overall verification time is around 85 seconds on a Sun UltraSparc.

7.4 Verifying the FIR Filter

The high-level design of a filter computation (benchmark 5.) together with a proof of its main invariant has already been described in Sect. 4.2. It is modeled as a collection of uninterpreted signals over reals, and the state transition of the machine are modeled as equations that describe the value of the signal at a certain point of time in terms of previous values. The proof of the main invariant uses a simplified version of the core strategy in Sect. 5.7: quantifier elimination followed by combined rewriting and call to the decision procedures. The verification time of the filter design of depth $N = 5$ in Fig. 4.3 is around 2 seconds on a Sun UltraSparc. No attempt has been made to generalize this formalization of the filter design to arbitrary depth N.

8. Conclusions

Automated theorem proving technology clearly has a great deal to contribute to hardware verification since hardware proofs tend to fall into certain systematic patterns. Our contention is that if theorem provers are to be effective in hardware verification, we must employ powerful and efficient deductive components within high-level strategies that capture the patterns of hardware proof.

We have illustrated the use of PVS with largely automatic correctness proofs of microprocessors, various arithmetic circuits including an SRT division circuit, and the benchmark circuits used throughout this book. The proof scripts for all the examples in this chapter are rather short and constructed using a combination of a small set of primitives (like assert, ground or grind) after a few initial decompositions (such as induction) are made. These examples demonstrate the effectiveness of the automation of hardware proofs in PVS, which derives mainly from PVS's tight integration of rewriting with decision procedures for equality, linear arithmetic over integers and rationals, propositional logic, and support for specialized logics suited for hardware verification like CTL.

The hardware examples in this chapter also demonstrate the value of an expressive specification language in mechanized verification. In particular, generic hardware components support modular proof development and have the advantage that some variations of a particular circuit design can be verified by just redoing one part of the proof. Moreover, the use of dependent types together with predicate subtypes allows the writing of specifications that are clear and concise. Such high-level descriptions not only minimize the pitfalls of introducing errors in initial design specification but also open the door to using these specifications as design documents.

Acknowledgement. Work on mechanized formal verification by our collegues S. Owre, F. von Henke, S. Rajan, J. Rushby, and N. Shankar has strongly influenced the exposition of this chapter. Some parts of the text are based on [CRSS94], and Sect. 3.5 describing model checking in PVS has been taken, with the permission of the author, from [Shan96]. H. Pfeifer and E. Canver provided many useful comments. Updated information about PVS and the source files for the examples in this chapter can be obtained from the addresses below.

    http://www.csl.sri.com/pvs.html
    http://www.informatik.uni-ulm.de/ki/pvs.html

Verifying VHDL Designs with COSPAN

Kathi Fisler and Robert P. Kurshan

1. Introduction

COSPAN is a general-purpose software tool for *coordination specification analysis* [HaHK96]. Its premier application is for largely automated refinement verification and model-checking of finite-state systems, particularly control-intensive systems. These include hardware controllers such as those for busses and arbitration, memory management protocols such as those for cache consistency, communications protocols, distributed consensus algorithms, network protocols, and even the high-level behavior of circuits modelled by differential equations [KuMc91], [HaKu90b]. Less suitable applications, which often require additional unautomated *ad hoc* methods or restriction of the model's behavior in order to complete the verification, include data-path-intensive models such as multipliers. Non-verification applications include prototyping and implementing control-intensive programs, puzzle-solving, and counting combinatorial objects.

Refinement verification determines whether all behaviors of one design model are consistent with some other (usually more abstract) design model. *Model-checking* determines whether all the behaviors of a design model satisfy some given property. To implement these procedures, COSPAN uses the paradigm of an ω-automaton language containment check [Kurs94a]. For automata A and B modelling respective designs, refinement verification is implemented by the language containment test

$$\mathcal{L}(A) \subset \mathcal{L}(B) \ ;$$

model-checking is implemented by the same test when B represents a property to be checked on a design modelled by A.

Each such test constitutes a mathematical proof (or disproof) that the design modelled by A is correct– relative to the test. The outcome of the test is determined algorithmically through symbolic analysis (*not* through execution or simulation). If the language containment fails, COSPAN produces an *error track* consisting of a succession of states (represented by respective assignments of values to program variables) from an initial state to the failure. Since COSPAN's automaton model is based upon non-terminating behaviors, the failure is represented as a "Bad Cycle": a succession of states of A ending in a cycle, describing a behavior of A not accepted by B. In the case of a run-time syntax error, *e.g.*, a variable forced out of range, an error track to the error is produced and there is no final cycle.

COSPAN realizes its full potential for system development when it is used to design, develop and implement a system through a formal top-down procedure based upon successive refinement. Starting at a very abstract level of design, a system definition is redefined repeatedly, through a succession of design levels, adding more functional detail at each successive design level. Every design level is checked for properties or the performance of "tasks" germane to the particular level of abstraction of that design level model. The refinement process guarantees that properties or tasks verified at one level are inherited automatically by all successive levels. The target software or hardware implementation may be generated *automatically* by COSPAN, from the final (most detailed) design level, through a production-quality C-code generator and an interface with a hardware synthesis tool. COSPAN supports a variety of optimization routines for both cases.

This approach permits detection of design errors during the entire course of development, well in advance of the design stages at which simulation test is applied customarily. Applications have demonstrated that the ability to detect errors earlier in the design cycle can accelerate the design process dramatically [HaKu90b]. Moreover, model-checking is able to uncover a class of design errors likely to be missed by simulation testing. These are the so-called "livelock" errors in which a design fails to complete its assigned task, although it never enters an identifiably bad state. Finally, in any design for which the verification is computationally tractable, it can advance the design's reliability to a level unattainable with simulation test, since unlike simulation, model-checking is capable of analyzing exhaustively every possible behavior of a design. The increased reliability that comes with model-checking is preserved when the implementation is generated automatically from a verified model.

COSPAN's native input language is S/R [KaKu86], a data-flow automaton-language. S/R supports parameterized macros which may be used to construct a procedural base of abstract data types supporting declarative structures. For a given class of coordination problems, an *ad hoc* library of such macros may be assembled and used as a special purpose data-flow language. It is natural to use this approach for developing distributed applications such as integrated circuits and communication protocols, in which structure and behavior already are understood in terms of data flow.

Such a library of S/R macros is used to define the properties and constraints with respect to which model-checking is conducted. Named QRY.h, this library is constructed expressly to facilitate the specification of temporal properties. It is complete in the sense that all ω-regular properties and constraints may be expressed through the automata defined with macros from QRY.h. This is an important aspect governing the use of COSPAN: the user does not need to learn a logic as in other verification paradigms, nor to define the required automata directly. All automata are invoked automatically through a small set of intuitively named parameterized macros, such as

$$\text{After\_Eventually\_}(\ enabled,\ fulfilled\ )$$

which defines an automaton accepting all design model behaviors in which after each occurrence of *enabled*, there eventually follows an occurrence of *fulfilled*; *enabled* and *fulfilled* are arbitrary Boolean expressions in the variables of the design model. Thus, all the properties to be verified in the context of model-checking may be specified simply by choosing the appropriate parameterized macro from QRY.h and defining Boolean expressions to be substituted for its parameters.

COSPAN supports documentation, conformance testing, verification coverage analysis (what portion of a design did a particular property check), software maintenance, libraries of abstract data types and reusable pre-tested components, debugging and simulation tools for pinpointing the source of logical failures, and timing verification, used to check behavior in the context of assumptions on timing. It uses algorithms based both on explicit state enumeration and symbolic state enumeration (BDD's), both separately and together.

Decision procedures for verification entail an exhaustive search of the design model state space. Central to the tractability of COSPAN's analysis algorithms are formal *reduction procedures* [Kurs94a] for coping with the large model sizes associated with virtually all production-oriented coordination problems. Reduction, used in conjunction with top-down development, sometimes (in a heuristic sense) can render the effective computational complexity of the analysis of a system design as constant throughout the refinement process. The reduction algorithms reduce the time and space complexity of the computations used for each verification, by tailoring the analysis to the property being verified. These reduction algorithms, which comprise a major portion of COSPAN's internal routines, are of four main types: *localization reduction* which automatically abstracts portions of the design irrelevant to the property being checked; *task decomposition* which automatically verifies a decomposition of a global property into local properties (in order to enhance the extent of localization reduction); *homomorphic reduction* which verifies the soundness of a user-provided abstraction (as well as the consistency of a refinement); and *symmetry reduction* which simplifies the verification of symmetric properties. All four types are demonstrated here.

A version of the COSPAN verification system is available without charge to universities for research and educational purposes: inquire of the second author at k@research.bell-labs.com. A commercial verification tool named FormalCheck[tm] which supports verification of both VHDL and Verilog design models is available from Lucent Technologies. FormalCheck operates through a graphical platform and uses COSPAN as its verification engine.

This paper focuses on using COSPAN as a verification tool for VHDL designs. In this mode, it is assumed that a VHDL model exists and verification consists of checking it against a specification described in terms of the performance of a variety of tasks (properties). Since this type of verification is performed commonly by hardware designers in a commercial setting, emphasis is placed on automation. The techniques that facilitate design development through successive refinement also are applicable to VHDL verification. Moreover, these techniques may be used for abstraction to deal with designs too complex to verify directly. However, on account of the need for user intervention to define the most general abstractions, in most cases of VHDL verification, abstractions are limited to those which can be generated automatically (algorithmically). Nonetheless, for illustrative purposes, one example of a verification based upon a used-defined abstraction is given: verification of a car seat controller, described in Section 6.5.

2. Related Work

There is an extensive literature on the theoretical background of automata-theoretic verification, as well as a variety of tools which perform this and related types of analysis, as detailed in [Kurs94a]. E. M. Clarke and his students at CMU have focused upon CTL (branching-time temporal logic) model-checking, including VHDL programs [CMCH96], [McMi93a]. There is at least one commercial CTL model-checker with a VHDL interface: CheckOff-M[tm], developed at Siemens [FSSS94] and marketed by Abstract Hardware Ltd. (UK). R. K. Brayton and his students at UC Berkeley have developed an automata-theoretic verification platform called HSIS [ABCH94] modelled after COSPAN, with an interface to Verilog (but not VHDL). (HSIS has evolved into a system called VIS, but VIS does not currently support automata-theoretic verification, although it may in the future.) COSPAN is the only automata-theoretic verification tool known to the authors with an interface to VHDL. An essential difference between automata-theoretic verification and CTL model-checking is that abstraction, refinement and localization reduction which form the basis in COSPAN for dealing with large designs, are not in general possible with CTL model-checking on account of the non-conservative nature of the existential path quantification in CTL. Although there exist a number of model-checkers for linear-time temporal logic, these do not appear to have been developed to the same extent as model-checkers for CTL or ω-automata. Moreover, ω-automata are strictly more expressive than linear-time temporal logic. For a general overview of the current state of hardware verification, see [ClKu96].

3. The Program Semantics

The semantic basis of an S/R program is founded on the notion of a *program state* which evolves in time. The program state simply is the simultaneous values of all program variables. This semantics applies to many programming languages. Since hardware at the register transfer level has a well-defined notion of state, it certainly applies to hardware description languages for this level of abstraction.

In the case of synthesizable VHDL programs, all variables have a finite range and thus the program evolves sequentially in discrete time: for any "current" state, there is a finite set of possible "next" states, depending upon program inputs and program nondeterminism, if any. Although not a part of VHDL, nondeterminism easily may be introduced into VHDL for purposes of abstraction. Primary inputs are assigned values nondeterministically, although this is only implicit in a VHDL program.

Asynchronous programs, although not synthesizable, may be modelled through the use of nondeterminism as well [KMOS95]. In fact, an asynchronous program is a special case of a synchronous program: a synchronous program in which only one component changes its state at any one time; the choice of which component, is nondeterministic, with all the other components updating their state to their current state.

In many programming languages, program variables may be partitioned (not necessarily uniquely) into a set of "independent" variables which hold the program's **sequential state** (in hardware terminology: *latches*) and the rest of the program variables which are then said to hold the program's **combinational state** (in hardware terminology: *logic*). The combinational state is given as a relation in terms of the sequential state: for each global assignment to all the sequential (latch) variables, the combinational (logic) variables assume some fixed value or set of values. The values of variables may be inter-related: for example, a pair of combinational variables (x, y) may be defined to assume nondeterministically exactly the values in the set $\{(0,0), (1,1)\}$, for some particular set of values of sequential variables.

Primary input values are part of the combinational state. In general, these are assigned nondeterministically– an assignment which is implicit in VHDL. However, for purposes of verification, they may need to be constrained according to some model of assumed behavior of offered inputs. Such constraints may be expressed implicitly in the specification of the property being verified, or explicitly in terms of a "driver" or "environment" model which defines the way in which inputs are assumed to be offered to the design model. Such an environment model has its own variables. At the semantic level of the S/R program which contains both the design model and such an environment model, the variables of the environment model need not be distinguished from the variables of the design model. Verification merely refers to relationships among all the program variables. When the environment model defines all of the inputs to a design model, the program represents a "closed" system.

We use the term *program* when we wish not to distinguish between the variables of the design model and those of the environment model or of automata which may be part of the program for the purpose of defining model constraints or properties to be verified.

Once there is a partition of the program variables into sequential and combinational, the combinational variables assume a secondary role to the sequential variables. Accordingly, the sequential state variables henceforth will be called merely the *state* variables. The *global state* will refer to a simultaneous evaluation of all (sequential) state variables and the *global selection* will refer to a simultaneous evaluation of all combinational variables, relative to some fixed global state. The terminology "selection" comes from the role of combinational variables in nondeterministically "selecting" each possible primary input, or more generally, selecting among a variety of nondeterministic choices, when abstraction is involved. In the latter case, nondeterministic values of combinational output variables may be "selected" as well.

Combinational variables (with nondeterministic valuation) are called *selection* variables in S/R. Thus, at the programming level, an S/R program defines a system model with some partition of the program variables into selection variables and state variables. All state variables must be initialized. Selection variables are not initialized since they are defined as relations of the current state.

The sequential state is updated in terms of its current value and the value of the combinational state (which in turn is determined as a relation of the current value of the sequential state). As the combinational state is updated in terms of the sequential state alone, there can be no dependency cycles ("combinational cycles") among the combinational variables.

Under the customary "discrete event system" interpretation, a program advances from each successive "current" program state to the respective "next" program state with some nonzero (possibly varying) delay, wherein the combinational state is determined in terms of the current sequential state with zero delay. Thus, a model can be viewed as progressing from one value of the global (sequential) state to a next value, with some global selection current at each global state. The global selection includes a valuation of all primary inputs and all combinational outputs– including any dependent upon the inputs. Given a "current" value of the global state, a next value is determined by a *transition relation* expressed in terms of the current state and selection. In general, the next state is a nondeterministic relation of the current state and selection.

A particular evolution of the design model, such as one that leads to a design error found in the course of verification, is depicted by COSPAN as a succession of global states. Each global state is followed by a global selection which enables the transition from the given global state to the succeeding one. Although combinational variables may be assigned iteratively, say x in terms of y in terms of z, the "discrete event" semantics is that they all are assigned in zero time as a relation of the current state, so their values at an operationally intermediate stage when some have been updated and others not, has no semantic meaning and is left undefined. Only the final (necessarily stable) valuation of such an iteration is depicted. Such an iteration is guaranteed to have a final stable value since combinational variables are forbidden to contain dependency cycles.

At a semantic level, S/R represents both a system model and its required properties in terms of automata, which succinctly capture the definition of model behaviors in terms of "accepted" global state/selection evolutions. Automata may be built up compositionally in terms of component automata called *processes* which can be interpreted as finite state machines which read input and write output and at the same time have the built-in notion of automaton acceptance [Kurs94a].

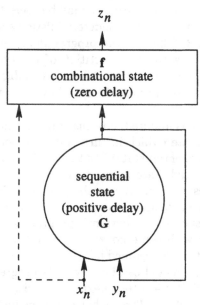

Fig. 3.1. Classical Sequential Machine Model. Given a sequence (x_n) of input values, a (zero-delay) combinational output function f and a (non-zero-delay) sequential transform function G, the state is given by $y_{n+1} = G(x_n, y_n)$. The output of a **Moore Machine** is $z_n = f(y_n)$ associated with the state, while that of a **Mealy Machine** is $z_n = f(x_n, y_n)$, associated with the state transition (y_n, y_{n+1}). Obviously every Moore machine is a Mealy machine. Given a Mealy machine which defines the set of input/output sequences $\{(x_n, z_n)\}$, a Moore machine whose state space is the set of state/output pairs of that Mealy machine can define the sequences $\{(x_n, z_{n-1})\}$; clearly no Moore Machine can define the unshifted sequences, in general. Moreover, the Mealy Machine supports a more efficient sequential state encoding.

Each process has its own set of *(local) states*. Each process, at its "current" state, has one or more possible outputs or *(local) selections* defined; these are the possible values of the combinational variables of the process at the given process state. Process selections differ from classical finite state machine outputs in that the latter are assumed to be a function of the state for Moore machines or a function of the state and input for Mealy machines (Figure 3.1), whereas selections may be a nondeterministic relation of the state and input (and other selections, as long as the dependency is acyclic).

The joint behavior of several processes evolving concurrently in time captures the coordination among these processes. Such a "parallel composition" of processes itself is a process, so the semantics of process is compositional, supporting hierarchical design development. The state of a parallel composition of processes is simply the vector of the respective local states of the components, and the selection of the parallel composition is the vector of the local selections.

Each process has a *state transition relation* which determines its next state as a "resolution" of its current state and the current global selection. In a parallel composition of processes, each component process has some (local) state and selection current. Each process *resolves* the current global selection by moving from its current (local) state to a next state through a state transition enabled by its transition relation. The global next-state is the vector of the local next-states. A parallel composition of processes thus evolves in time through a synchronous parallel succession of "selection" and "resolution".

The semantics of this evolution through repeated selection and resolution is independent of the order in which the processes are declared syntactically in the S/R program. If one selection variable is dependent upon another, then the dependent one is assigned after the one on which it is dependent, irrespective of their order in the program. This is well-defined since there are no dependency cycles among combinational variables. State variables may have dependency cycles as each state variable is assigned a next value in terms of all the current values of all the state variables and all the selection variables. The state transition relation of a parallel composition of processes is simply the conjunction of the transition relations of the respective components.

This non-deterministic automaton-based version of the classical finite state machine representation depicted in terms of a selection/resolution evolution gave the name to the s/r semantic model for processes depicted in Figure 3.2 and the S/R language used to program it.

The partition of program variables into sequential and combinational is prevalent especially in languages which define digital hardware at the register transfer level. Different partitions may define the same behavior, but for a program which will be implemented as hardware, the choice of partition can have an enormous significance for the time and space complexity of the hardware which will be synthesized from the program. For example, although Mealy and Moore machines can implement essentially the same function, a Mealy machine can make a more efficient use of space by using combinational

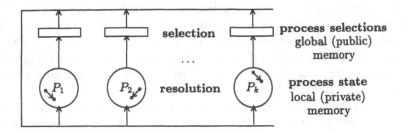

Fig. 3.2. The s/r Model. Each process $P_1, \ldots, P_k$ updates its state by resolving the current global selection and then (instantaneously) making a new selection from the new state.

output variables in places where a Moore machine must use sequential variables (*cf.* Figure 3.1). The same choice of partition mediates the tractability of the program's verification: partitions good for hardware tend to be good for verification, and *vice versa*. Latches, which for hardware synthesis require expensive space and retard execution, have precisely the same effect for verification; thus, the ability to distinguish syntactically between sequential and combinational state is a priority for both. It is no coincidence therefore that the subset of VHDL supported for hardware synthesis also is the subset which is efficient for verification.

It is this common interest in a good variable partition which has created a synergy between hardware and its verification not presently found in software verification, and goes a long way to explain why verification has made greater inroads in hardware. Nonetheless, the techniques for hardware verification described here generally could apply to software verification as well, and it is widely held that in the future there may be little distinction between hardware verification and software verification. Currently, the missing link for software verification is a venue like VHDL where software and verification can meet. The advent of "advanced" languages which distance themselves from data-path operations such as pointer manipulations could advance the opportunity for computer-aided software verification of the sort described here, in much the same way as hardware synthesis and hence synthesizable languages have advanced the opportunity of hardware verification.

4. The Semantics of Verification

The input to the verification process consists of the provided VHDL design model (in this paper, each respective benchmark design model), an environment or constraint model for the primary inputs to the design model, and a property to be verified. Automatic translation of the design model from VHDL to S/R preserves its original VHDL semantics.

The default constraint model is trivial: all primary inputs are uncon-strained and thus nondeterministically assigned. However, most properties require some constraints on primary inputs for the proper functioning of the design. In fact, constraints may be applied to any aspect of model behavior. The general effect of a constraint is to except the constrained behavior from the scope of the verification. This is implemented as follows: the state space search conducted in the course of verification is truncated at any state at which a constraint is violated; nondeterministic assignments (such as primary input combinations) which violate a constraint are omitted; and eventualities which contravene fairness assumptions are exempted from the property being verified.

Both the property to be verified and the constraint or environment model, including all fairness constraints, are represented by ω-automata [Kurs94a]. An ω-automaton is an edge-labelled finite-state transition system in which certain states or sets of states and edges are designated as the automaton *acceptance structure*. The acceptance structure defines which successions of transitions determine the *language* of the automaton.

In the case of COSPAN, the edge labels are Boolean expressions in the design model variables. When the design model is in a state s and the values of its respective variables cause some automaton edge label λ to have the value *true*, then the automaton transition lablelled by λ is *enabled*, and the design model and the automaton update their respective states accordingly. Thus, if the design model resolves its current state s to a next-state s' (Figure 3.2) and λ labels the edge (t, t') of the automaton, then the joint current state (s, t) of the design model and the automaton makes a transition to the joint next-state (s', t'). In fact, the design model too may be construed as a finite-state transition system in which each transition is labelled by the Boolean expression in the design model variables which describes all possible valuations of those variables for which the transition will be taken. If M and N are two such labelled transition systems with respective current states s and t, if $M(s, s')$ is the Boolean condition labelling the transition of M from s to s' and $N(t, t')$ is the Boolean condition labelling the transition of N from t to t', then the Boolean conjunction of the two conditions, $M(s, s') * N(t, t')$ is the Boolean condition under which both the transition of M from s to s' and the transition of N from t to t' are enabled. The joint system, denoted $M \otimes N$, has among its states (s, t) and (s', t'), and the edge $((s, t), (s', t'))$ is labelled with the joint enabling condition $M(s, s') * N(t, t')$. Each of M and N may be represented as a matrix over the Boolean algebra of Boolean expressions in the design model variables, wherein if the rows and columns of M are labelled by its states, the matrix element $M(s, s')$ is the label on the edge (s, s'), and likewise for N. The joint system then is represented by the tensor product [Kurs94a] $M \otimes N$ of the component matrices, given by

$$(M \otimes N)((s, t), (s', t')) = M(s, s') * N(t, t') .$$

The *language* $\mathcal{L}(A)$ of an automaton A is the set of behaviors accepted by A. A *behavior* is a sequence of global assignments to the design model variables (which set of variables is assumed to be fixed – one may take the largest set of variables ever to be considered). A behavior is in $\mathcal{L}(A)$ if it enables a succession of transitions of A which is accepted by the acceptance structure of A.

With COSPAN, two types of automata are used – one type to represent the design model together with the environment or constraint model, and another type to represent the property to be verified. In each type of automata, the acceptance conditions are given in terms of two types of acceptance structures: a set of *cycle sets* of states and a set of *recurring* edges. In the class of automata called *processes*, used to specify the design, its environment and constraints, the acceptance structures are interpreted negatively: a behavior is accepted if it neither eventually stays within a cycle set nor crosses a recurring edge infinitely often. Thus, acceptance corresponds intuitively to satisfying fairness constraints which bar "livelock" or more generally, bar violation of a "strong fairness constraint" (the most general type of automaton fairness constraint). This negative interpretation conforms to the need to have such automata represent *constraints*. In the automata used to specify the property to be verified, the acceptance structures are interpreted positively: a behavior is accepted if it either eventually stays within a cycle set or crosses a recurring edge infinitely often. This conforms with the usual representations of temporal properties in terms of eventualities and recurrant behaviors. It follows that for deterministic automata, a property and its corresponding constraint are represented by respective automata which define complementary languages. This has an important implementational consequence: in order to verify that a design modelled by a process D (with concomitant constraints) satisfies a property represented by the automaton P, that is, in order to verify the language containment

$$\mathcal{L}(D) \subset \mathcal{L}(P)$$

it is necessary and sufficient to verify the process language emptiness of $\mathcal{L}(D \otimes P^{\#})$ where $P^{\#}$ is the process obtained by construing the acceptance structure of the automaton P in the negative sense of a process, as just described. In particular, there is no ensuing computational cost associated with complementing the automaton P, as there is with automata-theoretic verification in which the same class of automata is used both for the constrained design model and the property.

Different properties may require different constraints. We refer to a property/constraint pair as a *query*. Each design model is verified relative to a set of queries. All the benchmark designs are verified with respect to queries whose properties and constraints are expressed through the QRY.h library of parameterized macros which define the underlying automata. If several queries pertaining to a single design model contain several different respective constraints, one could view the conjunction of those constraints as the environment model, against which the actual environment implementation could be verified.

Not all constraints may be considered to form a natural part of the environment model. Some constraints may cover special modes of operation (for example, "interrupts off") which would be inappropriate as a component of a general model of the environment. In this case, the constraint is interpreted as a precondition: "if the environment behaves like this, then the design model will behave like ... [*some property*]". In such a case, the constraint really is a part of the property being checked, and would not be used as a component of a specification of the environment.

Still other constraints are *restrictions* intended to focus the verification on problematic or interesting subcases of the design model's functionality. Since constraints allow COSPAN to simplify its analysis, restriction is a standard method for coping with computational complexity, when full verification is infeasible. As with all constraints, restrictions are designated through QRY.h elements, without any need to alter the design model source code (a very important attribute for commercial verification).

In FormalCheck there is a way to designate a constraint as *default*. This designation has the effect of including it automatically in all queries for the model in question. The conjunction of the default constraints defines a specification against which the actual environment may be checked, if there is a VHDL design model for it. Non-default constraints are inferred as a part of the property being checked or as restrictions, rather than a part of the environment specification.

Any constraint on offered inputs may be viewed as an abstract specification model of the environment, since in general, the environment mediates and hence constrains the inputs. In fact, any model, abstract or detailed, can be used as the environment or input constraint model of another, as long as circularity is avoided. The danger is a potential for circular reasoning of the sort: "M is verified under an assumption about N which will be true under an assumption about M", which may be vacuously true on account of the failure of both assumptions. This may be understood more clearly in terms of notation.

Letting $\otimes$ denote the parallel composition of models, we make use of the "language intersection property" for processes [Kurs94a]:

$$\mathcal{L}(M \otimes N) = \mathcal{L}(M) \cap \mathcal{L}(N) .$$

(Recall that the models are automata and their respective *languages* define their "behaviors" or accepted (*i.e.*, valid) non-terminating executions.)

We will say M' and N' are specification or abstraction models of implementation models M and N respectively, if

$$\mathcal{L}(M) \subset \mathcal{L}(M'), \ \mathcal{L}(N) \subset \mathcal{L}(N') .$$

Thus, we may use N' as the abstract environment model for verification of M and M' as the abstract environment model for the verification of N.

Checking M for a property defined by the automaton T in the context of the abstract environment model N' corresponds to verifying the automaton language containment

$$\mathcal{L}(M \otimes N') \subset \mathcal{L}(T) .$$

It clearly follows from the language intersection property that

$$\mathcal{L}(M \otimes N) \subset \mathcal{L}(T)$$

or in other words, that substituting the abstract environment model N' for the actual environment model in the verification of M for the property T is conservative and hence legitimate.

We also may proceed in the reverse manner: we may determine a constraint on the primary inputs of M consistent with the behavior of its environment model N, which facilitates the verification of M for the property defined by T. (Note as above that a constraint on the inputs of M consistent with N in fact is an abstraction N' of N, and viewing N' as a constraint on the inputs of M or as a model of the environment of M are simply two ways of saying the same thing.) Having verified M for T under a constraint N' as above, we next seek to verify that N' in fact is an abstraction of N. We may verify this by checking that $\mathcal{L}(N) \subset \mathcal{L}(N')$ and if this is true, we are done. On the other hand, perhaps this is not true "in general", but only under the assumption of a certain precondition M' on M:

$$\mathcal{L}(M' \otimes N) \subset \mathcal{L}(N') .$$

This in fact is sufficient if we know that $\mathcal{L}(M) \subset \mathcal{L}(M')$, as then by the language intersection property,

$$\mathcal{L}(M \otimes N) \subset \mathcal{L}(M' \otimes N) \subset \mathcal{L}(N')$$

and thus

$$\mathcal{L}(M \otimes N) \subset \mathcal{L}(M \otimes N') \subset \mathcal{L}(T)$$

(since then $\mathcal{L}(M \otimes N) = \mathcal{L}(M \otimes N) \cap \mathcal{L}(M) \subset \mathcal{L}(N') \cap \mathcal{L}(M) = \mathcal{L}(M \otimes N')$ $\subset \mathcal{L}(T)$).

However, if instead of $\mathcal{L}(M) \subset \mathcal{L}(M')$, all we know is the mirror assertion

$$\mathcal{L}(M \otimes N') \subset \mathcal{L}(M') ,$$

then we are in the circular situation where although we may verify

$$\mathcal{L}(M \otimes N') \subset \mathcal{L}(T) ,$$

$$\mathcal{L}(M' \otimes N) \subset \mathcal{L}(T) ,$$

and even

$$\mathcal{L}(M' \otimes N') \subset \mathcal{L}(T) ,$$

nonetheless it is entirely possible that

$$\mathcal{L}(M \otimes N) \subset \mathcal{L}(T) \text{ fails}$$

as, for example, is the case if $\mathcal{L}(M') = \mathcal{L}(N') = \emptyset$, $\mathcal{L}(M \otimes N)$ is the universal language and $\mathcal{L}(T)$ is not.

On the other hand, resolution of the precondition may be deferred for any number of such steps. For example, if:

$$\mathcal{L}(M \otimes N') \subset \mathcal{L}(T) ,$$

$$\mathcal{L}(M' \otimes N) \subset \mathcal{L}(N') ,$$

$$\mathcal{L}(M \otimes N'') \subset \mathcal{L}(M') ,$$

$$\mathcal{L}(N) \subset \mathcal{L}(N'')$$

then it is easily checked that in this case indeed the environment substitutions are legitimate and

$$\mathcal{L}(M \otimes N) \subset \mathcal{L}(T) .$$

In the final analysis, it is stylistically cleanest to use the simplest paradigm: for each model, its environment has an abstraction (without preconditions) which may be used in the verification of that model. This avoids all possibility of circularity, is easiest to understand and appears to be both sufficient and most natural in practice.

When this is done methodically, verifying each of several design model components with respect to some corresponding environment model, the procedure is called *compositional* (or *modular*) model-checking. For each design model component, one must verify that its respective environment model indeed abstracts all the other design model components. In the case of a set of properties each of which applies only to one or another component of a design model, compositional model-checking is an efficient and natural way to limit the complexity of verification.

5. Operational Verification

Operationally, the verification specification and proof paradigm is as follows. Given a VHDL design model, it is translated into S/R using the Formal-Check front-end. An environment model and properties are specified by the user in accordance with the natural language and diagrammatic descriptions provided with the design. One way to determine an appropriate environment model is to begin with the default (unconstrained) environment model, and adjust that by adding constraints in the course of verification, until the design model is sufficiently constrained to allow the specified properties to be verified.

Properties and constraints are specified through the automaton macro library QRY.h. Since these specifications are the result of interpreting a natural language document, there may be variations in the results of verification, from one practitioner to another. There always is a potential that an improperly formulated property or constraint will result in a spurious verification. Although faulty property formulation is unusual, overconstraining a design model is a common error among less experienced verifiers. A typical source of overconstraint is to make an unwarranted assumption about the behavior of a design, in lieu of a possibly more cumbersome assumption about offered inputs. For example, it may be that a flow control protocol on inputs is intended to prevent a design model buffer from overflowing. However, simply assuming that the buffer never will overflow is dangerous: a flaw in the design model could undercut the flow control. If buffer overflow were inevitable in certain cases on account of the design flaw, then constraining the buffer not to overflow would truncate the search of the design model state space at each point of overflow. Thus, any design errors past the point of overflow (including the overflow itself) would be missed, and one could erroneously conclude that the design were error-free. (The correct approach is to model the flow control that will govern the offered inputs, and *verify* that the buffer never will overflow, before using that as an assumption.) If all constraints restrict primary inputs only, overconstraint usually is not an issue. However, frequently it is useful and warranted to place assumptions on the internal operation of a design, as when the operation has been verified previously. COSPAN contains several "sanity" checks for overconstraint. Among these are checking that all properties with an enabling condition get enabled, and that the model is "ergodic" in the sense that every state is reachable from every other (this is the CTL formula AGEFinit, where init is the initial or reset state). Ergodicity must hold for reinitializable models and hence most hardware. Since ergodicity involves existential path quantification (the E in AGEF), it is not preserved under localization reduction (see below). However, the context of this check in COSPAN is not that of a "property" but rather a sanity check of an already-reduced, constrained model, so the non-conservative nature of ergodicity is not an issue.

Each verification run of COSPAN is based upon a *query* which consists of a property/constraint pair. If M represents the automaton which is the S/R translation of the VHDL design model, N' represents the automaton which is the given input constraint or model of the environment of M, and T represents the automaton which defines the property or task which M is intended to perform, then the verification check is the language containment test

$$\mathcal{L}(M \otimes N') \subset \mathcal{L}(T) .$$

Operationally, the property may be expressed as a conjunction of subproperties, and likewise the constraint may be expressed as a conjunction of subconstraints. For example, if $N_1, \ldots, N_k$ are the component automata which comprise N', then $N' = N_1 \otimes \cdots \otimes N_k$ and

$$\mathcal{L}(N') = \mathcal{L}(N_1) \cap \cdots \cap \mathcal{L}(N_k) .$$

All component properties which together comprise T and all component constraints which together comprise N' are expressed in terms of automata defined through the macro library QRY.h.

COSPAN is applied to a pair of files, say model.sr and query.sr, containing the S/R code for the design model and the query, respectively. Properties and constraints are constructed from automata with dual respective acceptance conditions as already described, so they may be grouped together. The automata in the macro library QRY.h which define constraints are distinguished from the automata which define properties by the tag *Assume* which appears in their macro names. Other than that, the two sets of automata are completely dual and syntactically indistinguishable. The syntactic descriptions of some representative automata from this macro library are presented in the Appendix.

After choosing run time options indicated by flags, COSPAN is run from the UNIX™ command line:

```
cospan [flags] model.sr query.sr
```

The major COSPAN run time options are listed in [HaHK96]. The main options offer a choice between explicit or symbolic state enumeration, and a choice of reduction algorithm. Often, explicit state enumeration is faster for "green" designs with many design errors, as well as designs with a considerable number of arithmetic operations. Symbolic enumeration usually is essential to verify designs once the errors have been removed, since without applying manual abstraction, virtually all commercial designs end up with a reduced state space whose explicit state enumeration space requirement exceeds the size of available memory. The effective limit of explicit state enumeration with 1GB of available RAM is, to give a rough sense, around 10 million states, depending upon the number and size of the state components. This is too small by far to hold most automatically reduced commercial models, which commonly have anywhere from 10^{10} to 10^{500} reachable states after reduction. The variations among the reduction algorithms trade off space *vs.* time.

The main reduction algorithm in COSPAN (and a default in Formal-Check) is *localization reduction*. In this algorithm the design model and its constraints are reduced (abstracted) conservatively relative to the property being checked [Kurs94a]. Conservative abstraction means that the reduction may not be bi-simulation-equivalent to the unreduced model. For this reason, localization reduction is not compatible with CTL. If the reduced model verifies, the unreduced model would too. Moreover, if the reduced model fails to verify, the ensuing error track is checked against the unreduced model and if it is found to be an artifact of the reduction, the reduced model is adjusted dynamically, and the verification analysis continues.

Localization reduction follows the topology of the design model variable dependency graph. Some design model variables become designated as *free* and are treated as primary inputs. The design model variables on which the variables defining the property thus no longer depend, are *pruned* (eliminated from the model) and the retained variables are *resized* to their relevant sub-ranges by discarding values relevant only to pruned variables. Expressions are *clipped* by lifting constants derived from pruning, resizing and constraints.

Localization reduction enables the verification or debugging of design models with as many as 5K latches and 100K combinational variables. Nonetheless, it is no panecea, and it is not uncommon that designs with many fewer variables and latches are intractable.

Additionally, COSPAN supports two user-assisted reductions: decomposition and abstraction. The user may propose a decomposition of a property or task into several subtasks, each of which may have a better localization reduction than the original task, and which together imply performance of the original task. The user proposes a decomposition and COSPAN then checks its validity.

Finally, COSPAN supports abstraction verification by checking the validity of a homomorphism between a model and its abstraction. Once the abstraction has been verified, it may be used in place of the original model for verification. The user proposes an abstraction to use in place of the model it purports to abstract; COSPAN checks the validity of the abstraction, and can automatically insert it in place of the model it abstracts.

The user specifies an abstraction by assigning an expression over variables and values from the detailed model to each output variable in the abstract model. COSPAN can verify whether these definitions form a homomorphism. The homomorphism check requires an exhaustive search of the state-space. This appears problematic given that abstractions are used in cases where the state-space is too large to be explored exhaustively. Fortunately, it is sufficient to verify a homomorphism on each respective component of a design, separately. Formally, let P be a process that is the product of processes $P_1, \ldots, P_n$. Let P' be the product of processes $P'_1, \ldots, P'_n$. P' is a valid abstraction of P if for all i, P'_i is a valid abstraction of P_i [Kurs94a].

Some abstractions are correct by construction (and hence need not be verified). FormalCheck supports a utility which permits the user to abstract complex blocks to the functionality of simpler sub-blocks, through a few mouse-clicks on a hierarchical rendering of the design model. The user designates the fundamental sub-blocks of a given block. FormalCheck translates these to ascii control strings in a file read by COSPAN, which then conservatively abstracts the block to the functionality of the designated sub-blocks, automatically. Since the abstraction is conservative, if the model with these abstractions verifies, so would the original model.

6. Experimental Results

The design models presented here were translated from VHDL to S/R using the FormalCheck front-end and then verified using COSPAN. The VHDL interface to COSPAN supports a commonly inferred synthesizable subset of VHDL.

This section presents the results of five such VHDL design model verifications: the four benchmark circuits and a fifth design model, a car seat controller, that demonstrates the use of abstraction and task decomposition. Verifications were performed on a Silicon Graphics IP19.

The verification procedure for each design consisted of three stages: translate the VHDL design model into S/R; formally define a set of queries through the macro library QRY.h, corresponding to the given benchmark specifications; run the query verification in COSPAN.

At the time of these experiments, the VHDL front-end to FormalCheck was under development, and the four following minor types of alterations had to be made to the VHDL source code to accommodate syntax that was not supported in the translator. The first two of these are requirements of the synthesizable subset of VHDL supported by the tool; the last two were deficiencies which have since been corrected:

– Clock events may not be tested in a nested "if" statement. These were easily fixed by reordering the conditional tests in the appropriate definitions.
– Several synthesizer keywords, such as "high" and "low" appeared in the VHDL code as signal names. These names were altered so as not to conflict with the synthesizer keywords.
– Top-level GENERICS were not yet fully supported. Dummy entities were declared that bound top-level generics to their default values. (This has been fixed subsequently.)
– Biputs (INOUT signals) left unconnected by using the *open* statement were not yet supported. Dummy signals were defined to pass in place of "open" in such instances. (This has been fixed subsequently.)

The following sections discuss the verification of each example in turn.

6.1 Single Pulser

The properties to be verified are as follows:

1. If there is a rising edge on the input, eventually the output is high.
2. The output is never high in two consecutive clock cycles.
3. There is only one output pulse per input pulse.

The first property requires a means of detecting a rising edge on the input. This is straightforward by saving the previous value of the input in a state variable and setting another variable to true whenever the current value on the input is 1 and the previous value is 0. The macro library QRY.h provides a macro state variable constructor to facilitate this. Assuming that a Boolean variable named InputRise stores this information, the first property is expressed as follows:

Property1: After_Eventually_(InputRise = 1, PULSE_OUT = 1)

where After_Eventually_(*enabled*, *fulfilled*) is as described in the Introduction. The formal definitions of the QRY.h macros used here are given in the Appendix.

In order to check this property, we put the S/R code generated by the VHDL translator in a file, say `pulser.sr` and the property in another, say `query.sr` and execute COSPAN on this pair of files from the UNIX[tm] command line, say, using symbolic state enumeration.

As a result of this execution, COSPAN produces the following output:

```
0 data variables with width >= -#databits=4
15 selection/local variables
6 bounded state variables: 64 states
0 unbounded state variables
2 free selection/local variables: 4 selections/state
0 non-deterministic (non-free) selection/local variables
        1 selections/state (maximum)
4 total selections/state (maximum)
```

```
query.sr: Synchronous model
1 initial states.
26 states reached.
Bad cycle.
Bounded stvar range coverage: 6 variables, 100.00% average coverage
    6 enumerated and boolean: 0 values of 0 variables unreached
    0 integer: 0 variables with unreached values; 100.00% average
            coverage
            worst coverage: 100.00% for 0 variables
222 bdd nodes, 0.12 seconds, 0 megabytes
query.sr: Task failed...
```

The error track corresponding to this task is written to a file in the working directory. The contents of the file for this particular run appear below. Each global state is indexed by a pair of numbers (state-index(search-index)) indicating the order in which the states were searched. Within the data for a particular global state, the values of component state variables are given, followed by the values of the combinational variables below the demarcation denoted "selections". For each global state, only those variables that have changed value since the previous global state are listed.

The "Post Mortem Track" marker indicates the global state at which the erroneous behavior begins, and the type of error. There are two types of errors: transition errors, which occur when a state invariant property is violated (at a specific state), and bad cycle errors, which occur when a cycle is found in the system that corresponds to a behavior not accepted by the automaton which defines the property. Such errors sometimes are called "livelocks" and indicate the potential of the design to thrash and thus fail to fulfill the stated eventuality property. Bad cycle errors are marked with a "C" after the "Post Mortem Track" marker, as shown in this example. Tracks without the "C"

denote transition errors. Given a bad cycle error, the offending infinite behavior is obtained by cycling from the end of the error track back to the state
appearing immediately following the "Post Mortem Track" line.

```
0(0)
.SINGLEPULSERSTRUC.SYNCHRONIZER._IO.SR_ST_Q=0
.SINGLEPULSERSTRUC.SYNCHRONIZER._IO._OLDCLKVAL_CLK=0
.SINGLEPULSERSTRUC.FINDER._IO.SR_ST_Q=0
.SINGLEPULSERSTRUC.FINDER._IO._OLDCLKVAL_CLK=0
.InputRise.PrevSig.$=0
.Property1.$=_INIT
++selections+++++++++++++++
.SINGLEPULSERSTRUC_ENV.PULSE_IN=1
.SINGLEPULSERSTRUC_ENV.CLK=0
.SINGLEPULSERSTRUC.SYNCHRONIZER._IO.Q=0
.SINGLEPULSERSTRUC.SYNCHRONIZER.Q=0
.SINGLEPULSERSTRUC.FINDER._IO.Q=0
.SINGLEPULSERSTRUC.FINDER.Q=0
.SINGLEPULSERSTRUC.INVERTER._IO._TO_=1
.SINGLEPULSERSTRUC.INVERTER._IO.Y=1
.SINGLEPULSERSTRUC.INVERTER.Y=1
.SINGLEPULSERSTRUC.LOGIC._IO._TO_=0
.SINGLEPULSERSTRUC.LOGIC._IO.Y=0
.SINGLEPULSERSTRUC.LOGIC.Y=0
.SINGLEPULSERSTRUC.PULSE_OUT=0
.InputRise.PrevSig.#=0
.InputRise.#=1
1(1) -------------------
.InputRise.PrevSig.$=1
.Property1.$=_ENABLED
Post Mortem Track: C::::::::::::::::::::::::::::::::::::::
1(1) -------------------
.SINGLEPULSERSTRUC.SYNCHRONIZER._IO.SR_ST_Q=0
.SINGLEPULSERSTRUC.SYNCHRONIZER._IO._OLDCLKVAL_CLK=0
.SINGLEPULSERSTRUC.FINDER._IO.SR_ST_Q=0
.SINGLEPULSERSTRUC.FINDER._IO._OLDCLKVAL_CLK=0
.InputRise.PrevSig.$=1
.Property1.$=_ENABLED
++selections+++++++++++++++
.SINGLEPULSERSTRUC_ENV.PULSE_IN=0
.SINGLEPULSERSTRUC_ENV.CLK=0
.SINGLEPULSERSTRUC.SYNCHRONIZER._IO.Q=0
.SINGLEPULSERSTRUC.SYNCHRONIZER.Q=0
.SINGLEPULSERSTRUC.FINDER._IO.Q=0
.SINGLEPULSERSTRUC.FINDER.Q=0
.SINGLEPULSERSTRUC.INVERTER._IO._TO_=1
.SINGLEPULSERSTRUC.INVERTER._IO.Y=1
.SINGLEPULSERSTRUC.INVERTER.Y=1
.SINGLEPULSERSTRUC.LOGIC._IO._TO_=0
.SINGLEPULSERSTRUC.LOGIC._IO.Y=0
.SINGLEPULSERSTRUC.LOGIC.Y=0
.SINGLEPULSERSTRUC.PULSE_OUT=0
.InputRise.PrevSig.#=1
```

```
.InputRise.#=0
2(2) --------------------
.InputRise.PrevSig.$=0
++selections++++++++++++++++
.SINGLEPULSERSTRUC_ENV.PULSE_IN=1
.InputRise.PrevSig.#=0
.InputRise.#=1
```

This error track presents a case in which the input rises but the clock signal CLK remains low indefinitely. This merely demonstrates the futility of trying to verify a totally unconstrained design. The clock can be constrained to toggle at each successive global state by using the following definition, where OldClk contains the value of CLK from the previous global state:

$$\text{MakeClk: AssumeAlways\_}(CLK = 1 - OldClk)$$

Even with the clock constraint however, COSPAN still finds a counter-example. The new problem is abstracted from the error track in tabular form:

PULSE_IN	0	1	0	1
CLK	1	0	1	0
PULSE_OUT	0	0	0	0
SYNCHRONIZER.Q	0	0	0	0
	0	1	2*	3

The numbers across the bottom of the table index the global state and the asterisk denotes the state following the "Post Mortem" designation.

The problem here relates to the signal debouncing done by the synchronizer. According to the design, an output pulse is generated based on the output of the synchronizer, not the input value. The error reports a case where the synchronizer failed to catch an input pulse that fell between rising edges of the clock. The easiest fix would appear to be a change in the enabling condition of the property, testing only for those times when the input is true on a rising edge of the clock.

Property1:
   After_Eventually_((OldClk = 0) * (CLK = 1) * (PULSE_IN = 1),
      (PULSE_OUT = 1))

Here, * denotes Boolean AND. Even this version yields a counterexample. The problem now is one of timing granularity as the input fluctuates with the clock. The property treats each of these as distinct input pulses to be tested for corresponding output pulses. The synchronizer, however, sees only one long input pulse since it does not catch the falling edges of the input landing in between the rising edges on the clock.

PULSE_IN	1	0	1	0	1	1	1
CLK	1	0	1	0	1	0	1
PULSE_OUT	0	1	1	0	0	0	0
SYNCHRONIZER.Q	0	1	1	1	1	1	1
	0	1	2	3	4	5*	6

This case can be addressed by adding the assumption that the clock runs sufficiently fast as not to be alignable with the input in this manner. This constraint is formed by slowing down the rate of change on the input signal. Assume that *slowinput* is a process with boolean-valued selection variable # that can only change value every five global states; we chose to use an odd number to prevent the environment model from inadvertently debouncing the clock signal. Such a process is defined by creating a state variable that acts as a modulo-5 counter and enabling the change of value on # only when the counter is at value 0. The constraint is then introduced as

SlowInput: AssumeAlways(PULSE_IN = slowinput.#)

Under this slow input assumption, the first property verifies, stated in its original form,

Property1: After_Eventually_(InputRise = 1, PULSE_OUT = 1)

with the corresponding COSPAN output:

```
0 data variables with width >= -#databits=4
18 selection/local variables
9 bounded state variables: 1280 states
0 unbounded state variables
1 free selection/local variables: 2 selections/state
1 kill/free optimization actions
1 variable assignments driven by kills
0 non-deterministic (non-free) selection/local variables
        1 selections/state (maximum)
2 total selections/state (maximum)

query.sr: Synchronous model
1 initial states.
34 states reached.
Bounded stvar range coverage: 9 variables, 100.00% average coverage
    8 enumerated and boolean: 0 values of 0 variables unreached
    1 integer: 0 variables with unreached values; 100.00% average
                coverage
            worst coverage: 100.00% for 0 variables
566 bdd nodes, 0.18 seconds, 0 megabytes
query.sr: Task performed!
```

The second property checks that the output is never true in two consecutive clock cycles. Assuming that OldOutput reflects the value on PULSE_OUT at the last rising edge on the clock, this property now verifies without further intervention.

Property2: Never_((OldOutput = 1)*(PULSE_OUT = 1))

The third property, which also verifies immediately, establishes that there is only one output pulse per input pulse. Intuitively, the required property says that if a new pulse starts, a subsequent new pulse does not occur unless the output is true in the interim. The property uses "oldpulse" to hold the previous value of PULSE_IN.

Property3: After_Never_Unless_((oldpulse = 0) * (PULSE_IN = 1),
                             (oldpulse = 0) * (PULSE_IN = 1),
                             (PULSE_OUT = 1))

6.2 Arbiter

The properties to be verified are as follows:

1. No two distinct cells are ever simultaneously acknowledged.
2. Every persistent request is eventually acknowledged.
3. No acknowledge is asserted without a current request.

Properties 2 and 3 verified easily with no constraints. Each instance of the second property is an example of an After_Eventually_Unless_() automaton, which takes parameters *enabled*, *fulfilled*, and *discharged*, and yields a bad cycle if *enabled* ever is true with no subsequent global state in which either *fulfilled* or *discharged* is true. Each instance of the third property is a Never_() condition. In each property definition, the [i < 4] denotes an array constructor producing one copy of the automaton for each value of i less than 4:

Property2[i < 4]:
    After_Eventually_Unless_((REQ_IN[i]=1),
                             (ACK_OUT[i]=1),
                             (REQ_IN[i]=0)+(RESET=0))

Property3[i < 4]: Never_((ACK_OUT[i]=1)*((REQ_IN[i] ≠ 1)))

Here, + denotes Boolean OR.

Verification of the first property required constraints on the primary inputs.

Property1[i < 4][j < 4]:
    Never_((i ≠ j)*(ACK_OUT[i]=1)*(ACK_OUT[j]=1))

For this property, COSPAN produced a counterexample in which cell 1 and cell 2 acknowledge simultaneously. The first 12 states of the error track reflect a problem in passing the token:

TokenOut[0]	1	1	0	0	0	0	0	1	1	1	0	0
TokenOut[1]	0	0	1	1	0	0	0	0	1	1	1	1
TokenOut[2]	0	0	0	0	1	1	0	0	0	0	1	1
TokenOut[3]	0	0	0	0	0	0	1	0	0	0	0	0
RESET	0	1	1	1	1	1	0	1	1	1	1	1
	0	1	2	3	4	5	6	7	8	9	10	11

The token appears to pass normally until global state 7, at which point it jumps back to cell 0; thereafter, there are two tokens working through the system, which explains how two cells acknowledged simultaneously. The error track shows that a reset occurred in global state 6. This reset caused the token to leave cell 3 and return to cell 0.

According to the architectural layout provided in the documentation, TokenOut[3] should pass directly into the input TokenIn[0], so the output of the inverter taking input TokenIn[0] should therefore contain the inverse of the value on TokenOut[3]. Expanding the word to include the value on that inverter, this is apparently not the case (see global state 6 and 7).

TokenOut[0]	1	1	0	0	0	0	0	1	1	1	0	0
TokenOut[1]	0	0	1	1	0	0	0	0	1	1	1	1
TokenOut[2]	0	0	0	0	1	1	0	0	0	0	1	1
TokenOut[3]	0	0	0	0	0	0	1	0	0	0	0	0
RESET	0	1	1	1	1	1	0	1	1	1	1	1
TokenIn[0]Inv	1	1	1	1	1	1	1	0	1	1	1	1
	0	1	2	3	4	5	6	7	8	9	10	11

This suggests an inconsistency between the architectural layouts and the VHDL source code. The arbiter architecture code indicates that components are missing from the layout diagram in the documentation. The token is passed around in a five stage ring, not a four stage ring, as evidenced by a "token(0) ⇐ token(N)" line, where N is set to 4. This line introduces a one clock delay between the time the token passes from cell 3 back to cell 0. The passing of the token to the intermediate variable, however, does not take the reset signal into account. If a reset arrives when the clock is low, the intermediate variable will still grab the token in the next clock, but the reset will have sent the token back to cell 0.

There are two ways to eliminate this problem. The VHDL design can be changed to read "token(0)⇐ token(N) AND reset". Alternatively, the problem can be constrained out of the original design by restricting resets to occur only when the clock is high. Introducing this restriction with an AssumeNever_() constraint, the property verifies:

ResetRestrict: AssumeNever_((RESET=0)*(CLK=0))

6.2.1 Verification Statistics. These properties were tested with symbolic state enumeration, but in view of the small models could just as well have been verified using explicit state enumeration. The two sets of statistics for Property 1 cover the two possible ways of handling the error. The number of reached states varied: for Property1, it was 134, for Property2 it was 1855 and for Property3 it was 262145 (illustrating the lack of correlation between state space size and required time or space for verification using symbolic state enumeration). For such small models, the relatively small complexity of the reduction nonetheless can swamp the complexity of the verification.

	BDD nodes	CPU seconds	Meg Memory
Property 1 (fix model)	1705	3.45	1.1
Property 1 (restrict env)	1701	2.45	1.1
Property 2	11326	14.33	1.0
Property 3	261	.76	0

6.3 Systolic Array Element

Of the benchmark examples, this was the only one which provided a challenge, and unfortunately, in the short time available to the first author who ran the verification, the verification was not completed. Moreover, on account of the intrinsic complexity of this model, it seems highly likely that it could not be fully verified using COSPAN in a straight-forward manner. Nonetheless, a substantial portion of the circuit was verified by reducing the width of the data-path.

The computational complexity problems associated with this design derive from the fact that it is mainly a "data-path" model which moreover contains large-integer multiplications. Multiplication is provably intractable for conventional BDD-based verification. Word-level model-checking [ClKZ96] also is BDD-based and can deal with multiplication, but appears not to be well-suited to automated general verification and localization reduction as implemented in COSPAN. Furthermore, although explicit state enumeration had no difficulty searching parts of the state space (up to about 5 million states with .5GB RAM), the state space of this model, after reduction, is around 10^{45} states for just a single stage.

As noted in the Introduction, COSPAN is best suited for verifying control-intensive models. Verification of data-path-intensive models such as this one generally requires application of *ad hoc* methods which go beyond the fully automated algorithms in COSPAN. Thus, this is a good example of a design model *not well suited* to the automated use of COSPAN. Nonetheless, through application of a straight-forward semi-automated decomposition of the model, a partial verification was possible within a short amount of time, as now reported.

For the verification of this model, the unconstrained environment model is too general. Accordingly, we developed an environment model consistent with the timing diagram provided in the benchmark documentation. The model contains a state for each stage of inputs required: *start* serves as an initial state for the environment, a system-wide reset is issued in *streset*, the weights are loaded during *ldwgt*, the weights are stored during *stwgt*, the inputs are provided during *stinput*, and results are stored and produced during *stresult*. The model passes through these states sequentially but with a nondeterministic choice whether to change state or not, modelling an indeterminate delay between state transitions. The inputs to the systolic array are assigned values based on these states. Using the QRY.h library macros, the environment model is as expressed in terms of the following constraints. Here,

$$STR(\ initial,\ expression\ )$$

defines a state variable $ of type *string* (holding an enumerated type), and

$$NoDet(\ limit\ )$$

defines a nondeterministic combinational variable # with range $0, \ldots, limit$. The symbols $ and # are merely convenient, with no special meaning attached. We use the latter to define a fully general nondeterministic delay between the state transitions of the environment model. With *limit* = 1, # = 1 denotes the end of the delay. The syntax

$$x\ ?\ P\ |\ y$$

denotes the if-then-else expression:

$$x\ if\ P\ else\ y$$

similar to the language C.

ENVIRONMENT MODEL

```
ENV_STATE: STR( "start",
                 ( "streset"   ? $ = start
                 | "ldwgt"     ? $ = streset
                 | "stwgt"     ? $ = ldwgt
                 | "stinput"   ? $ = stwgt
                 | "stresult"  ? $ = stinput
                 | start )     ? DELAY.# = 1 | $
                 )
DELAY: NoDet(1)
E1: AssumeAlways( SELECTWGTSTR = 0 ? ENV_STATE.$=stwgt | 1 )
E2: AssumeAlways( STOREWGT = 1 ? ENV_STATE.$=stwgt | 0 )
E3: AssumeAlways( STORESTR = 0 ? ENV_STATE.$=stwgt | 1 )
E4: AssumeAlways( STORERES = 1 ? ENV_STATE.$=stresult | 0 )
E5: AssumeAlways( RESET_N = 0 ? ENV_STATE.$=streset | 1 )
```

Our partial verification consisted of three phases. First, we verified that a single stage of the array computed the expected function: when STORERES is asserted, the weight and input registers are multiplied together and added to RESULT_IN. Second, we used this result to compose a larger, general result about the function computed by three stages. Finally, we attempted to use this larger result to prove that the circuit computed the desired function over streams of input values. Only the first two phases were verified in the full model; verifying the third required that we limit ourselves to a narrower data-path. These phases are detailed in the remainder of the section.

To verify a portion of a larger design such as a single stage of the systolic array, it may be helpful to use a utility in COSPAN which automatically restricts the model to a designated submodel. In a commercial setting, it is of considerable importance that all such semantic modifications of the original design model be accomplished automatically– through COSPAN directives– *without touching the design model source code*. While COSPAN's localization reduction algorithm applied relative to a property of a single stage in theory should restrict the analysis to that stage, sometimes spurious dependencies can contaminate the algorithm and cause it to fail to produce as small a reduction as it apparently should. In the case of this circuit model, we initially attempted to verify the behavior of a single stage using the entire circuit, but COSPAN failed to reduce the analysis to that stage, instead retaining the entire model. Explicitly restricting the analysis to a single stage, the check went through. In the process, we discovered a violation of the property we were trying to verify, and also the reason the localization reduction was not succeeding to isolate that stage: the value on RESULT_IN must be smaller than 2^{10} in order for the property to hold. Without this, spurious dependencies on the entire circuit were created. With this assumption, the property verified both in the single stage model and in the full model through a now-successful localization reduction. The RESULT_IN-bound assumption later was verified to hold for each stage of the full model. The following is the formal definition of the property verified for each single stage.

RESULT_IN_bound: AssumeAlways(RESULT_IN $< 2^{10}$)
SingleStageProperty: Always((STORERES $\neq$ 1) +
    (ENV_STATE.$ = stresult)*(oldstres = 0) +
    (STAGERESULTOUT =
        STAGEWEIGHT_REGISTER*inputreg_t + resin_t1))

The auxiliary variables *inputreg_t* and *resin_t1* store input-stream size integers and result-stream size integers, respectively, on the rising clock edge of CLK1 and CLK2 respectively. Each clears its value when RESET_N = 0 . The auxiliary variable *oldstres* holds the value of the Boolean expression

(ENV_STATE.$ = stresult)*(CLK2Rise=1) .

All auxiliary variables may be defined through the QRY.h macro library variable constructors in the same fashion as ENV_STATE was defined.

Using COSPAN's task decomposition verification algorithm and the result that each stage satisfies the SingleStageProperty, we verified that the entire circuit model satisfies the AllStages property given below, but only for 2-bit input registers and 4-bit output registers.

AllStages: Always((STORERES $\neq$ 1) +
$\qquad$ (ENV_STATE.$ = stresult)*(oldstres = 0) +
$\qquad$ (result_delay < 5) +
$\qquad$ (STAGE0RESULTOUT =
$\qquad\qquad$ STAGE0WEIGHT_REGISTER * input0reg_t +
$\qquad\qquad$ STAGE1WEIGHT_REGISTER * input1reg_t2 +
$\qquad\qquad$ STAGE2WEIGHT_REGISTER * input2reg_t3 +
$\qquad\qquad$ resin2_t13))

Here, *result_delay* is an auxiliary variable that counts the number of rises on CLK2 since the STORERES signal was asserted. Each of *input0reg_t*, *input1reg_t2*, and *input2reg_t3* is an instance of *inputreg_t* described for a single stage. The suffixes indicate the storage delay in with respect to the number of clock cycles: for example, *input2reg_t3* holds the value that was on the input stream to stage 2 three clock cycles previously. These auxiliary variables are defined as a succession of stages, each of which delays the value by one clock cycle.

This verification succeeded after exploring 9.53915e+15 states, using 561378 BDD nodes, 63 megabytes of memory, and taking approximately half an hour. COSPAN's symbolic state enumeration algorithms utilize dynamic variable reordering, which typically dramatically reduces BDD size. Sometimes, especially with data paths, hand reordering can improve further on the automatic reordering. In this example, when the BDD variable ordering was modified by hand, only 51137 nodes, 36.8 megabytes of memory, and just over ten minutes of elapsed time were required. We ordered the variables roughly in a left to right sweep over the signals in the layout diagram: the stage 2 variables, then the stage 1 variables, then the stage 0 variables. In order to specify the BDD ordering by hand, one runs COSPAN twice: once to dump its internally determined order to a file; then a second time, reading in that file, after it has been reordered by the user.

We used the same compositional technique as just reported to extend AllStages to the required result over the previous input values. We defined an array of five processes called *PrevInputs* such that *PrevInputs[i]* contains the value on the input stream *i* clock cycles earlier; each process updates its value on the rising edge of CLK1. By composing the *PrevInputs* definitions with the AllStages result and the definitions of *input0reg_t*, *input1reg_t2* and *input2reg_t3*, we were able to prove the following result on the narrower data-path through a symbolic enumeration of 1.6e+12 states.

Output0:
$\qquad$ Always((STORERES $\neq$ 1) +
$\qquad\qquad$ (ENV_STATE.$ = stresult)*(oldstres = 0) +

$$(result\_delay < 5) +$$
$$(STAGE0RESULTOUT =$$
$$STAGE0WEIGHT\_REGISTER * PrevInputs[4] +$$
$$STAGE1WEIGHT\_REGISTER * PrevInputs[2] +$$
$$STAGE2WEIGHT\_REGISTER * PrevInputs[0]))$$

The input values used in this property are two apart, rather than consecutive, to reflect the requirement given in the documentation that the input values should hold for two full clock cycles. Even in the reduced-width model, we did not succeed to verify the full input behavior in which the *PrevInputs* are extended far enough to capture the desired weight register values.

6.3.1 Verification Statistics.

	BDD nodes	CPU seconds	Meg Memory
OneStage	57103	531.6	41.0
AllStages	561378	1863.7	63.0
AllStages (with custom variable ordering)	51137	641.7	36.8
Output0	27606	184.0	11.6

6.4 BlackJack Dealer

The properties to be verified are as follows:

1. If the score is greater than 16 but less than 21, the Stand indicator is lit.
2. If the score is greater than 21, the Broke indicator is lit.

Both of these properties verified quickly and easily without any special treatment. As with the Single Pulser, the clock had to be constrained to toggle at each successive global state. The same OldClk and MakeClk processes from the Single Pulser section were used in verifying these properties.

Property1:
$$After\_Eventually\_Unless((SCOREGT16 = 1)*(SCOREGT21 = 0)$$
$$*(RESET = 1),$$
$$STAND\_IND = 1,$$
$$RESET = 0)$$

Property2:
$$After\_Eventually\_Unless((SCOREGT21 = 1)*(RESET = 1),$$
$$BROKE\_IND = 1,$$
$$(RESET = 0) + (ACE11FLAG = 1))$$

6.4.1 Verification Statistics.

	BDD nodes	CPU seconds	Meg Memory
Property1	3863	6.41	0
Property2	4341	6.64	0

6.5 Car Seat Controller

This additional VHDL example demonstrates how COSPAN's various complexity reduction routines can be utilized in concert in a verification methodology. The car seat has three axes of movement: the seat bottom can be raised or lowered, the seat back can be raised or lowered, and the seat can be moved closer to or further away from the steering wheel. Preferred settings on each axis can be stored and recalled for up to three people. The controller is responsible for producing two signals for each axis: a pulse to drive the motor on that axis and a signal indicating which direction the seat should move.

The main component of the architecture is the axis module, which is instantiated once for each of the three axes; the axis module architecture appears in Figure 6.1. We exploited this modular design style to take advantage of the symmetry-based reduction methods in COSPAN to reduce the verification effort. The inputs to the module are *up* and *dn* buttons for moving the seat while in manual mode, a toggle (*memsw*) indicating memory mode when on and manual mode when off, a system-wide *reset* signal, a *store* button for indicating that values should be saved to memory, a button for each person (*person1*, *person2*, *person3*) to control which registers to access on loads and stores, and the system clock *clk*. The module uses an eight-bit data-path and contains the following major components:

- The *Actual Position Counter* (APC) for storing the current seat position.
- The *Desired Position Counter* (DPC), a loadable counter containing the desired position of the seat when in manual mode.
- Three registers, for storing the preferred setting for each of three persons.
- A multiplexer (I4) that determines the target position (POS) depending upon whether the controller is in memory or manual mode; if the former, POS is also determined by the person whose settings are being recalled.
- A comparator that indicates the relationship between POS and the APC.
- A motor (MTR) that produces the output pulses and the input signals to the APC, depending upon the output of the comparator.

The remaining portions of the design in Figure 6.1 are combinational logic blocks, with their functionality defined as follows:

- I0 sets *nsmem* to the current value of *memsw*. The output to *ldenb* is asserted exactly when *csmem* is asserted and *memsw* is not.
- I2 produces the disjunction of *cenbup* and *cenbdn*.

- I3 sets *udc* to true if *up* is asserted and to false if *dn* is asserted. *udc* will hold its previous value if neither is asserted. The environment guarantees that the *up* and *dn* are never asserted simultaneously.
- I5 asserts *loada* if *store* and *person1* are true; *loadb* and *loadc* are similarly defined for *person2* and *person3*.
- I6 sets *csmem* to the value of *nsmem* on the rising edge of *clk*; at all other times, *csmem* holds its previous value.

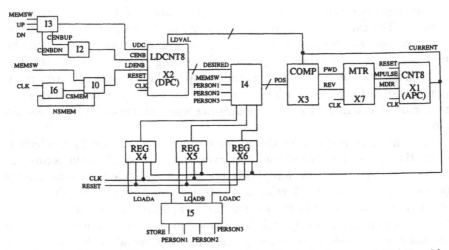

Fig. 6.1. Architectural layout of the axis module. Slashed wires are eight bits wide.

The axis design is straightforward, with the possible exception of the wire for loading the DPC from the APC. When the controller is in manual mode, the APC is designed to follow the movements of the DPC. Once the controller loads a value from memory, the APC and the DPC can lose the alignment that makes this behavior possible. As a result, the DPC must be explicitly aligned with the APC when the controller switches from memory mode to manual mode.

The car seat environment model introduces the following constraints:

- The *up* and *dn* buttons for an axis may not be asserted simultaneously.
- Storage or recall may be selected for at most one person at a time.
- *Reset* remains high for a period of time (indicating initial reset) then remains low thereafter.
- A period of time must be allowed for the controller to initially reset.
- The controller is in exactly one of manual mode or memory mode at all times.

Three copies of the axis module comprise the design. Since the axes do not communicate with one another, all properties of interest are relevant to the individual modules. We are interested in verifying the following properties of each axis:

1. Pushing the *up* button in manual mode causes the APC to increase.
2. Pushing the *dn* button in manual mode causes the APC to decrease.
3. Settings are properly stored to memory when store is activated.
4. Settings are properly recalled from memory when in memory mode.
5. The seat never attempts to move past the maximum and minimum positions.

Statements of the properties relative to the first axis appear below. Each of Property3 and Property4 needs to be duplicated for *person2* and *person3*. OldAPC and OldDPC store the respective values of the APC and DPC in the immediately preceding global state. Signals *fthtup* and *fthtdn* are the *up* and *dn* signals for the first axis. Signal *motorRot* indicates that the motor should move; *motorDir* indicates the direction in which it should move: 0 indicates backwards and 1 indicates forwards.

Property1: After_Eventually ($\neg reset$ * (APC < 255) * *fthtup*,
$\qquad\qquad\qquad$ *memsw* + (APC > OldAPC))
Property2: After_Eventually ($\neg reset$ * (APC > 0) * *fthtdn*,
$\qquad\qquad\qquad$ *memsw* + (APC < OldAPC))
Property3: After_Eventually (*store* * *person1* * $\neg reset$ * (APC = POS),
$\qquad\qquad\qquad$ (REG1 = APC) + $\neg person1$ +
$\qquad\qquad\qquad$ $\neg$(APC = POS))
Property4: After_Eventually(*memsw* * $\neg reset$ * *person1*,
$\qquad\qquad\qquad$ $\neg person1$ + $\neg memsw$ + (APC = REG1))
Property5min: Never($\neg reset$ * (APC = 0) * *motorRot* * $\neg motorDir$)
Property5max: Never($\neg reset$ * (APC = 255) * *motorRot* * *motorDir*)

The controller model contains 5.14e62 states. None of the properties verified in a reasonable amount of time when tested with the default reduction algorithm (localization reduction), even with symbolic state enumeration. To accomplish the verification, a combination of localization reduction, abstraction, symmetry reduction and task decomposition was used.

The first two properties were the most difficult to verify. We discuss the verification of the first in detail. Attempting to verify the first property using localization reduction alone yielded a model containing 6.04e24 states; the localization reduction removed the other two axes and their respective *up* and *dn* inputs. This model was still too large to handle in a reasonable amount of time, even with symbolic state enumeration.

6.5.1 Abstraction. This design appears to be a good candidate for abstraction, as described in Section 5. Its data-path is wide, but largely irrelevant to the desired properties; only two data-path values (0 and 255, the minimum and maximum values) are mentioned explicitly in the properties. This suggests an abstraction to a data-path with only three distinct values: one corresponding to 0, one to 255, and one for everything in between. Our abstraction maps 0 to 0, 255 to 2, and all intermediate values to 1, thus reducing the data-path to two bits as opposed to eight.

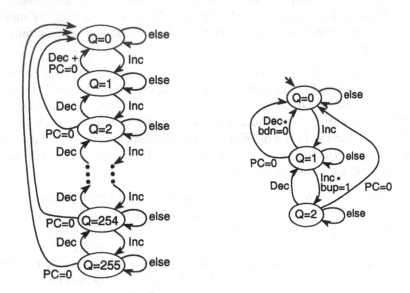

Fig. 6.2. State machine depictions of the original and abstract counters. Q is the counter output. Inc and Dec represent the input values requesting the counter to increment and decrement, respectively.

Consider the abstraction of the APC. Attempting to abstract the APC to a deterministic two-bit counter would not result in a homomorphism. Let ϕ be the desired mapping from concrete values to abstract values: $\phi(0) = 0$, $\phi(255) = 2$, and ϕ applied to any other eight-bit value is 1. For ϕ to be a homomorphism,

$$\phi(a) + \phi(b) = \phi(a + b)$$

must be satisfied for all concrete values a and b. Assume b carries value 1. If a is 254, the requirement is satisfied. If a is 253, the requirement is not satisfied, since $\phi(253) + \phi(1) = 1 + 1 \neq 1 = \phi(254)$. The abstract counter must therefore be non-deterministic when it holds value 1.

State machine depictions of the original counter and the non-deterministic abstract counter appear in Figure 6.2. The signals *bup* and *bdn* appear only in the abstract counter and are non-deterministically assigned either zero or one at each clock cycle; the machine transitions based upon these values. The homomorphism relationship is specified with statement:

hom *bup* := 1 ? (APC $\geq$ 254) | 0,

$$bdn := 0 ? (APC \leq 1) \mid 1$$

Under this mapping, COSPAN verifies that the non-deterministic two-bit model is a valid abstraction of the eight-bit counter. After performing similar abstractions on each component, we have an abstract model defined on the same variable names as the original model; this allows us to use the original properties file unchanged except for replacing uses of 255 with 2 (this replacement is valid under the homomorphism). However, the use of nondeterminism in the model introduces new problems that are discussed in the next section.

One practical note is in order. While attempting to verify a previous version of the controller, we found it quite useful to simply reduce the controller to having a two-bit, deterministic data-path. This allowed us to locate potential bugs in the control quickly and easily. Although the verifications done on the two-bit deterministic model were not mathematically guaranteed to apply to the eight-bit model, we did locate several control bugs in this manner, all of which turned out to apply to both the two-bit and eight-bit models.

6.5.2 Handling Nondeterminism. Attempting to verify Property1 uncovers a problem with using the abstract model. Pressing the *fthtup* button in manual mode should result in the DPC being incremented. The DPC should now be larger than the APC, so the APC should also be incremented. The DPC and the APC are non-deterministic counters in the abstract model. It is therefore possible for the DPC to choose to increment and for the APC to choose not to increment. Since the APC is never forced to increment from value 1, COSPAN detects a cycle in which the desired property is not satisfied.

A first thought was to impose a fairness constraint on the nondeterminism; this would prohibit the APC from ignoring the increment requests indefinitely. Unfortunately, this solution is too general to be useful in this situation. The APC could choose not to increment, then the DPC could start to decrement before the fairness constraint took effect. The automaton for Property1 waits for its *fulfilled* condition to be satisfied, even after the DPC begins to decrement. If the DPC never incremented again, the eventually condition would never be satisfied, resulting in an error. Clearly, a different approach is required.

The problem is that the non-deterministic behavior of the two counters is not aligned. The two counters should take the same action when dealing with the same flow of the input. This is more subtle than requiring that the two counters always make the same non-deterministic choices. The architectural layout indicates that data reaching the DPC at time t doesn't reach the APC until time $t+2$, where time is measured relative to clock ticks. The APC must therefore follow the non-deterministic choice of the DPC two time steps later. These constraints are introduced using state variables bup_t+2 and bdn_t+2 to store the appropriate previous values of bup and bdn.

AlignNDup: AssumeNever($\neg$(APC.bup_t+2 = DPC.bup))

AlignNDdn: AssumeNever($\neg$(APC.bdn_t+2 = DPC.bdn))

The homomorphism must be re-verified once these restrictions are introduced.

6.5.3 Property Decomposition. Aligning the nondeterminism is still insufficient to verify the property because the property assumes that the DPC is deterministic. Aligning the nondeterminism requires that the APC follows the movements of the DPC, but it does not force the DPC to increment on each press of *fthtup*. A task decomposition is useful here. Intuitively, Property1 can be decomposed into two subproperties: one that pressing *fthtup* causes the DPC to increment and one that the APC increments whenever the DPC increments (provided the controller is in manual mode).

Our initial attempt at a decomposition for Property1 split the circuit around the POS signal (see Figure 6.1). POS indicates the target position for the seat taking into account memory or manual mode, and any respective recall requests. The subproperties appeared as

DecompProperty1: After_Eventually($\neg$*reset* * (APC < 255) * *fthtup*,
$\qquad\qquad\qquad\qquad$ *memsw* + (POS > OldAPC))
DecompProperty2: After_Eventually((POS > OldAPC), (APC > OldAPC))

The decomposition check tests that the language of the original automaton contains the language of the intersection of all the subproperty automata.

Surprisingly, this is not a valid decomposition. COSPAN locates a sequence of values that result in Property1 being starved while both decomposed properties are satisfied. The error is summarized in the table below. "I" indicates that a process is in its initial state. "E" indicates that the process has been enabled and is waiting to be fulfilled. Abbreviating the above Boolean expressions by

$$b0 = \neg reset * (APC < 255) * fthtup$$
$$b1 = memsw$$
$$b2 = (POS > OldAPC)$$
$$b3 = (APC > OldAPC)$$

the status of a process in a given row is determined by its status and the Boolean values in the previous row.

P1	I	E	I	E
DP1	I	E	E	I
DP2	I	E	I	I
b0	1	0	1	0
b1	0	0	1	0
b2	1	0	0	0
b3	0	1	0	0
	0	1	2	3*

Examining this table indicates that the problem is a failure to accurately depict an important feature of the circuit. There is no way for POS to be larger than APC without either the DPC incrementing or memory mode being turned on. It is possible to capture this condition by introducing another automaton to the list of subproperties. In fact, an automaton that just places this restriction is sufficient to prove the decomposition.

Unfortunately, DecompProperty2 itself failed on the circuit, because it did not take the system-wide *reset* signal into account. Adding ¬*reset* to the *enabled* condition in DecompProperty2 suppressed this error, but the verification attempt failed to terminate in a reasonable amount of time.

We concluded that having the APC in the enable condition of Property1 was contributing extensive complexity to the decomposition process. As a result, we rewrote the original Property1 (named NewProperty1 in the discussions that follow) so that the enable condition tested whether the DPC, rather than the APC, was at less than maximum value. An additional property was added to check that the APC reached the maximum value whenever the DPC was at maximum, the APC was below maximum, and the system was in manual mode.

In addition, when we wrote the decomposition for NewProperty1, we chose to split the circuit at the DPC, rather than at POS. Any property requiring deterministic behavior of the system must be verified in the original rather than the abstract model. Since the original model is so large, we needed to localize as much as possible the portion of the model that had to be considered. NewProperty1 only required that the DPC be checked deterministically; the portion of the circuit from the DPC to POS did not require deterministic behavior. Decomposing around the DPC reduced the required state space for the deterministic test. NewProperty1 and the new decomposition are provided below:

DecompProperty1:
      After_Eventually(¬*memsw* * (DPC > OldDPC) * (Mode : man),
          (APC > OldAPC))

DecompProperty2:
      After_Eventually(¬*memsw* * ¬*reset* * *fthtup* * (DPC < max)*
          (Mode : man), *memsw* + (DPC > OldDPC))

NewProperty1:
      After_Eventually(¬*memsw* * ¬*reset* * *fthtup* * (DPC < max)*
          (Mode : man), *memsw* + (APC > OldAPC))

These properties test the state of an auxiliary variable called Mode, depicted in Figure 6.3. Mode tracks when the system is fully operating in manual mode. Immediately after a switch from memory to manual mode, the APC and the DPC need to be synchronized, as described earlier. In synchronizing,

242   Kathi Fisler and Robert P. Kurshan

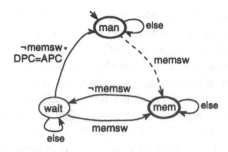

Fig. 6.3. The Mode state machine.

the DPC may increment from its previous value. By design, this increment of DPC will not trigger a subsequent increment in the APC; as a result, it must be eliminated from the *enabled* condition of DecompProperty1. Mode is in its state *man* only after the synchronization is complete. The state of **Mode** therefore eliminates this case from DecompProperty1. The *man* and *mem* states in Figure 6.3 appear in bold. A fairness constraint is imposed on each, to constrain the model from remaining in either state forever, through the QRY.h query library macro

AssumeEventually_(not(**state**))

applied (separately) to each of the two states. Another fairness constraint prevents the transition from *man* to *mem* (dashed line) from repeating indefinitely, in order to discard the case in which synchronization never completes.

After aligning the nondeterminism as discussed in Section 6.5.2, DecompProperty1 verified quickly in the abstract model. However, attempts to verify DecompProperty2 in the full model failed to terminate in a reasonable amount of time. Examination of COSPAN's status information indicated that the reduction algorithm was failing to find a suitable localization for DecompProperty2. Looking at the architecture (Figure 6.1), we were able to define a seed for the localization algorithm to start it from a reasonable set of variables. (Layout diagrams were extremely useful in helping to locate all signals on the boundary of a localizable property.) The seed included the variables appearing in DecompProperty2, plus those forming the boundary around the DPC. Our seed therefore contained OldDPC, OldAPC, DecompProperty2, DPC, *reset, fthtup, memsw* and the output of component I0 in Figure 6.1; it turned out that the reduction algorithm had tried to exclude the I0 component, thereby causing it to blow up. This precluded the algorithm from subsequently including this variable. Once this variable was forced into the model via the seed to the algorithm, the localization succeeded and the property was verified in the full model. Execution times appear in Section 6.5.4.

The remaining properties of the designated axis were verified without much difficulty. Property2, being similar to Property1, was verified in a similar manner. The remaining three properties each verified without decomposition in the abstract model. Verifying Property4, however, required a fairness constraint that the abstract APC does not ignore the increment and decrement signals indefinitely. The QRY.h query library macro

IfRepeatedly_AssumeEventually_(*enabled, fulfilled*)

defines such a fairness constraint (known as "strong fairness"). Including this restriction therefore was straightforward.

The properties stated in this paper are given relative to the first axis. There are two options for verifying the properties of the remaining two axes. One is to simply rewrite the properties using the associated variable names from the other axes and verify them explicitly. Another is to exploit the symmetry of the axes. Just as homomorphisms can be used to show properties hold between abstract and detailed models, they can be used to establish symmetry. In this case, we define the homomorphism to exchange the variables feeding into the first and second axes. Verifying this homomorphism is sufficient for us to conclude that the properties proved of the first axis also hold for the second. A similar procedure verifies the behavior of the third axis.

6.5.4 Verification Statistics.
The initial VHDL specification contained 447 lines of code. The FormalCheck translator produced a 927 line S/R specification. The following table summarizes running time and memory usage statistics for the discussed properties:

	CPU seconds	Meg Memory
DecompProperty1	21.37	1.06
DecompProperty2	437.46	8.42
Property3	41.32	5.27
Property4	53.71	5.27
Property5min	14.48	1.05
Property5max	13.61	1.05

The most expensive property, DecompProperty2, is the one verifying the behavior of the counter in the full model. This is the only one of the properties that checks predominantly data-path, as opposed to control; that its verification is so much more expensive emphasizes the need for reduction techniques for automated verification of data-path-intensive designs (*cf.* Section 6.3).

In conclusion, we succeeded to verify this controller using localization reduction, task decomposition, homomorphic reduction (abstraction) and symmetry reduction. In spite of the difficulties and false starts reported (and the fact that at the time the entire verification methodology was new to the first author, who conducted this verification essentially on her own), the verification was completed in the course of a summer visit (under three months), including the time to learn how to use COSPAN, understand the controller design and redesign the original buggy controller. We expect a similar, subsequent project would take considerably less time. Finally, parameterizing the data-path of a design is a very advantageous design practice from the perspective of verification. Setting a data-path to 1-2 bits often permits full verification without recourse to *ad hoc* reductions, and may uncover most of the bugs present in the full design. In the case of this controller, all the design errors could have been uncovered using a 2-bit data-path.

7. Summary

We have described the utilization and operation of COSPAN through some VHDL examples. COSPAN is a software tool whose primary application is the verification of finite-state programs through an ω-automaton language containment check. While VHDL programs typically are synchronous, COSPAN is capable of verifying both synchronous and asynchronous models, through the use of nondeterminism. The backbone of COSPAN is its reduction algorithms through which design models with as many as 5K latches and 100K combinational variables may be checked. The central reduction algorithm is *localization reduction* through which a design model is automatically replaced with a conservative (non-bi-simulation-equivalent) model, reduced relative to the property being checked.

Acknowledgement. We thank Arthur Glaser for his help in understanding the Car Seat Controller design, and in correcting the bugs found in the original design.

A. Appendix: Automata Definitions

We give here some representative definitions of the **proctypes** (S/R parameterized process macros) used to define the properties and constraints in the foregoing examples. All of these proctype definitions are drawn from QRY.h library of proctypes used to define queries. In general, COSPAN model-checking may be based exclusively upon queries defined through this library, without recourse to any other user-defined automata whatsoever.

All the library elements are built upon the paradigm of an *enabling condition* which triggers a check, a *fulfilling condition* with respect to which the check is conducted and a *discharging condition* which terminates the check. The enabling and discharging conditions are optional. The fulfilling condition has two "safety" formats: Always_ and Never_, and two "liveness" formats: Eventually_ and EventuallyAlways_ (not illustrated here). The enabling condition also has two formats: After_ and IfRepeatedly_ (not illustrated here), as does the discharging condition: Unless_ and UnlessAfter_ (not illustrated here).

Proctypes whose names contain the tag "Assume" are used to define constraints, while those without "Assume" are used to define properties. Automata for properties and constraints thus are paired; the automaton for a given constraint defines the formal language complementary to the one defined for the corresponding property.

Each definition below consists of the S/R syntactic definition augmented by an explanation and a pictorial representation of the automaton it defines. Automata which accept infinite words, such as are used with COSPAN, require acceptance conditions defined in terms of states and edges that are visited infinitely often; states and edges in the acceptance conditions are denoted by shading and dashed lines, respectively. Property automata (those without "Assume" in their names), accept words for which all of the states visited infinitely often are shaded or for which the dashed edges are crossed infinitely often. Constraint automata (those with "Assume" in their names) use these notations in a dual fashion to denote word exclusion: words that visit shaded states or cross dashed edges infinitely often are not in the languages of such automata. More complete details on these relationships are provided elsewhere [Kurs94a].

A.1 AssumeAlways_()

Used to define constraints, AssumeAlways_(*fulfilled*) excludes from the verification check any executions in which *fulfilled* is not always true. For efficiency of verification, this proctype is implemented using the **kill** primitive, which truncates the search of behaviors inconsistent with the constraint.

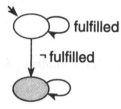

fulfilled

¬ fulfilled

proctype AssumeAlways_(fulfilled : boolean)
    import fulfilled
    kill ~fulfilled
end AssumeAlways()

A.2 Never_()

Used to define properties, Never_(*fulfilled*) accepts all executions in which *fulfilled* never is true at any reachable state. For efficiency of verification, this proctype is implemented so as to to force a run-time syntax error (out of range variable) if the associated property fails. This causes COSPAN to stop its analysis as soon as the property is violated, and to pinpoint the error more precisely than it would if the proctype were implemented as a syntactically proper ω-automaton.

proctype Never_(fulfilled : boolean)
   import fulfilled
   stvar $: (1)
   init $:=1
   trans true → 1 : ~fulfilled
end Never_()

A.3 After_Eventually_()

After_Eventually(*enabled*, *fulfilled*) accepts all executions in which after each occurrence of *enabled*, an occurrence of *fulfilled* eventually follows.

proctype After_Eventually_(enabled, fulfilled : boolean)
   import enabled, fulfilled
   stvar $: (INIT, ENABLED)
   recur $: *→ INIT
   init $:= INIT
   asgn $ → ENABLED ? ($=INIT)*enabled | INIT ? fulfilled | $
end After_Eventually_()

A.4 PHASE

PHASE(*set*, *clear*) is a bit that goes high whenever *set* is true and low whenever *clear* is true. It is used as an auxiliary variable in the following two definitions.

proctype PHASE(set, clear: boolean)
   import set, clear
   stvar $: boolean
   init $:= 0
   asgn $ → 1 ? set*~$ | 0 ? clear | $
end PHASE()

A.5 After_Never_Unless_()

After_Never_Unless_(*enabled*, *fulfilled*, *discharged*) accepts all executions in which after each occurrence of *enabled*, *fulfilled* never is true unless *discharged* becomes true.

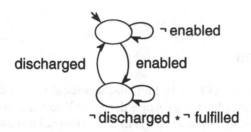

```
proctype After_Never_Unless_(enabled, fulfilled, discharged : boolean)
    import enabled, fulfilled, discharged
    monitor ENABLED: PHASE(enabled, discharged)
    monitor G: Always(~fulfilled + discharged + ~ENABLED.$)
end After_Never_Unless_()
```

A.6 After_Eventually_Unless_()

After_Eventually_Unless_(*enabled*, *fulfilled*, *discharged*) accepts all executions in which after each occurrence of *enabled*, an occurrence of *fulfilled* or *discharged* eventually follows.

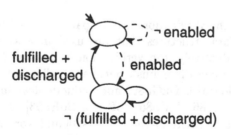

```
proctype After_Eventually_Unless_(enabled, fulfilled, discharged : boolean)
    import enabled, fulfilled, discharged
    recur  ~ENABLED.$
    monitor ENABLED: PHASE(enabled, fulfilled+discharged)
end After_Eventually_Unless_()
```

1

The C@S System: Combining Proof Strategies for System Verification

Klaus Schneider and Thomas Kropf

1. Introduction

Formal verification is the task to find mathematical proofs that ensure that a given specification holds for a certain design. Although more powerful than traditional simulation, a breakthrough in the industrial use has been achieved only after the introduction of binary decision diagrams and symbolic state traversal algorithms [Brya86, CoBM89a], avoiding the explicit enumeration of states of a finite state machine model. These methods lead to powerful techniques like symbolic model checking [BCMD92], symbolic trajectory evaluation or language containment based on ω-automata (the latter two can be found in this book on the pages 3 and 206, respectively). Using these approaches, a fully automated verification of significantly large systems has become possible. Common to all of these automated approaches is the view that the system to be verified has a finite number of states such that the verification problem is decidable. However, the drawback of these formalisms it that the size of the systems that can be verified with acceptable resources (in terms of memory and runtime) is limited due to the so-called state explosion problem that arises in most cases when the system has a nontrivial data path.

For this reason, the verification of data paths is often not done by these methods. As the data structures that are used in a system can in most cases be defined inductively, the verification is usually done by induction based methods. One can distinguish between explict induction approaches [Burs69, Aubi79, BoMo79, GaGu89] and inductionless induction methods [Muss80, HuHu82, Frib86, JoKo86, JoKo89, BoRu93]. The latter ones are closely related to term rewrite methods [Ders90] and work more or less automatically (guided by some user interactions). However, they can only be applied to verification goals that are given as equation systems, possibly extended by conditional equations. Explicit induction can be applied to a broader class of goals, but requires a higher degree of manual interaction, as e.g. the induction variables, the induction schema and the induction hypotheses have to be chosen manually.

Hence, almost all formalisms used for formal verification can be classified into two categories: on the one hand there are decidable formalisms, unable to

---

* This work has been funded by a German research grant (Deutsche Forschungs-gemeinschaft, SFB 358, TP C2.)

reason about abstract data types[1], and on the other hand there are undecidable formalisms, able to reason about recursive data type by induction proofs. While these two classes of formalisms are limited to either control or data path dominated systems, this restriction does not hold for theorem provers that are based on higher-order logic [Gord86, ORSS94]. Unfortunately, these approaches require a considerable amount of manual interaction. Thus various approaches have been presented to partially automate the verification by incorporating automated reasoning procedures [KuSK93a, ORSS94] or by adding abstraction and compositional verification techniques to allow larger systems to be verified by finite state approaches (see [ClLM89a, ClGL92, Long93, Schn96a] or many of the other approaches presented in this book).

Clearly, it is useful to separate control and data path oriented parts of a verification goal prior to verification [ScKK94a, LaCe94, HoBr95, HuGD95, ClZh95]. Proceeding this way, finite state approaches may be used for the controller part 'guiding' the verification of the data path, whereas the latter often requires theorem proving techniques. However, these approaches are also not able to provide full automation for systems with heavy interaction between control and data.

In this chapter, a new approach is presented that allows the combination of different proof techniques. These proof procedures and the methods to combine them have been implemented in the verification system C@S[2] of the University of Karlsruhe. C@S is based on a linear temporal predicate logic that is a superset of both linear temporal (propositional) logic [Pnue77] and first order predicate logic with (inductive) abstract data types. To be precise, the underlying formalism of C@S is a higher order logic as the induction principles of the abstract data types are not expressible in first order predicate logic. However, the logic is far less powerful than the higher order logics used in HOL [GoMe93] or PVS [ORSS94]. Based on our experiments, we claim however that the logic is powerful enough to express almost all hardware verification problems and due to its special form of specifications, it allows a degree of automation that can not be reached by a general higher order theorem prover. In case of pure control-based systems, a full automation can be achieved as C@S provides state-of-the art model checking procedures for temporal logic. Also, if a pure data path problem is to be verified, C@S provides induction methods that can be invoked to prove these problems semiautomatically, i.e. by giving some information in form of lemmata and term orderings, that can be seen as 'milestones' of the proof.

The integration of various decision procedure in interactive theorem provers is not a new idea. E.g. also the idea of the PVS system [ORSS94] is to obtain a more powerful higher order theorem prover by extending the

---

[1] Reasoning about abstract data types leads to undecidable problems due to Gödel's theorem [MaYo78]. Hence, each method that is able to reason about abstract data types can no longer work fully automatically.

[2] speak: cats

interactive prover by decision procedures, and the HOL system is also currently extended by automated proof procedures. The added value of C@S is however, that these decision procedures do not only work isolated of each other. Instead, due to the restriction of the underlying logic, C@S provides some special proof tactics that can be applied to split general proof goals such that finally various decision procedures can be used together for the remaining proof. Currently, C@S has interfaces to the SMV [McMi93a] model checker implemented at the Carnegie Mellon University by Clarke et al. and the rewrite rule laboratory RRL [KaZh88] implemented by Kapur and Zhang. Furthermore, we have implemented a front end for SMV that enables SMV to check not only CTL formulas, but also to check and to prove LTL formulas (see section 5.3). As a result, C@S currently offers the following decision procedures and proof methods:

1. decision procedures for verifying temporal properties (control dominated systems):
   a) CTL model checking
   b) linear time temporal logic (LTL) theorem proving and model checking
   c) ω-automata
2. semi-automated proof procedures for proving lemmata about abstract data types
   a) explicit induction like cover set induction as implemented in RRL
   b) inductionless induction as implemented in RRL
3. interactive proof rules as e.g. invariant rules for breaking up interactions between control and data paths

Additionally, we are currently implementing additional proof procedures like an arithmetic decision procedures for Pressburger arithmetic and Büchi's monadic second-order theory of one successor. These arithmetic decision procedures can be used for proving lemmata that can be used for data path verification or for the verification of temporal properties.

In case of temporal properties, we suggest to use LTL instead of branching time temporal logic such as CTL for two reasons: Firstly, LTL temporal specifications can directly express facts where chains of events have to be considered, and secondly, LTL directly supports hierarchical reasoning in contrast to CTL in form of the assume-guarantee paradigm. The drawback of LTL is its principal non-linear time complexity for model checking and the insufficient support for efficiently implementable proof algorithms as compared to well-established symbolic CTL model checking implementations [BCMD90a]. However, the theoretical disadvantage of LTL does not necessarily lead to much larger runtimes in practice. Our approach contains new and efficient methods for LTL theorem proving and model checking. They are essentially based on transforming LTL verification problems into finite-state machine problems that can finally be checked by an arbitrary CTL model checker as SMV. Proceeding this way, we use ω-automata as an intermediate formalism. Of course, it is also possible to use these automata directly for specifying temporal behavior. In our opinion, there are some examples as e.g. protocols, where it is more natural to use automata based specifications than temporal logics. This has also the advantage to directly visualize the specification as a finite-state transition system.

On the other hand, if a module is data-driven and does contain only a simple control flow, e.g. systolic architectures or signal processors, then neither temporal logics nor ω-automata are well suited to solve the problem due to their lack of reasoning about data types other than booleans. Based on our experiments, we claim that in most cases a first-order formalization together with possibilities to perform inductive proofs is sufficient. Moreover, data paths can be well described by data-assigning equations; hence, the best-suited proof methods are term rewrite methods together with inductionless induction capabilities. In many cases, only a few interactions are necessary and even completely automatic proofs can be obtained for some systems.

If a system contains both a complex control and a data path that interact with each other, then neither temporal logics nor first order predicate logics are well suited for performing hardware verification. In such cases, CQS offers two solutions: if the data path is small, abstract data types are transformed to a purely propositional level. Then temporal logics or ω-automata are used to automatically verify the system. Obviously, this fails for the well-known problems, i.e. if the data path is too large (in terms of resulting propositional variables). Moreover, this approach does also not allow to reason about generic modules such as n-bit components. In these cases, CQS offers methods to eliminate the temporal part of the specification: it is often possible to translate the temporal logic part into an ω-automaton with only safety properties (see section 3.2). Such automata can be interpreted directly as rewrite systems, and can hence can be tackled by term rewrite methods and inductionless induction techniques, even when they are extended by abstract data types. If such a translation is not possible, then CQS offers invariant rules for each temporal operator. These invariant rules allow to eliminate a temporal operator such that the remaining temporal part can be translated into an ω-automaton with safety properities, i.e. into a rewrite system.

Moreover, the possibility of interactive rule applications in CQS allows to give the system 'proof hints' that also may significantly reduce the effort of a model checking proof. Proceeding this way it is sometimes possible to verify a system with a large bit width by model checking techniques, not being possible before. A simple, but powerful algorithm for heuristically choosing the suitable combination of proof procedures for a given system leads to a further automatioa.

There are two ways to read this article: on the one hand, the reader who is interested in the theory and the algorithms used in CQS will find in section 3 the basic formalisms, i.e. the specification language of CQS and in section

5 the new proof procedures of C@S. For example, we explain in detail how linear temporal logic is translated into special ω-automata. A reader more interested in practical applications will find in section 4 the definition of our system description language and in section 6 many case studies showing which verification procedure(s) of C@S are used to verify which circuit classes and specifications.

2. Related Work

2.1 Hardware Verification using Temporal Logics

In principle, for each logic $\mathcal{L}$, there is a temporal extension $\mathcal{L}_t$ of this logic $\mathcal{L}$, where the truth value of a formula depends on time. The semantics of the temporal logic $\mathcal{L}_t$ is given by a sequence $M_0, M_1, \ldots$ of models for $\mathcal{L}$, where it is assumed that M_i contains the truth values at time i and that each transition from one model M_i to the next one M_{i+1} in the sequence requires one unit of time. The syntax of $\mathcal{L}$ is extended by temporal operators [Emer90] that express temporal relations between formulas, as e.g. $\mathsf{G}\varphi$ which means that φ has to hold for all future models.

Using this extension for propositional logic, *linear temporal logic* (LTL) [Emer90] is obtained. Most temporal properties can be described in LTL in a very natural, concise and readable manner. Hence, LTL has been widely used for specifying temporal properties: Manna and Pnueli used LTL for specifying concurrent programs [MaPn81, MaPn82, MaPn83]; Bochmann was the first who used LTL for specifying hardware systems [Boch82]; Malachi and Owicki considered 'self-timed systems' [MaOw81] with LTL; Manna and Wolper used LTL for the description of the synchronization of communicating processes [MaWo84]. Lichtenstein and Pnueli developed an algorithm for checking a LTL formula φ in a given finite state model M with a runtime $O(\|M\|^2 \, 2^{\alpha\|\varphi\|})$ [LiPn85], where $\|\varphi\|$ means the length of the formula and $\|M\|$ the number of states of M. Using this algorithm for LTL model checking, it is possible to check small LTL formulas in big models. Moreover, LTL is well-suited for hierarchical verification [Lamp83, Pnue84].

In branching time temporal logics (BTL) [Emer90], a model is not a simple sequence of models of the base logic $\mathcal{L}$. Instead, at each 'point of time', there may exist more than one succeeding model which is chosen arbitrarily to model nondeterminism. Consequently, models of BTL form a tree instead of a linear sequence, where each node contains a traditional model for the logic $\mathcal{L}$. This tree is sometimes called a computation tree and therefore branching time temporal logics are sometimes called computation tree logics. If a system is described that consists of only a finite number of states, then it is possible to redraw the computation tree as a graph with only a finite number of nodes. In terms of modal logics, which are the general framework of temporal logics, this graph is called a *Kripke structure* [Krip75]. As each path through the

Kripke structure is a sequence of models, Kripke structures can also serve as an encoding of a set of models for LTL. Thus, BTL can be viewed as an extension of LTL and the complexity of the proof procedures cannot be better than the corresponding ones for LTL. On the other hand, Emerson and Lei showed that model checking problems[3] for branching time temporal logic have the same complexity as the corresponding linear ones [EmLe87].

A general branching time temporal propositional logic that corresponds to LTL is CTL* [Emer90]. Unfortunately, both the model checking problem for LTL and CTL* have a high complexity, i.e. they are both PSPACE-complete [Emer90]. For this reason, Clarke and Emerson [ClEm81] developed a restricted computation tree logic called CTL. CTL is not powerful enough to describe all facts that can be formalized in LTL or CTL* [Lamp80, LiPn85]. The advantage of CTL is however the efficient solvability of the model checking problem: given a Kripke structure M and a CTL formula φ, we can determine the sets of states in M where φ holds in time $O(\|M\|^2 \|\varphi\|)$, where $\|\varphi\|$ means the length of the formula and $\|M\|$ the number of states of M [ClES86]. Hence, CTL has been used for the verification of quite large systems as for example the alternating bit protocol [ClES86], the Gigamax cache coherence protocol [McMi93a], the Futurebus cache coherence protocol [CGHJ93a], a traffic light controller [BCDM86] or a DMA system [McMi93a]. A disadvantage of CTL is its weak expressiveness which does for example not allow to state that a property has to hold infinitely often. For this reason, CTL has been extended also by fairness constraints [EmLe85a, ClES86] leading to roughly the same complexity of the model checking problem (additionally, it is linear in the size of the fairness constraints), but that have still a limited expressiveness.

Another disadvantage of CTL is that it cannot be easily used for a hierarchical verification where the model is given as a composition of submodules which can be replaced by previously verified specifications. In order to use the given design hierarchy despite this fact, some approaches have been discussed: Clarke, Long and McMillan modeled in [ClLM89a, ClLM89b] the environment of a module by an interface process and stated conditions for the correctness of this approach. Grumberg and Long [GrLo91] developed an approach, which is based on an ordering of the models and allowed to state similar to [Pnue84] that temporal properties hold for all systems that are implemented by the considered component. However, this approach required a further restriction for CTL (the use of existential path quantifiers has to be forbidden) that leads to the temporal logic ACTL.

Apart from LTL and BTL, interval temporal logic (ITL) [HaMM83] has been proposed for the specification of temporal properties. In contrast to

---

[3] Verification problems are often modeled in such a manner that the system is mapped on a logical model and the specification on a logical formula. Hence, the verification problem becomes a model checking problem, where the truth value of a formula (the specification) is to be checked in a given model (the system).

LTL or BTL, the truth value of a formula is related to a sequence of states rather than to a single state. The most interesting operator of ITL is the 'chop operator ;' which has the following effect: $\varphi; \psi$ holds on a sequence of states iff this sequence can be split up in two parts, where φ hold on the first and ψ on the second part. The chop operator can be used to describe many interesting facts in ITL. However, the satisfiability of ITL formulas is not decidable [HaMM83], and hence, the logic has been used only in some rare cases [Mosz85, Mosz85a, Lees88a] on transistor level.

2.2 Hardware Verification Based on Predicate Logics

Propositional logics and their temporal extensions are not well suited for describing systems that process abstract data types, as the data types must first be mapped to the propositional level. In order to use abstract data types directly in the logic, i.e. to have variables of the corresponding type, predicate logics and their temporal extensions have to be used instead. The research in this area is mainly based on the Boyer-Moore logic and its prover and also to approaches based on higher order logic.

The Boyer-Moore logic is a quantifier-free first order predicate logic with the extension to formalize induction schemes [BoMo88]. The logic as well as the corresponding interactive prover have been developed by R.S. Boyer and J.S. Moore. The prover allows to define new inductive data types by declaring constructor and destructor functions for this data type. Of course, it is also possible to define new functions on the defined data types. In order to avoid inconsistent definitions, these definitions have to fulfill certain constraints. For example, the recursive definitions must follow a well-ordering such that a termination of the recursion is guaranteed. The Boyer-Moore prover has strong heuristics, but does not work automatically. Usually, the user has to provide appropriate lemmata which are useful for the proof. Boyer-Moore has already been used for verifying hardware: Hunt verified in 1985 the FM8501 microprocessor [Hunt85] and 1989 the 32 bit extension FM8502 of FM8501 [Hunt89]. The verification used generic n-bit wide descriptions in order to make use of the induction tactics of the Boyer-Moore prover.

Apart from the explicit induction methods implemented in the Boyer-Moore prover, there are other induction methods which are summarized under the category 'inductionless induction' [Muss80, HuHu82, Frib86, JoKo86, JoKo89, BoRu93] and are available in theorem provers as e.g. the rewrite rule laboratory RRL of Kapur and Zang [KaZh88], the ReDuX system [Bouh94, BeBR96] or SPIKE [BoRu93]. Inductionless induction methods are based on term rewriting, which is a general theorem proving method. Examples, where term rewrite methods have been used for formal verification are [NaSt88], where the verification of of the Sobel chip has been presented, [KaSu96] where multipliers have been verified and [ChPC88], where term rewriting has been used for proving propositional formulas. Term rewriting is also available in almost every interactive proof assistant, in particular in those for higher order logics. In contrast to explict induction methods, the advantage of these methods is that they are able to generate their own induction hypothesis and hence, offer in principle a higher degree of automation than explicit induction methods. However, it is in most cases necessary to formulate some lemmata that are useful for the later proof. These lemmata are then automatically proved first, and by their help, the remaining proof can also be done automatically.

In comparison to (temporal) propositional logic, a major advantage of first-order logic in combination with induction schemes is the use of abstract data types. On the other hand, the temporal behavior of a system can not be modeled appropriately in this logic. For this reason, C@S is based on a temporal extension of second-order predicate logic with induction schemes, as outlined section 3. From a formal point of view, this logic is a higher order logic, but it is strictly weaker than usual higher order logics as supported by higher order proof assistants as e.g. HOL [GoMe93], VERITAS [HaDa86], NUPRL [Cons86], ISABELLE [Paul94], LAMBDA [FoHa89] and PVS [ORSS94]. A disadvantage of this logic is its undecidability and its incompleteness that holds for each higher order logic: each decidable set of rules is not powerful enough to derive all theorems [MaYo78]. In practice, this incompleteness is however not a serious drawback. Usually, the set of inference rules to be applied manually is sufficient to prove all verification tasks. The advantage of our logic is that it contains exactly the constructs that are necessary from our point of view for the specification of systems with a complex temporal behaviour together with a complex data processing.

2.3 Combined Approaches

The key idea of C@S is the combination of different formalisms for performing hardware proofs. Compositional reasoning and abstraction play also important parts in our approach. Both aspects will be briefly related to other existing approaches.

Using combined approaches for hardware verification can be motivated twofold: first, not all formalisms are equally well suited for expressing different properties and abstractions. Thus, as it has been shown in the last section, temporal logics are better suited for reasoning about time, whereas predicate logics are well suited for dealing with data paths. If however, both kind of properties are to be verified then both aspects have to be treated simultaneously. The second reason for combining approaches is to enrich an interactive proof tool with decision procedures thus that at least the decidable parts of the verification can be automated, although an undecidable framework is used.

The idea of enriching interactive theorem provers for higher order logic with powerful decision procedures can also be found in the PVS system. However, PVS does not support a proof methodology which is based on a separation of control and data flow and no common formalism as available in

C@S is used. Our specification language is based on linear time temporal logic (LTL), enriched by abstract data types. This language is similar to the CTL extension used for word level model checking [ClZh95]. However,[ClZh95] aims at a full automation, hence only finite bitvectors are possible as abstract data types. The approach of Hojati and Brayton [HoBr95, HoDB97] can be seen as complementary to our approach as especially for systems without full control/data interaction valuable decidability results have been achieved, which can be directly used in C@S. By using invariants we are also able to cope with those systems, not treated by them – the price to be paid for that is interaction.

The approach of Hungar, Grumberg and Damm [HuGD95] is also based on the idea to separate control and data. They however use FO-ACTL (first order version of ACTL), whereas we support full LTL enhanced by data equation expressions. As they use a first-order variant of temporal logics, the reasoning process has to be performed using a dedicated calculus involving the construction of first-order semantic tableaux. Instead, C@S uses a more rigorous separation of control and data such that already existing proof tools for propositional temporal logics may be used in conjunction with e.g. term rewriting systems.

C@S is a direct descendant of MEPHISTO [ScKK91a, ScKK91b, KuSK93a]. MEPHISTO was a verification system that was implemented by our group on top of the HOL system [GoMe93]. Similar to C@S, MEPHISTO provided the user with tactics for structuring the proof goals such that a further automation of the proof had been achieved by a first order theorem prover called FAUST [ScKK92b, ScKK93a, ScKK94b]. A key idea of MEPHISTO was also to use so-called 'hardware formulas' as a restricted form for specifications in order to achieve a higher degree of automation of the proofs inside the HOL system. These hardware formulas have been extended in the C@S system to the fair prefix formulas, and new algorithms for translating linear temporal logic to these fair prefix formulas have been developed and implemented in C@S. C@S is no longer on top of HOL, but a complete system on its own that has interfaces to include other verification systems such as SMV or RRL.

2.4 Outline of the Chapter

In the next section, the basics of the underlying formalism of C@S are given. First, we define a second-order temporal logic that is the basis of the C@S logic and a special kind of ω-automata called fair prefix automata. The strategy of C@S is to split verification goals given in this logic such that they belong to a subset of the logic where automated proof procedures are available. In this book section, we mainly focus on two proof procedures of C@S: one for translating linear temporal propositional logic to fair prefix automata, shown in detail in section 3.2, and a proof procedure for equation systems, given in section 3.3. It is important to note here, that fair prefix formulas that have only safety properties can be directly interpreted as rewrite systems and hence, can be proved on the one hand by model checking procedures and on the other hand by rewrite based methods. The latter proof method allows also the extension to abstract data types. If the translation of a LTL formula does not yield in such an ω-automaton, then C@S allows to modify the formula by interactive proof rules such that the modified formulas can finally be translated to a rewrite system.

3. Basics

The description of the entire behavior of general systems requires one hand to express temporal behavior and on the other hand to reason about abstract data types defined by the user. The logic used in C@S is powerful enough to define new abstract data types and to express temporal behavior, but is on the other hand weak enough such that a high degree of proof automation can be achieved. For this reason, proof rules of C@S aim at reducing formulas to subsets where powerful proof procedures are available: currently, these are on the one hand linear temporal propositional logic together with finite state ω-automata and, on the other hand, rewrite systems with abstract data types. These logics are considered in detail in the following three subsections.

3.1 Underlying Formalism of C@S

The underlying formalism of C@S is a second order linear time temporal logic with inductively defined abstract data types. This logic is on the one hand a proper subset of higher order logic, and on the other hand a common extension of LTL and first order logic with abstract data types. The syntax of this logic is determined by a signature which fixes the function symbols, constants and variables.

Definition 3.1 (Signature). *A signature Σ is a tuple $(C_\Sigma, V_\Sigma, \mathcal{TP}_\Sigma, \mathsf{typ}_\Sigma)$, where C_Σ is the set of constant symbols, V_Σ is the set of variables, and $\mathcal{TP}_\Sigma$ is a set of basic types. typ_Σ is a function that assigns to symbols of $C_\Sigma \cup V_\Sigma$ a type of the form $\alpha_1 \times \ldots \times \alpha_n \to \beta$ where $\alpha_i, \beta \in \mathcal{TP}_\Sigma$. In case $n = 0$, simply a type $\beta \in \mathcal{TP}_\Sigma$ is assigned. It is assumed that $\mathcal{TP}_\Sigma$ contains at least the set of boolean values $\mathbb{B}$.*

Usually, we also have natural numbers $\mathbb{N}$ in the signature, but these can also be defined as shown in the following. Given a signature that provides basic types and symbols for constants and variables, we can define the sets of terms and the set of formulas.

Definition 3.2 (Terms $\mathcal{T}_\Sigma$). *The set of terms $\mathcal{T}_\Sigma$ over a signature $\Sigma = (C_\Sigma, V_\Sigma, \mathcal{TP}_\Sigma, \mathsf{typ}_\Sigma)$ is the smallest set that satisfies the following properties:*

1. $x \in \mathcal{T}_\Sigma$ for all $x \in V_\Sigma$

2. $f(\tau_1, \ldots, \tau_n) \in \mathcal{T}_\Sigma$ with $\mathrm{typ}_\Sigma(f(\tau_1, \ldots, \tau_n)) := \beta$ for all $f \in C_\Sigma$ provided that $\mathrm{typ}_\Sigma(f) = \alpha_1 \times \ldots \times \alpha_n \to \beta$, $\mathrm{typ}_\Sigma(\tau_i) = \alpha_i$ and $\tau_1, \ldots, \tau_n \in \mathcal{T}_\Sigma$.

3. Given that $\tau \in \mathcal{T}_\Sigma$ with $\mathrm{typ}_\Sigma(\tau) = \alpha$, then $\mathsf{X}\tau \in \mathcal{T}_\Sigma$ with $\mathrm{typ}_\Sigma(\mathsf{X}\tau) = \alpha$

It has to be noted here, that the set of terms is defined as usually for first-order predicate logic (see e.g. [Fitt90]). However, we also have an additional terms $\mathsf{X}\tau$ that represent the value of a term at the next point of time. These terms are necessary when temporal behavior is mixed with abstract data types. All other temporal operators work on formulas, i.e. of terms of type $\mathbb{B}$.

Definition 3.3 (Formulae $\mathcal{TL}_\Sigma$ of a signature).

The set of formulas $\mathcal{TL}_\Sigma$ over a given signature $\Sigma = (C_\Sigma, V_\Sigma, \mathcal{TP}_\Sigma, \mathrm{typ}_\Sigma)$ is the smallest set that satisfies the following properties (all formulas have type $\mathbb{B}$):

1. $q \in \mathcal{TL}_\Sigma$ for all $q \in V_\Sigma$ with $\mathrm{typ}_\Sigma(q) = \mathbb{B}$
2. $p(\tau_1, \ldots, \tau_n) \in \mathcal{T}_\Sigma$ for all $p \in C_\Sigma$ provided that $\mathrm{typ}_\Sigma(p) = \alpha_1 \times \ldots \times \alpha_n \to \mathbb{B}$, $\mathrm{typ}_\Sigma(\tau_i) = \alpha_i$ and $\tau_1, \ldots, \tau_n \in \mathcal{T}_\Sigma$.
3. $\tau_1 = \tau_2 \in \mathcal{TL}_\Sigma$, for all terms $\tau_1, \tau_2 \in \mathcal{T}_\Sigma$ with $\mathrm{typ}_\Sigma(\tau_1) = \mathrm{typ}_\Sigma(\tau_2)$
4. $\neg\varphi$, $\varphi \wedge \psi$, $\varphi \vee \psi$, $\varphi \to \psi$, $\varphi = \psi \in \mathcal{TL}_\Sigma$ for all $\varphi, \psi \in \mathcal{TL}_\Sigma$
5. $\mathsf{X}\varphi$, $\mathsf{G}\varphi$, $\mathsf{F}\varphi$, $[\varphi \mathrel{\mathsf{W}} \psi]$, $[\varphi \mathrel{\underline{\mathsf{W}}} \psi]$, $[\varphi \mathrel{\mathsf{U}} \psi]$, $[\varphi \mathrel{\underline{\mathsf{U}}} \psi] \in \mathcal{TL}_\Sigma$, for all $\varphi, \psi \in \mathcal{TL}_\Sigma$
6. $\exists x.\varphi \in \mathcal{TL}_\Sigma$ and $\forall x.\varphi \in \mathcal{TL}_\Sigma$, for all $x \in V_\Sigma$
7. abstype α with $c_1, \ldots, c_n$ in $\varphi \in \mathcal{TL}_\Sigma$ for all $\varphi \in \mathcal{TL}_\Sigma$ provided that $c_1, \ldots, c_n \in C_\Sigma$ and $\alpha \in \mathcal{TP}_\Sigma$

Moreover, let $x = \tau$ in φ end is used as syntactic sugar that corresponds to the formula that is obtained by replacing every occurrence of the variable x in φ by τ.

The first three items introduce *atomic formulas*, while the fourth and fifth state the boolean closure and the closure under temporal operators, respectively. Item 6 introduces first and second order quantification. The second order quantification, in particular the quantification over variables of type $\mathbb{B}$, is important for our formalism as it allows us to formalise finite automata in a very convenient way (see next subsection). The possibility to define new abstract data types is added by item 7 in the above definition, where the constants $c_1, \ldots, c_n \in C_\Sigma$ are defined as the *constructors* of the new data type α according to [Melh88d]. The definition of α is locally declared for the formula φ. We assume that the signature Σ has enough types for each user defined type definition.

Note that all constructs of the C@S logic are viewed as depended on time, i.e. if we have a term of type α, then it is always assumed that the value of the term may change during time. Hence, it would be more precise to assign to the term the type $\mathbb{N} \to \alpha$, but this would lead to an overhead of syntax that can be avoided once we have made the convention that all types are interpreted as dependent on time.

After defining the syntax of the CQS logic, we define now its semantics. First, we need the notion of a variable assignment, interpretations and models.

Definition 3.4 (Domains, Interpretations, Assignments).
Given a signature $\Sigma = (C_\Sigma, V_\Sigma, \mathcal{TP}_\Sigma, \mathsf{typ}_\Sigma)$, *each set* $\mathcal{D} := \bigcup_{\alpha \in \mathcal{TP}_\Sigma} D_\alpha$ *is called a domain for* Σ. *A function* $I : C_\Sigma \to \bigcup_{\alpha \in \mathcal{TP}_\Sigma} D_\alpha$ *that maps a constant of type* α *to a function* $\mathbb{N} \to D_\alpha$ *is called an interpretation for* Σ *and* $\mathcal{D}$. *Similarly, a function* $\xi : V_\Sigma \to \bigcup_{\alpha \in \mathcal{TP}_\Sigma} D_\alpha$ *that maps a variable of type* α *to a function* $\mathbb{N} \to D_\alpha$ *is called a variable assignment for* Σ *and* $\mathcal{D}$.

Given a signature, a domain, an interpretation and a variable assignment, we can define the semantics of arbitrary formulas and terms. This is done recursively on the structure of the corresponding term or formula. Note again, that each term and formula is mapped to a function that depends on time, where time is modeled by the natural numbers $\mathbb{N}$. Functions are given in the following in a λ-calculus syntax, and the application of such a function f on a point of time t is written as $f^{(t)}$.

Definition 3.5 (Semantics). *Given a signature* $\Sigma = (C_\Sigma, V_\Sigma, \mathcal{TP}_\Sigma, \mathsf{typ}_\Sigma)$, *and a domain* $\mathcal{D}$ *for* Σ, *an interpretation* I *for* Σ *and* $\mathcal{D}$, *and a variable assignment* ξ *for* Σ *and* $\mathcal{D}$, *the semantics of terms and formulas is defined as follows:*

1. $\mathcal{V}_\xi^{D,I}(x) := \xi(x)$ *for all* $x \in V_\Sigma$

2. $\mathcal{V}_\xi^{D,I}(f(\tau_1, \ldots, \tau_n)) :=$
$$\lambda t. \left[\mathcal{V}_\xi^{D,I}(f)\right]^{(t)} \left(\left[\mathcal{V}_\xi^{D,I}(\tau_1)\right]^{(t)}, \ldots, \left[\mathcal{V}_\xi^{D,I}(\tau_n)\right]^{(t)}\right)$$

3. $\mathcal{V}_\xi^{D,I}(\mathsf{X}\tau) := \lambda t. \left[\mathcal{V}_\xi^{D,I}(\tau)\right]^{(t+1)}$

4. $\mathcal{V}_\xi^{D,I}(\tau_1 = \tau_2) := \lambda t. \left(\left[\mathcal{V}_\xi^{D,I}(\tau_1)\right]^{(t)} = \left[\mathcal{V}_\xi^{D,I}(\tau_2)\right]^{(t)}\right)$

5. $\mathcal{V}_\xi^{D,I}(\neg\varphi) := \lambda t. \neg \left[\varphi^{(t)}\right]$
$\mathcal{V}_\xi^{D,I}(\varphi \wedge \psi) := \lambda t. \left[\varphi^{(t)}\right] \wedge \left[\psi^{(t)}\right]$
$\mathcal{V}_\xi^{D,I}(\varphi \vee \psi) := \lambda t. \left[\varphi^{(t)}\right] \vee \left[\psi^{(t)}\right]$

6. $\mathcal{V}_\xi^{D,I}(\mathsf{X}\varphi) := \lambda t. \left[\mathcal{V}_\xi^{D,I}(\tau)\right]^{(t+1)}$

$\mathcal{V}_\xi^{D,I}(\mathsf{G}\varphi) := \lambda t. \begin{cases} 1 & \mathcal{V}_\xi^{D,I}(\varphi)^{(t+\delta)} = 1 \text{ holds for all } \delta \in \mathbb{N} \\ 0 & \text{otherwise} \end{cases}$

$\mathcal{V}_\xi^{D,I}(\mathsf{F}\varphi) := \lambda t. \begin{cases} 1 & \mathcal{V}_\xi^{D,I}(\varphi)^{(t+\delta)} = 1 \text{ holds at least for one } \delta \in \mathbb{N} \\ 0 & \text{otherwise} \end{cases}$

$$\mathcal{V}_\xi^{D,I}([\varphi \; \mathsf{W} \; \psi]) := \lambda t. \begin{cases} 1 & \text{if } \mathcal{V}_\xi^{D,I}(\psi) = \lambda t.0 \\ \mathcal{V}_\xi^{D,I}(\varphi)^{(t+\delta)} & \text{otherwise, where } t+\delta \text{ is the} \\ & \text{first point of time after } t, \\ & \text{where } \mathcal{V}_\xi^{D,I}(\psi)^{(\delta)} \text{ holds} \end{cases}$$

$$\mathcal{V}_\xi^{D,I}([\varphi \; \underline{\mathsf{W}} \; \psi]) := \mathcal{V}_\xi^{D,I}([\varphi \; \mathsf{W} \; \psi] \wedge \mathsf{F}\psi)$$
$$\mathcal{V}_\xi^{D,I}([\varphi \; \mathsf{U} \; \psi]) := \mathcal{V}_\xi^{D,I}([\psi \; \mathsf{W} \; (\varphi \to \psi)])$$
$$\mathcal{V}_\xi^{D,I}([\varphi \; \underline{\mathsf{U}} \; \psi]) := \mathcal{V}_\xi^{D,I}([\varphi \; \mathsf{U} \; \psi] \wedge \mathsf{F}\psi)$$

7. For a given assignment ξ, a variable x and an element $D_{\mathrm{typ}_\Sigma(x)}$, the x-modified assignment ξ_x^d is defined as follows:

$$\xi_x^d := \begin{cases} \xi(y) & \text{if } y \neq x \\ d & \text{if } y = x \end{cases}$$

This allows to define the semantics of quantified formulas:

$$- \; \mathcal{V}_\xi^{D,I}(\exists x.\varphi) := \lambda t. \begin{cases} 1 & \text{if there is a } d \in D_{\mathrm{typ}_\Sigma(x)} \\ & \text{such that } \mathcal{V}_{\xi_x^d}^{D,I}(\varphi)^{(t)} \text{ holds} \\ 0 & \text{otherwise} \end{cases}$$

$$- \; \mathcal{V}_\xi^{D,I}(\forall x.\varphi) := \mathcal{V}_\xi^{D,I}(\neg\exists x.\neg\varphi)$$

8. Provided that $\mathrm{typ}_\Sigma(c_i) = \alpha_{i,1} \times \ldots \times \alpha_{i,m_i} \to \beta_i$ holds for $i = 1, \ldots, n$, we define:

$$\mathcal{V}_\xi^{D,I}(\text{abstype } \alpha \text{ with } c_1, \ldots, c_n \text{ in } \varphi) :=$$

$$\mathcal{V}_\xi^{D,I} \left(\begin{array}{l} \forall g_1 \ldots g_n. \exists_1 f.\alpha \to \beta. \\ \bigwedge\limits_{i=1}^{n} \forall x_1 : \alpha_{i,1} \ldots x_{m_i}\alpha_{i,m_i}. \\ \qquad f(c_i(x_1, \ldots, x_{m_i})) = g(f(x_1), \ldots, f(x_{m_i}), x_1, \ldots, x_{m_i}) \\ \wedge\varphi \end{array} \right)$$

X is the usual next time operator, $\mathsf{G}\varphi$ means that φ has to hold for all future points of time, and $\mathsf{F}\varphi$ means that φ has to hold for at least one future points of time. $[\varphi \; \mathsf{W} \; \psi]$ means that φ must hold at the first point of time when ψ holds and $[\varphi \; \mathsf{U} \; \psi]$ means that φ must hold until the first point of time when ψ holds. There are strong $\underline{\mathsf{W}}$, $\underline{\mathsf{U}}$ and weak variants W, U of the binary temporal operators. Strong variants require that the event actually has to occur at a certain point of time, while weak variants also allow the absence of the event.

The semantics of the type definitions assures that functions over the type α can be defined recursively. This primitive recursion axiom assures also that the set D_α is isomorphic to the set of variable-free terms that can be built from the constants $c_1, \ldots, c_n$ and that the constructor functions c_i are one-to-one [Melh88d]. This means on the one hand, that none of the terms $c_i(x_1, \ldots, x_{m_i})$ and $c_j(y_1, \ldots, y_{m_j})$ are mapped to the same element in D_α and on the other hand, that for each element $d \in D_\alpha$ there is a variable-free term $c_i(x_1, \ldots, x_{m_i})$ such that $\mathcal{V}_\xi^{D,I}(c_i(x_1, \ldots, x_{m_i})) = d$ holds.

As an example, consider the definition of the natural numbers:

abstype $\mathbb{N}$ with $0 : \mathbb{N}, \text{SUC} : \mathbb{N} \rightarrow \mathbb{N}$ in φ

Here, we have $\text{typ}_\Sigma(0) = \mathbb{N}$ and $\text{typ}_\Sigma(\text{SUC}) = \mathbb{N} \rightarrow \mathbb{N}$. The definition of the semantics now assures that Peano's axioms hold, i.e. for each element $d \in D_{\mathbb{N}}$ we have a term of the form $\text{SUC}^n(0)$ such that $\mathcal{V}_\xi^{D,I}(\text{SUC}^n(0)) = d$ holds, and of course the induction principle holds for $\mathbb{N}$. This axiomatic definition of the new data types introduces directly an induction scheme that allows to reason inductively on the new data type. For more details, see section 3.3.

Clearly, also linear temporal propositional logic is a subset of the above defined logic. This logic is obtained when the set of basic types $\mathcal{TP}_\Sigma$ of the signature Σ contains only the type $\mathbb{B}$ and neither quantification over variables nor the definition of abstract data types is allowed. In the following, we consider two other important subsets of $\mathcal{TL}_\Sigma$ which are important for specifying hardware modules: fair prefix formulas and first order abstract data types.

3.2 Fair Prefix Automata

In general, an ω-automaton [Thom90a] is a finite state system for accepting infinite words according to an acceptance condition that is usually a simple LTL formula. For example, Büchi automata use as acceptance condition formulas of the form $\text{GF}\varphi$ while Rabin automata use formulas of the form $\bigwedge_{i=0}^f \text{GF}\varphi_i \vee \text{FG}\psi_i$, where φ, φ_i and ψ_i are propositional formulas.

An important proof strategy of the C@S system is to translate the temporal logic part of a verification goal into an ω-automaton that has an acceptance condition of the form $\text{G}\varphi$ where φ is a propositional formula. As already mentioned, these automata can also be viewed as equation systems and can hence be simply proved with term rewriting methods hence they allow to verify temporal properties with abstract data types. The kind of ω-automata that is considered in this section is strongly related to the class of ω-regular languages which can be viewed as the boolean closure of safety (rsp. liveness) properties [StWa74]. In order to capture full LTL, we extend these automata by additional fairness constraints.

Throughout this chapter, we describe ω-automata as logical formulas similar to [Sief70]. For this reason, suppose that the states and the input alphabet is encoded by boolean tuples. In the following, these tuples are written as vectors $\vec{b}$ and for $\vec{b} = (b_1, \ldots, b_n)$, the length of $\vec{b}$ is defined as $\|\vec{b}\| := n$. An infinite word $\vec{i}$ over an alphabet $\Sigma =$ can be modeled as a function from natural numbers $\mathbb{N}$ to $\mathbb{B}^n$, where $\vec{i}^{(k)} \in \Sigma$ is a boolean tuple that encodes the k-th symbol of the sequence $\vec{i}$. Formulas that contain exclusively the free variables $\vec{b} = (b_1, \ldots, b_n)$ are written as function applications $\Phi(\vec{b})$ on $\vec{b}$.

In general, each ω-automaton can be described by a logical formula of the form $\exists \vec{q}.\mathcal{T}(\vec{i}, \vec{q}) \wedge \Theta(\vec{i}, \vec{q})$, where $\vec{q}$ are the state variables, $\vec{i}$ are the input variables, $\mathcal{T}(\vec{i}, \vec{q})$ is the transition relation, and $\Theta(\vec{i}, \vec{q})$ is the acceptance condition (specific to each kind of ω-automaton). A deterministic transition relation can always be given as a conjunction of initialisation equations of the form $q_k = \omega_k$ with $\omega_k \in \{1, 0\}$ and transition equations of the form $G\left(Xq_k = \Omega_k(\vec{i}, \vec{q})\right)$. Fair prefix formulas are now defined in this style:

Definition 3.6 (Fair Prefix Formulas (FPF)).
Let $\Omega_k(\vec{i}, \vec{q})$ for $k \in \{0, \dots, s\}$, $\xi_l(\vec{i}, \vec{q})$ for $l \in \{0, \dots, f\}$, and $\Phi_m(\vec{i}, \vec{q})$ and $\Psi_m(\vec{i}, \vec{q})$ for $m \in \{0, \dots, a\}$ be propositional formulas with the free variables $\vec{i}$ and $\vec{q}$. Moreover, let $\omega_k \in \{1, 0\}$ for $k \in \{0, \dots, s\}$, then the following formula $\mathcal{P}(\vec{i})$ is a (fair) prefix formula:

$$
\mathcal{P}(\vec{i}) := \left(
\begin{array}{c}
\exists q_1 \dots q_s. \\
\bigwedge_{k=0}^{s} \left[(q_k = \omega_k) \wedge G\left(Xq_k = \Omega_k(\vec{i}, \vec{q})\right)\right] \wedge \\
\left(\bigwedge_{l=0}^{f} GF\xi_l(\vec{i}, \vec{q})\right) \rightarrow \left(\bigwedge_{m=0}^{a} \left[G\Phi_m(\vec{i}, \vec{q})\right] \vee \left[F\Psi_m(\vec{i}, \vec{q})\right]\right)
\end{array}
\right)
$$

The formulas $\Phi_m(\vec{i}, \vec{q})$ and $\Psi_m(\vec{i}, \vec{q})$ are called the safety properties and liveness properties, respectively. $\xi_l(\vec{i}, \vec{q})$ is called a fairness constraint.

Note that the definition includes that each FPF is deterministic and that each propositional formula over the variables $\vec{i}$ and $\vec{q}$ is the characteristic function of a set of edges in the state transition diagram of the automaton. The run $\vec{q}$ of an input sequence $\vec{i}$ is fair with respect to a set of edges $\xi_l(\vec{i}, \vec{q})$, iff the run traverses infinitely often at least one of the edges of $\xi_l(\vec{i}, \vec{q})$. An infinite word $\vec{i}$ is accepted by a FPF $\mathcal{P}(\vec{i})$, iff the corresponding run $\vec{q}$ is fair with respect to all fairness constraints $\xi_l(\vec{i}, \vec{q})$ and satisfies all acceptance pairs $(\Phi_m(\vec{i}, \vec{q}), \Psi_m(\vec{i}, \vec{q}))$ of $\mathcal{P}(\vec{i})$ as follows: either it stays exclusively in the set $\Phi_m(\vec{i}, \vec{q})$ or it reaches at least once an edge of $\Psi_m(\vec{i}, \vec{q})$.

Languages that can be accepted by a FPF are exactly the languages that can be accepted by a deterministic Büchi automaton. We are however mainly interested in simple prefix formulas, i.e. FPFs without fairness constraints. In contrast to FPFs and deterministic Büchi automata, simple prefix formulas are closed under boolean operations. This is important for the translation procedure given in section 5.3.

3.3 Rewrite-Based Theorem Proving for Abstract Data Types

Formulae 'abstype α with $c_1, \dots, c_n$ in φ' of the C@S logic define α to be a type that is isomorphic to the set of ground terms that can be obtained by the constructor functions $c_1, \dots, c_n$. Most abstract data types can be described in this way; for example, consider again the definition of the natural numbers:

$$\text{abstype } \mathbb{N} \text{ with } 0 : \mathbb{N}, \text{SUC} : \mathbb{N} \rightarrow \mathbb{N} \text{ in } \varphi$$

$\mathbb{N}$ is now a type visible in φ that is isomorphic to the set 0, SUC(0), SUC(SUC(0)), ... and new operations can be defined on $\mathbb{N}$ by primitive recursion over the constructors 0 and SUC. As another example, consider abstype *Nat_List* with NIL : *Nat_List*, CONS : $\mathbb{N} \times$ *Nat_List* $\rightarrow$ *Nat_List* in φ_ℓ. In this formula, a type of the name *Nat_List* that is isomorphic to the set of lists containing natural numbers is declared for the formula φ.

Of course, it is possible to define also new operations on this new data type. Usually, φ is an implication, where the premise contains such definitions in form of a recursive definition over the constructors $c_1, \ldots, c_n$. For example, φ_ℓ could have the following form:

$$\left(\begin{array}{l} \text{TL (NIL)} = \text{NIL} \wedge \\ \forall x\, y . \text{TL (CONS)}\, (x, y) = y \end{array} \right) \rightarrow \varphi_1$$

The premise defines a new function TL that removes the first element of a list if the list is not empty. TL can be used with this semantics in the remaining formula φ_1.

The reason for this form of defining new data types is that there are special proof methods for these data types that belong to the class of 'inductionless induction' proof methods [Muss80, HuHu82, Frib86, JoKo86, JoKo89, BoRu93]. Of course, the special form of definition directly leads to an induction scheme as we know all members of the set D_α (these are the variable free terms built from the constructor constants of the data type). In the given examples, the induction schemes are as follows:

$$\frac{P(0) \qquad \forall x : \mathbb{N} . P(x) \rightarrow P(\text{SUC}(x))}{\forall x : \mathbb{N} . P(x)}$$

$$\frac{P(\text{NIL}) \qquad \forall x : \mathbb{N}\, y : \textit{Nat\_List} . P(x) \rightarrow P(\text{CONS}(x, y))}{\forall x : \textit{Nat\_List} . P(x)}$$

The advantage of inductionless induction methods is that they are able to directly use and derive complex induction schemes from the definition of the abstract data types similar to the ones given above. The methods are also able to automatically generate intermediate lemmata for making a complex proof. However, in practice it is usually necessary to guide the proof search by providing the prover by some lemmata that one expects to be of some help for the current proof. On the other hand, these methods are limited to oriented equation systems, to be more precise to term rewrite systems [Ders90].

Definition 3.7 (Term Rewrite System). *A term rewrite system is a set of equations of the form $l = r$, where each equation is also called a rewrite rule. A rule $l = r$ can be applied to a term t if this term t contains a subterm τ such that there is a substitution σ of the variables occurring in l such that $\sigma(l) = \tau$. The application of the rule $l = r$ on the term t is then performed by replacing some occurrences of the subterm τ in t by $\sigma(r)$.*

A term rewrite system is called terminating iff for each term t there is no infinite number of rule applications. A term rewrite system is confluent iff all possible rule applications finally yield in the same term. Finally, a rewrite system is called canonical, iff the system is terminating and confluent. Two rewrite systems $R_1 = \{l_{1,1} = r_{1,1}, \ldots, l_{n,1} = r_{n,1}\}$ and $R_2 = \{l_{1,2} = r_{1,2}, \ldots, l_{m,2} = r_{m,2}\}$ are equivalent iff $(\bigwedge_{i=1}^{n} l_{i,1} = r_{i,1}) = (\bigwedge_{i=1}^{m} l_{i,2} = r_{i,2})$ holds.

If the rewrite system contains more than one rule, it is undecidable whether the system is terminating. For this reason, usually term orderings are used [Ders87] to assure the termination property. The idea is to order the set of terms by a well-founded ordering $\succ$ such that the lefthand sides of the rewrite rules are larger than the righthand sides. As each chain of terms must have a minimal term, it is clear that the corresponding rewrite system must be terminating.

If it is known that a rewrite system is terminating, the confluence property is decidable. However, rewrite systems are often not confluent, although there is an equivalent confluent rewrite system that can be computed by the completion algorithm due to Knuth and Bendix [KnBe70]. This algorithm is also the basis of the inductionless induction proof methods. The basic idea of these methods is the following theorem [Ders83]:

Theorem 3.1. *Given a a canonical rewrite system R_1 and a set of equations $E = \{e_1, \ldots, e_n\}$. Let R_2 be the canonical rewrite system that has been obtained by applying the completion algorithm on $R_1 \cup E$. Then $R_1 \vdash e_i$ holds iff the set of irreducible variable free terms of R_1 is also the set of irreducible variable free terms of R_2.*

The computation of the irreducible variable free terms of a rewrite system is however undecidable. For this reason, some more specialized algorithms have been developed. For our special form of data type definitions, the algorithm of Huet and Hullot [HuHu82] is of particular use. This algorithm is based on the fact that we know for each data type the set of constructors, that the primitive recursion theorem holds for these constructors (see the definition of the semantics of our logic).

A function $f \in \Sigma_D$ is completely defined iff for all constructors $c_1, \ldots, c_n$ there is a reduction rule $f(c_1(\ldots), \ldots, c_n(\ldots)) = r$. If all function symbols are completely defined in a rewrite system R, and R is canonical, and no lefthand side of any equation of R starts with a constructor, then the proof procedure of [HuHu82] is given in figure 3.1. The algorithm has the following properties:

Theorem 3.2. *Let $\succ$ be a term ordering $\succ$ and E_1 be a rewrite system with completely defined function symbols. If $\mathsf{HuHu}(\{\}, E_1 \cup E_2, \succ)$ terminates with* PROOF, *then $E_1 \vdash E_2$ holds. The thereby obtained rewrite system R does also follow the definition principle and is equivalent to E_1. If $\mathsf{HuHu}(\{\}, E_1 \cup$*

```
(* INPUT rewrite system R, set of equations E, term ordering ≻ *)
(* OUTPUT PROOF, DISPROOF oder FAILURE *)
PROCEDURE HuHu(R,E,≻);
    IF E = {} THEN PROOF
    ELSE
        Choose l = r ∈ E, E := E \ {l = r}
        l₁ := rewrite(R,l);
        r₁ := rewrite(R,r);
        CASE
            l₁ ≡ r₁ : HuHu(R,E,≻)
            l₁ ≡ f(t₁,... ,tₙ) ∧ f ∈ Σ_C :
                CASE
                    r₁ ≡ f(t₁,... ,tₙ) : HuHu(R,E ∪ {τᵢ = tᵢ | i ≤ n},≻);
                    r₁ ≡ g(t₁,... ,tₘ) : DISPROOF
                    r₁ ∈ V_Σ : DISPROOF
                ENDCASE
            l₁ ∈ V_Σ ∧ r₁ ≡ f(t₁,... ,tₙ) ∧ f ∈ Σ_C : DISPROOF
            l₁ ≻ r₁ : l₂ := l₁; r₂ := r₁;
            r₁ ≻ l₁ : l₂ := r₁; r₂ := l₁;
            OTHERWISE FAILURE; (* nicht-orientierbare Gleichung *)
            (R',E') := SIMPLIFY(R,E,l₂ = r₂);
            C := CRITICAL(R' ∪ {l₂ = r₂});
            HuHu(R'∪{l₂ = r₂},E' ∪ C,≻)
    ENDIF
END PROCEDURE HuHu
```

Fig. 3.1. Inductionless Induction according to Huet and Hullot

$E_2, \succ$) *terminates with* DISPROOF, *then at least one equation of* E_2 *does not hold under* E_1. *Moreover, if an equation of* E_2 *does not hold under* E_1, *then* HuHu($\{\}, E_1 \cup E_2, \succ$) *terminates with* DISPROOF *or will fail.*

The above algorithm is specialized to abstract data types that are given by constructors as it is the case in the logic of C@S. There are lots of refinements of the algorithm, some of them also implemented in the RRL system, that is interfaced with C@S.

4. Modelling Hardware

In this section, we describe the simplest version of the input language of C@S that is called SHDL. SHDL is an acronym for 'synchronous hardware description language' as it focuses on the description of synchronous systems. SHDL allows to describe systems in a modular way and it offers also the possibility to describe generic structures. In contrast to the logic as given in definition 3.3, abstract data types are defined in SHDL globally, i.e. SHDL has a construct abstype α with $c_1, \ldots, c_n$ that introduces a new data type α that is globally visible. Similarly, SHDL allows to define new functions on these defined abstract data types according to the primitive recursion theorem that is given by the definition of the data types according to definition 3.5.

4.1 Describing Nongeneric Structures

There are two kinds of (nongeneric) modules in SHDL: one for describing the behavior of base modules and another one for describing the wiring of composed modules.

```
BASIC Module_Name(i₁, ... , iₙ)
    init(q₁) := ω₁;
    next(q₁) := Ω₁((⃗i), (⃗q));
    ⋮
    init(qₛ) := ωₛ;
    next(qₛ) := Ωₛ((⃗i), (⃗q));
OUTPUT
    o₁ := Φ₁((⃗i), (⃗q));
    ⋮
    oₜ := Φₜ((⃗i), (⃗q));
END;
```

```
BASIC D_FlipFlop(inp)
    init(q) := 0;
    next(q) := inp;
OUTPUT
    o₁ := q;
    o₂ := ¬q;
END;

BASIC And(i₁, i₂)
OUTPUT
    o := i₁ ∧ i₂;
END;
```

Fig. 4.1. Structure of basic modules (left) and two examples (right)

The overall structure of a base module is given in figure 4.1. The lefthand side shows the overall structure of a basic module and the righthand side the description of a D-Flipflop and an AND gate as an example. The module has a name *Module_Name* and the inputs $i_1, \ldots, i_n$. A base module may have internal states, which are encoded by boolean state variables $q_1, \ldots, q_s$. This means that each tuple of $\{1,0\}^s$ is potentially an internal state of the above base module, However it may be the case that some of the potential states are not reachable. The reachability is given by the initial state that is in turn given by the initialization equations $\mathsf{init}(q_k) := \omega_k$, where $\omega_k \in \{1,0\}$. Hence, the initial state of the module is $(\omega_1, \ldots, \omega_s)$. The transition relation is given by the equations $\mathsf{next}(q_k) := \Omega_k(\vec{i}, \vec{q})$ where $\Omega_k((\vec{i}), (\vec{q}))$ is an arbitrary propositional formula over the variables $i_1, \ldots, i_n$ and $q_1, \ldots, q_s$.

Clearly, the semantics of a base module as given in figure 4.1 is the following simple prefix formula, that has as acceptance condition only a single safety property:

$$\mathcal{P}(\vec{i}) := \left(\begin{array}{l} \exists q_1 \ldots q_s. \\ \displaystyle\bigwedge_{k=0}^{s} \left[(q_k = \omega_k) \wedge \mathsf{G}\left(\mathsf{X}q_k = \Omega_k(\vec{i}, \vec{q}) \right) \right] \wedge \\ \displaystyle\mathsf{G}\left[\bigwedge_{k=0}^{t} o_k = \Phi_m(\vec{i}, \vec{q}) \right] \end{array} \right)$$

We can interpret the above prefix formula either as a rewrite system or as an ω-automaton. Hence, SHDL descriptions for nongeneric structures are independent on the proof procedures to verify them. However, it has to be mentioned, that model checking procedures use module descriptions to compute a Kripke structure and hence, SHDL modules are never transformed into prefix formulas. The above prefix formula is just given for defining the semantics of a basic SHDL module.

Note also that the transition relation for basic modules is always deterministic: it is required that for each state variable a transition equation, i.e. an assignment $\text{next}(q_i) := \varphi_i$ exists. It is allowed that a basic module has no inputs and it is also allowed that a basic module has no internal state variables. In particular, the description of boolean basic gates such as AND-gates do not require internal states. Furthermore, initialisation equations are optional, such that there may be more than one initial state. On the other hand, it is required that each base module has at least one output, otherwise the module does not make sense at all.

```
MODULE M(inp₁, . . . , inpₙ)
   C₁ := M₁(v₁,₁, . . . , v₁,ₙ₁);
   ⋮
   Cₛ := Mₛ(vₛ,₁, . . . , vₛ,ₙₛ);
   OUTPUT
   o₁ := ψ₁;
   ⋮
   oₜ := ψₜ;
   SPEC
   Spec_Name₁  φ₁;
   ⋮
   Spec_Nameᵣ  φᵣ;
END;
```

```
MODULE Mux(sel, inp₁, inp₂)
   C₁ : And(inp₁, inp₂);
   C₂ : Not(sel);
   C₃ : And(C₂.o, inp₂);
   C₄ : Or(C₁.o, C3.o);
   OUTPUT
   o := C₄.o;
   SPEC
   S₁  G[o = (sel ⇒ inp₁ | inp₂)];
END;
```

Fig. 4.2. Structure of composed modules (left) and an example (right)

The structure of a composed module and an example of a composed module is given in figure 4.2. SHDL distinguishes between the *definition* of a module and the *use* of a module, such as imperative languages distinguish between the definition of a function with formal parameters and the call of the function with some actual parameters. This can also be understood as an object-oriented paradigm: the definition of a module defines a *class* and uses of the modules define some objects of that class. In the following, the notions of class and object are used, instead of definition and use. The access of an output o_i of an object C_j is done by the 'access operator (.)' and is written as $C_j.o_i$.

The definition of a composed module starts with a module name *Module_Name* and a formal parameter list $inp_1, \ldots, inp_n$. Then a list of objects $C_1, \ldots, C_s$ follows that belong to the classes *Module_Name$_1$*, ..., *Module_Name$_s$*. It is assumed that these classes are already defined with argument lists of lengths n_1, ..., n_s, respectively. In the definition of the objects $C_1, \ldots, C_s$, the module classes are instantiated with actual parameters $v_{i,j}$ that can be either an input variable $inp_1, \ldots, inp_n$ or an access expression $C_j.o_i$. In the latter case it is required that (i) the module class *Module_Name$_j$* has an output o_i and that (ii) no cycles occur. This means that if in the actual parameter list of the definition of C_j an output of C_j is accessed, then it is required that at least one delay element (with state variables) must be on this cycle.

The definition of the used objects C_j for the implementation of a composed module implicitly gives the wiring of these modules by the actual parameters that are access expression. In contrast to other hardware description languages, this is done object-oriented, that is by considering the underlying systems instead of the wires between them.

Analogously to basic modules, the outputs of a composed module are given by an equation $o_i := \psi_i$. However, ψ_i is now restricted to be either an input variable $inp_1, \ldots, inp_n$ or an access expression $C_j.o_i$ where $j \in \{1, \ldots, s\}$. In the latter case it is required that the definition of the module class *Module_Name$_j$* has an output o_i. Note that the right hand sides ψ_i of output equations in basic modules can be arbitrary propositional formulas.

In contrast to basic modules, composed modules can have some specifications. Each specification has a name to identify it and a formula of the specification language. In figure 4.2, the specification of the multiplexer is given in temporal logic, and means that for each computation it will always hold that if *sel* holds then output o equals to inp_1, otherwise to inp_2.

The semantics of a composed module is given by the product automaton of the used objects. Using access expressions, the module hierarchy can be flattened such that the composed module becomes a basic module. This is the same as computing the product automaton of the ω-automata of the used basic modules.

4.2 Describing Regular Structures

Basic modules and composed modules as described in the last section, allow to describe arbitrary systems. However, regular structures can often be described more compact, if the underlying implementation scheme is given instead of a particular fixed implementation. For example, the implementation of a carry-ripple adder follows a simple cascadic implementation scheme that is even more intuitive than an implementation of a 32-bit version as a composed module. SHDL has therefore language constructs for describing regular implementation schemes. These are given in this section.

```
RECURSIVE Module_Name(p, p₁ ... , pₓ; inp₁ : α₁, ... , inpₙ : αₙ)
    CASE p = 0:
        B₁ := B_Class₁ (v₁,₁, ... , v₁,ₙ₁);
            ⋮
        Bₛ := B_Classₛ (vₛ,₁, ... , vₛ,ₙₛ);
    OUTPUT
        o₁ := ψ₁;
            ⋮
        oₜ := ψₜ;
    CASE p > 0:
        R₁ := R_Class₁ (w₁,₁, ... , w₁,ₗ₁);
            ⋮
        R_q := R_Class_q (v_w,₁, ... , v_w,ₗ_q);
    OUTPUT
        o₁ := Ψ₁;
            ⋮
        oₜ := Ψₜ;
    SPEC
        Spec_Name₁ φ₁;
            ⋮
        Spec_Nameᵣ φᵣ;
    END;
```

Fig. 4.3. Structure of recursive module definitions

Figure 4.3 shows the structure of a recursive module definition with the recursion parameter $p \in \mathbb{N}$. Beneath the inputs $inp_1, \ldots , inp_n$, an additional parameter list $p, p_1, \ldots , p_x$ is added to the formal argument list. The first parameter of this list, i.e. p, is the one on which the recursive definition is done. Additionally, it may be the case that there are also other parameters $p_1, \ldots , p_x$ that are used by components which are required for the definition of the considered module. The list of parameters and the list of inputs are separated by a semicolon. The inputs can be constrained by terms $\alpha_1, \ldots , \alpha_n$ in order to specify that inp_i is a list of length α_i. If no constraint is given, the input is supposed to be a boolean valued signal. The constraints $\alpha_1, \ldots , \alpha_n$ usually depend on the recursion parameter p.

A recursive module defines for each $p \in \mathbb{N}$ a module that can be described for a fixed $p \in \mathbb{N}$ by the constructs of the last section, if not other parameters have to be instantiated. The definition of a recursive module is split up in the definition of a base case $p = 0$ and the definition of a recursion case $p > 0$. Each of these cases is in principle the definition of a composed module. However, in the recursion case it is allowed to use the module class also for instantiation with a parameter smaller than p, usually with $p - 1$. It is mandatory that the names, the types and the number of outputs are the same in each of the two cases.

Recursive modules can be 'unroled' for a fixed parameter $p \in \mathbb{N}$, i.e. the recursion case of the implementation description is eliminated by successive calls and interpretations of the definition for the fixed parameter. Each instantiation of a parameter of a recursive module with a fixed value $p \in \mathbb{N}$ is again a module class that can be used for generating objects of this class. If a module contains a variable in the parameter list, it is called a *parameterized* module.

Recursive modules define also an induction scheme, which is important for the verification of all modules that can be obtained by instantiating the recursion parameter. In particular, one has to show the correctness for the base case $p = 0$ and, by using the induction hypothesis, that the module is correct for $x < p$. It is then necessary to prove that the module is correct for the parameter p. This involves also the elimination of hierarchy, in so far as the objects belonging to the currently considered class and that are used in the definition of the recursion case have to be replaced by one or more of the specifications of the module.

RECURSIVE $Carry\_Ripple(n; inp_1 : n + 1, inp_2 : n + 1, carry)$
    CASE $n = 0$:
        $C_1 : Full\_Add(\text{HD}(inp_1), \text{HD}(inp_2), carry)$;
      OUTPUT
        $sum := [C_1.sum]$;
        $cout := C_1.cout$;
    CASE $n > 0$:
        $C_1 : Full\_Add(\text{HD}(inp_1), \text{HD}(inp_2), C_2.cout)$;
        $C_2 : Carry\_Ripple(n - 1; \text{TL}(inp_1), \text{TL}(inp_2), carry)$;
      OUTPUT
        $sum := (C_1.sum \triangleright C_2.sum)$;
        $cout := C_1.cout$;
    SPEC
      S_1 $\text{AG}[\neg carry \rightarrow ((cout \triangleright sum) \sim_\mathbb{N} (inp_1 \bar{+} inp_2))]$;
      S_2 $\text{AG}[carry \rightarrow ((cout \triangleright sum) \sim_\mathbb{N} \text{INC}((inp_1 \bar{+} inp_2)))]$;
  END;

Fig. 4.4. Example of a recursive module

Figure 4.4 gives the recursive definition of a carry-ripple adder in SHDL. The module is parameterized with a single parameter n that is the number of bits of the inputs that are to be added. Hence, the inputs inp_1 and inp_2 are lists of bits of the length $n + 1$ and the third input $carry$ is a simple boolean valued signal. The outputs are a list of length n that has the name sum and a boolean valued output $cout$. The latter is high exactly when an overflow occurred (it may also be viewed as the most significant bit of the sum) and *sum* contains the first n bits of the sum of the inputs.

In the base case $(n = 0)$, the system consists simply of a full adder, whose inputs are $inp_1[0]$ and $inp_2[0]$ (note that in this case $inp_1 = [inp_1[0]]$ and $inp_2 = [inp_2[0]]$ holds). Note that the output $C_1.sum$ of the full adder is a boolean valued signal that has to be put in a list such that the types of the outputs are correct. The recursion case the inputs $inp_1 = [inp_1[0], \dots, inp_1[p-1]]$ and $inp_2 = [inp_2[0], \dots, inp_2[p - 1]]$ are split up. The lower $p - 1$ bits $\mathsf{TL}(inp_1) = [inp_1[1], \dots, inp_1[p-1]]$ and $\mathsf{TL}(inp_2) = [inp_2[1], \dots, inp_2[p-1]]$ are fed into a carry-ripple adder C_2 of size $p-1$ and the most significant bits $\mathsf{HD}(inp_1) = inp_1[0]$ and $\mathsf{HD}(inp_2) = inp_2[0]$ are fed into a full adder C_1. The overflow output of C_1 is wired to the carry-in input of C_2. The overflow output is then the overflow output of C_1 and the sum is the sum computed by C_2 with the extension of the sum of C_1 at the left hand side of $C_2.sum$.

Additionally, some specifications are given for the system that are not considered in detail here. S_1 states that the systems output *sum* and *cout* can be viewed as the sum of the inputs when *carry* is low, otherwise it can be viewed as the increment of the sum of the inputs.

It is important to assure the consistency of the definition of a recursive module. This involves a type checking together with the constraints on list lengths given by the recursion parameter p. For example, in figure 4.4 the inputs of the carry-ripple adder inp_1 and inp_2 are declared in the header as lists of length p (length constraint). The given component lists make then clear that these lists are lists of boolean values (type constraint). In the component declaration of the module object C_2 in the recursion case, it is necessary that the conditions $\|\mathsf{TL}(inp_1)\| = \|inp_1\| - 1$ and $\|\mathsf{TL}(inp_2)\| = \|inp_2\| - 1$ are fulfilled. Otherwise, the definition of the module can run into the risk of being inconsistent and hence of making no sense at all. These consistency conditions can be generated automatically for each definition of a recursive module, but need to be proved by hand (usually it requires interactive proofs, but can be done with a high degree of automation).

Figures 4.5 and 4.6 show the definition of an Omega-network as a more complicated example of a generic structure. Figure 4.5 defines the perfect shuffle function and a switching circuit. The latter one has two inputs inp_1, inp_2 and an additional control input *sel*. The outputs o_1, o_2 directly correspond to inp_1, inp_2 if *sel* is high, or to inp_2, inp_1, otherwise. $Switch\_N(n; s, a)$ is a column of $n + 1$ switches, hence there are $n + 1$ control inputs s for mapping the $2(n + 1)$ inputs to the $2(n + 1)$ outputs. The modules given in figure 4.6 combine these modules to the Omega-network. $Net\_Help(m, n; s, a)$ has two parameters m and n and as inputs a list s that contains $m + 1$ boolean lists and a boolean list a of length $2(n + 1)$. The type of s is firstly determined by the length restriction $s : m + 1$ in the parameter definition of the module Net_Help. The module instantiation $S : Switch\_N(n; \mathsf{HD}(s), a)$ requires that $\mathsf{HD}(s)$ is a boolean list of length $n + 1$.

```
RECURSIVE Perfect_Shuffle(n; a : n + 1, b : n + 1)
    CASE n = 0:
        OUTPUT
            out := [EL₀(a), EL₀(b)];
    CASE n > 0:
        P : Perfect_Shuffle(n − 1; TL (a) TL (b));
        OUTPUT
            out := (EL₀(a) ▷ (EL₀(b) ▷ P.out));
END;

MODULE Switch(s, inp₁, inp₂)
        C₁ : Mux(sel, inp₁, inp₂);
        C₂ : Not(sel);
        C₃ : Mux(C₂.o, inp₁, inp₂);
    OUTPUT
        o₁ := C₁.o;
        o₂ := C₂.o;
END;

RECURSIVE Switch_N(n; s : n + 1, a : 2(n + 1))
    CASE n = 0:
        C₁ : Switch(HD (s), EL₀(a), EL₁(a));
        OUTPUT
            out := [C₁.o₁, C₁.o₂];
    CASE n > 0:
        C₁ : Switch(HD (s), EL₀(a), EL₁(a));
        C₂ : Switch_N(n − 1; TL (s), TL (TL (a)));
        OUTPUT
            out := (C₁.o₁ ▷ (C1.o₂ ▷ C₂.out));
END;
```

Fig. 4.5. Perfect shuffle function and a column of switches as recursive modules

$Net\_Help(m, n; s, a)$ simply consists of $m + 1$ columns of $Switch\_N$ and Perfect_Shuffle modules. This is in principle the implementation of the Omega-network. However, as Omega-networks are defined to consist of $m+1$ columns of perfect shuffles with switching elements of length 2^{m+1}, we have to constrain the parameter n to $2^{m+1} - 1$. This is the task of the module $Omega\_Net$. Hence, the argument list a in the module $Omega\_Net$ has now the length $2(2^{m+1} - 1 + 1) = 2^{m+2}$. Note that the definition of the module $Omega\_Net$ is not done recursively. A correctness proof of such generic modules requires usually to prove some lemmata about its components that are combined by some interactive rules to obtain the final correctness result.

RECURSIVE $Net\_Help(m, n; s : m + 1, a : 2(n + 1))$
   CASE $m = 0$:
        S : $Switch\_N(n; \text{HD}(s), a)$;
        P : $Perfect\_Shuffle(n; \text{FIRSTN}_{n+1}(S.out), \text{LASTN}_{n+1}(S.out))$;
     OUTPUT
        $out := P.out$;
   CASE $n > 0$:
        S : $Switch\_N(n; \text{HD}(s), a)$;
        P : $Perfect\_Shuffle(n; \text{FIRSTN}_{n+1}(S.out), \text{LASTN}_{n+1}(S.out))$;
        N : $Net\_Help(m - 1, n; \text{TL}(s), P.out)$;
     OUTPUT
        $out := N.out$;
END;

GENERIC $Omega\_Net(m; s : m + 1, a : 2^{m+2})$
   N : $Net\_Help(m, 2^{m+1} - 1; s, a)$;
   OUTPUT
        $out := N.out$;
END;

Fig. 4.6. Omega-network as generic module

5. Specification and Proof

5.1 The Unifying Principle

In general, a specification in C@S may consist of a combination of temporal operators and abstract data type expressions as presented in section 3.1. Clearly, it is useful to separate the temporal and the data part such that they can be tackled with different approaches. For example, the specification $[(C = A + 1) \text{ W } (B = 0)]$ (A, B, C are natural numbers and $+$ is a defined function symbol for the usual addition) is transformed into $[\varphi_1 \text{ W } \varphi_2]$ where φ_1 and φ_2 are new variables of type $\mathbb{B}$ with the meaning $\varphi_1 := (C = A + 1)$ and $\varphi_2 := (B = 0)$. These new variables are added by let-expressions similar to those of functional programming languages as follows:

$$\text{let} \left[\begin{pmatrix} \varphi_1 \\ \varphi_2 \end{pmatrix} := \begin{pmatrix} C = A + 1 \\ B = 0 \end{pmatrix} \right] \text{ in } [\varphi_1 \text{ W } \varphi_2] \text{ end}$$

In general, all specifications in C@S follow the template given below:

$$\text{C\_SPEC}(\vec{i}, \vec{o}) := \begin{pmatrix} \text{let} \\ \quad \begin{pmatrix} \varphi_1 \\ \vdots \\ \varphi_n \end{pmatrix} := \begin{pmatrix} \Theta_1(\vec{i}, \vec{o}) \\ \vdots \\ \Theta_n(\vec{i}, \vec{o}) \end{pmatrix} \\ \text{in} \\ \quad \Phi(\varphi_1, \ldots, \varphi_n) \\ \text{end} \end{pmatrix}$$

The above specification consists of a *temporal abstraction* Φ and the *data equations* $\Theta_1(\vec{i}, \vec{o}), \ldots, \Theta_n(\vec{i}, \vec{o})$. Φ is a propositional linear temporal logic formula containing only the propositional variables $\varphi_1, \ldots, \varphi_n$, i.e. no data expressions may occur. The above separation between temporal and data abstraction parts follows the usual distinction of digital systems into a controller and a data path. In some way, $\Phi(\varphi_1, \ldots, \varphi_n)$ corresponds to the specification of the control flow, while the data equations correspond to the specification of the data flow. Of course, it is usually not the case that these specification can be verified independently from each other.

As it has been shown in section 3., there exist proof procedures for temporal propositional logic and for proving the consistency of rewrite systems with abstract data types. However, it is difficult to combine both classes of proof procedures. The next subsection explains some strategies how this is done in the C@S system.

5.2 Strategies for Verifying Temporal Behavior

In this section, the W operator is used to illustrate different translation procedures of C@S. All presented procedures are able to capture full linear temporal logic as defined in definition 3.3. The main strategy is Φ_{PF} for translating verification goals into prefix formulas. If the corresponding prefix formula consists only of a safety property, then a simple combination with proof procedures for term rewrite systems can be achieved by interpreting the prefix formula as a term rewrite system.

If, on the other hand, the corresponding prefix formula has also liveness or fairness properties in its acceptance condition, then the strategy Φ_{IV} can be used to eliminate some temporal operators. Φ_{IV} can also be used to reduce the complexity of the result of Φ_{PF} by providing more proof information by additional invariants. C@S has also other proof procedures for temporal reasoning as Φ_{N}, that translates temporal logic to arithmetic and Φ_{μ} that translates temporal logic to fixpoint equations. These proof procedures are however not given in this article.

5.2.1 Strategy Φ_{PF}. It is straightforward to define various temporal operators by prefix formulas. For example, the W-operator can be defined as follows:

$$\vdash_{def} [x \, \mathsf{W} \, b] := \left(\begin{array}{c} \exists q. \, (q = 0) \wedge \mathsf{G} \, (\mathsf{X}q = q \vee b) \wedge \\ \mathsf{G} \neg b \vee x \vee q \end{array} \right)$$

The strategy Φ_{PF} uses these definitions to translate the temporal abstraction Φ into an equivalent prefix formula $\mathcal{H}_{\Phi}$. In order to capture full LTL, additional universal quantification over input variables and additional fairness constraints are required for prefix formulas [Schn96b]. Section 5.3 explains in detail how strategy Φ_{PF} works.

5.2.2 Strategy Φ_{IV}. Provided that invariants are given for some occurrences of temporal operators, this strategy eliminates the corresponding temporal operators by invariant rules of CℚS . For example, the invariant rule of the W operator is as follows [ScKK94a]:

$$[x \text{ W } b] := \left(\begin{array}{l} \exists J. \\ \quad J \wedge \\ \quad G\,(\neg b \wedge J \to XJ) \wedge \\ \quad G\,(b \wedge J \to x) \end{array} \right)$$

The bound variable J is called an invariant and the subformulas have the following meaning: The first one states that the invariant J holds at the beginning of the computation, the second one states that if the termination condition b does not hold, the invariant J implies that the invariant J holds also at the next point of time, and the third subformula states that if the termination condition b and the invariant J hold at a certain time, then the condition x holds also at this point of time.

As already mentioned, invariant rules are applied for the following purposes: first, they are used to cut data dependencies between control and data part of the verification goals. Second, they often eliminate the introduction of fairness constraints by Φ_{PF}, such that an application of Φ_{PF} yields in prefix formulas with only a safety property that can be checked by traditional rewrite methods. A third purpose is that even in case that model checking techniques are used, the complexity of these procedures can be significantly reduced as shown in the experimental results section.

5.3 Translating LTL Formulas into Fair Prefix Formulae

In this section, the translation method for LTL as defined in section 3.3 into universally quantified fair prefix formulas is presented. The translation of a LTL formula φ involves the following steps:

1. Computation of the negation normal form
2. Computation of the prenex next normal form φ_p with kernel ψ_p
3. Reduction of the kernel ψ_p to quantified W normal form
4. Computation of the X- and W-closures and generation of fairness constraints by a bottom-up traversal through the syntax tree.

The negation normal form has to be computed since negations of SPFs might cause an exponential blow-up. Nevertheless, SPFs are closed under arbitrary boolean operations as stated in the following theorem.

Theorem 5.1 (Boolean Closure of Simple Prefix Formulas). *For all SPFs* $\mathcal{P}_1(\vec{\imath}), \mathcal{P}_2(\vec{\imath})$ *there are SPFs* $\mathcal{P}_\neg(\vec{\imath})$, $\mathcal{P}_\wedge(\vec{\imath})$, $\mathcal{P}_\vee(\vec{\imath})$ *such that* $\models \mathcal{P}_\neg(\vec{\imath}) = \neg\mathcal{P}_1(\vec{\imath})$, $\models \mathcal{P}_1(\vec{\imath}) \wedge \mathcal{P}_2(\vec{\imath}) = \mathcal{P}_\wedge(\vec{\imath})$ *and* $\models \mathcal{P}_1(\vec{\imath}) \vee \mathcal{P}_2(\vec{\imath}) = \mathcal{P}_\vee(\vec{\imath})$ *holds.*

Proof. According to the closure lemma [Schn96a], it is sufficient to show that all boolean operations on SPF acceptance conditions can be expressed as SPF. The proof of the closure under conjunction is trivial. To show the closure under disjunction, an equivalent SPF for the following formula has to be found:

$$\left(\bigwedge_{j=0}^{a_1} [\mathsf{G}\phi_{1,j}] \vee [\mathsf{F}\psi_{1,j}] \right) \vee \left(\bigwedge_{k=0}^{a_2} [\mathsf{G}\phi_{2,k}] \vee [\mathsf{F}\psi_{2,k}] \right)$$

Using distributivity laws for $\vee$ and $\wedge$, the following equivalent formula is obtained:

$$\bigwedge_{j=0}^{a_1} \bigwedge_{k=0}^{a_2} ([\mathsf{G}\phi_{1,j}] \vee [\mathsf{G}\phi_{2,k}] \vee [\mathsf{F}\psi_{1,j}] \vee [\mathsf{F}\psi_{2,k}])$$

Disjunctions of liveness properties can be reduced to a single liveness property using the law $[\mathsf{F}\psi_{1,j}] \vee [\mathsf{F}\psi_{2,k}] = \mathsf{F}(\psi_{1,j} \vee \psi_{2,k})$. Reducing the disjunction of safety properties is not possible in this manner. This can be done by introducing new state variables p_j, q_k to watch the corresponding safety properties $\phi_{1,j}$ and $\phi_{2,k}$, respectively. These 'watchdog variables' are initially set to 0 and switch to 1 if the corresponding safety property is 0 at a certain point of time. After switching to 1 the watchdog variables stay on the value 1 regardless what happens. Using the watchdog variables, $[\mathsf{G}\phi_{1,j}] \vee [\mathsf{G}\phi_{2,k}]$ is equivalent to $\mathsf{G}(\neg p_j \vee \neg q_k)$. Thus the following SPF is obtained:

$$\left(\begin{array}{l} \exists p_0 \ldots p_{a_1} \, q_0 \ldots q_{a_2}. \\ \displaystyle\bigwedge_{j=0}^{a_1} [(p_j = 0) \wedge \mathsf{G}(\mathsf{X}p_j = p_j \vee \neg\phi_{1,j})] \wedge \\ \displaystyle\bigwedge_{k=0}^{a_2} [(q_k = 0) \wedge \mathsf{G}(\mathsf{X}q_k = q_k \vee \neg\phi_{2,k})] \wedge \\ \displaystyle\bigwedge_{j=0}^{a_1} \bigwedge_{k=0}^{a_2} ([\mathsf{G}(\neg p_j \vee \neg q_k)] \vee [\mathsf{F}(\psi_{1,j} \vee \psi_{2,k})]) \end{array} \right)$$

To show the closure under negation, the closure lemma is first used to shift the negation over the transition relation. Then negation is shifted inside the safety and liveness properties, such that the acceptance condition $\bigvee_{j=0}^{a} [\mathsf{F}\neg\Phi_j(\vec{q})] \wedge [\mathsf{G}\neg\Psi_j(\vec{q})]$ is obtained. As each safety and each liveness property is itself a SPF (without state transitions), this is just the disjunction of $a+1$ SPFs. As already proved, this can be expressed as an equivalent SPF. As each boolean operation can be expressed by $\wedge, \vee$ and $\neg$ the boolean closure of SPFs can be concluded in general.   ∎

Conjunctions of SPF can be performed in constant time, while performing disjunctions requires at most quadratic space and time. Negations, on the other hand, can blow up a SPF by an exponential factor. Thus, negations of SPF are avoided in the translation procedure by converting the given LTL formula first into negation normal form as follows:

Theorem 5.2 (Negation Normal Form (NNF)). *The application of the following theorems converts each LTL formula into an equivalent one where negation symbols only occur in front of variables:*

- $\neg\neg\varphi = \varphi, \ \neg(\varphi \wedge \psi) = \neg\varphi \vee \neg\psi$ *and* $\neg(\varphi \vee \psi) = \neg\varphi \wedge \neg\psi$
- $\neg X\varphi = X\neg\varphi, \ \neg G\varphi = F\neg\varphi$ *and* $\neg F\varphi = G\neg\varphi$
- $\neg[x \ W \ b] = [(\neg x) \ \underline{W} \ b] = [(\neg x) \ W \ b] \wedge [Fb]$
- $\neg[x \ U \ b] = \neg[b \ W \ (x \to b)] = [(\neg b) \ W \ (x \to b)] \wedge F(x \to b)$

The definition and existence of the prenex next normal form is based on the following theorem. The computation itself is straightforward and therefore not presented in detail.

Theorem 5.3 (Prenex Next Normal Form (PNNF)).
For every LTL formula Φ, there exists an equivalent formula of the following form, where Ψ is a LTL formula without X-operators:

$$\exists q_1 \ldots q_n. \left(\bigwedge_{k=1}^{n} G\left(Xq_k = x_k\right) \right) \wedge [X \ldots X\Psi]$$

The variables q_k must not occur in Φ and the variables x_k are either variables occurring in Φ or one of the variables q_k. The formula Ψ is called the kernel of the above PNNF.

The X-operator commutes with all other operators, i.e. $(Xx)\wedge(Xy) = X(x\wedge y)$, $[(Xy) \ W \ (Xx)] = X[y \ W \ x]$, etc. However, if only one argument of a binary operator has a leading X-operator, as e.g. in $[(Xy) \ W \ x]$, these laws cannot be used directly. In this case, a new variable q is introduced by defining $G(Xq = x)$, such that the formula $\exists q.G(Xq = x) \wedge [(Xy) \ W \ (Xq)]$ is obtained. After that the X-operators can be shifted outwards. For example, the PNNF of $X[b \ W \ Xa] \wedge c$ is

$$\exists q_1 \ q_2 \ q_3.$$
$$G(Xq_1 = b) \wedge G(Xq_2 = c) \wedge G(Xq_3 = q_2)\wedge$$
$$XX([q_1 \ W \ a] \wedge q_3)$$

As the initial values of the new variables q_k in the above theorem are not considered for the evaluation of the truth value of Ψ, they may be arbitrarily set to 1 or 0. The equations $G(Xq_k = x_k)$ can therefore be viewed as transition equations of a transition relation and hence, it is sufficient to translate the remaining formula $[X \ldots X\Psi]$ into a SPF. The following closure theorem for the X-operator shows how this can be done, if the Ψ can be translated into a FPF:

Theorem 5.4 (X-Closure of Fair Prefix Formulas).
Given the FPF $\mathcal{P}(\vec{i})$ of definition 3.6, the following FPF is equivalent to $X\mathcal{P}(\vec{i})$:

$$
\left(
\begin{array}{l}
\exists p\, q_1 \ldots q_s. \\
\quad [(p = 0) \wedge \mathsf{G}\,(\mathsf{X}p = 1)] \wedge \\
\quad \bigwedge_{k=0}^{s} \left[(q_k = \omega_k) \wedge \mathsf{G}\left(\mathsf{X}q_k = \left(p \Rightarrow \Omega_k(\vec{i}, \vec{q})\middle|\omega_k\right)\right)\right] \wedge \\
\quad \left(\bigwedge_{l=0}^{f} \mathsf{GF}\xi_l(\vec{i}, \vec{q})\right) \rightarrow \left(\bigwedge_{m=0}^{a} \left[\mathsf{G}(p \rightarrow \Phi_m(\vec{i}, \vec{q}))\right] \vee \left[\mathsf{F}(p \wedge \Psi_m(\vec{i}, \vec{q}))\right]\right)
\end{array}
\right)
$$

A proof can be found in [Schn96a]. Thus, it remains to show how LTL formulas without X-operators can be translated into FPF. This is based on the following closure theorem for the temporal operator W, which can also be adapted to $\underline{\mathsf{W}}$.

Theorem 5.5 (W-Closure of Fair Prefix Formulas). *Given the $\mathcal{P}(\vec{i})$ of definition 3.6 and a propositional formula b, the following FPF is equivalent to $\left[\mathcal{P}(\vec{i})\ \mathsf{W}\ b\right]$:*

$$
\left(
\begin{array}{l}
\exists p\, q_1 \ldots q_s. \\
\quad [(p = 0) \wedge \mathsf{G}\,(\mathsf{X}p = p \vee b)] \wedge \\
\quad \bigwedge_{k=0}^{s} \left[(q_k = \omega_k) \wedge \mathsf{G}\left(\mathsf{X}q_k = \left((b \vee p) \Rightarrow \Omega_k(\vec{i}, \vec{q})\middle|\omega_k\right)\right)\right] \wedge \\
\quad \left(\bigwedge_{l=0}^{f} \mathsf{GF}\xi_l(\vec{i}, \vec{q})\right) \rightarrow \bigwedge_{m=0}^{a} \left(\begin{array}{l}\left[\mathsf{G}((b \vee p) \rightarrow \Phi_m(\vec{i}, \vec{q}))\right] \vee \\ \left[\mathsf{F}((b \vee p) \wedge \Psi_m(\vec{i}, \vec{q}))\right]\end{array}\right)
\end{array}
\right)
$$

Again, a proof of the theorem can be found in [Schn96a]. However, these theorems do not allow to derive closure theorems for the remaining temporal operators from the above theorem, since the assumption that the event b has to be propositional is not fulfilled in general. In order to handle the closure for these operators, the following theorem is used to reduce these operators into equivalent quantified W expressions with a propositional event. The key of this theorem is the introduction of new signals as events, which may become true in a special interval.

Theorem 5.6 (Reduction to Quantified W expressions). *Given that b is propositional, the following equations hold (a is a new variable):*

1. $[\mathsf{G}\varphi] = \forall a.\,[\varphi\ \mathsf{W}\ a]$
2. $[\mathsf{F}\varphi] = \exists a.\,[\varphi\ \underline{\mathsf{W}}\ a] = \exists a.\mathsf{F}a \wedge [\varphi\ \mathsf{W}\ a]$
3. $[\varphi\ \mathsf{U}\ b] = \forall a.\,[(\neg a)\ \mathsf{U}\ b] \vee [\varphi\ \mathsf{W}\ a]$
4. $[(\forall a.\mathcal{P}(a))\ \mathsf{W}\ b] = \forall a.\,[(\mathcal{P}(a))\ \mathsf{W}\ b]$
5. $[(\exists a.\mathcal{P}(a))\ \mathsf{W}\ b] = \exists a.\,[(\mathcal{P}(a))\ \mathsf{W}\ b]$

The first three items of the above theorem eliminate the temporal operators G, F and U if the event of U is propositional. Finally, the last two equations allow to shift quantifiers outside W-operators such that a prenex normal form

```
VAL 𝒱 := {};

FUNCTION add_var(Θ) = {a := new_var; 𝒱 := 𝒱 ∪ {(a,Θ)}; return a; }

FUNCTION QW(φ) =
  CASE φ of q
    prop(φ)   :  return ∃p. [(p = 0) ∧ G (Xp = 1)] ∧ G (φ ∨ p)
    φ₁ ∧ φ₂  :  return SPF_CONJ(QW(φ₁),QW(φ₂))
    φ₁ ∨ φ₂  :  return SPF_DISJ(QW(φ₁),QW(φ₂))
    Xφ₁       :  return SPF_NEXT(QW(φ₁))
    Gφ₁       :  if prop(φ₁) then return Gφ₁
                 else {a = add_var(∀); return [QW(φ₁) W a]}
    Fφ₁       :  if prop(φ₁) then return Fφ₁
                 else {a = add_var(∃); return Fa ∧ [QW(φ₁) W a]}
    [φ₁ U b]  :  if prop(φ₁) then
                    return ∃q. [(q = 0) ∧ G (Xq = q ∨ b)] ∧ G (q ∨ b ∨ φ₁)
                 else {a = add_var(∀);
                    return QW([(¬a) U b] ∨ [φ₁ W a])}
    [φ₁ W b]  :  if prop(φ₁) then
                    return ∃q. [(q = F) ∧ G (Xq = q ∨ b)] ∧ G (q ∨ ¬b ∨ φ₁)
                 else return SPF_WHEN(b,QW(φ₁))

FUNCTION LTL2QWHEN(φ) =
  { 𝒱 := {}; 𝒫:= QW(φ); return mk_quantify(𝒱,𝒫); }
```

Fig. 5.1. Algorithm for translating a subset of LTL to quantified SPF

can be obtained similar to the construction of the prenex normal form for first order logic. A proof of the theorem can be found in [Schn96a]. This directly leads to the following theorem:

Theorem 5.7. *Given a LTL formula $\Phi(\vec{i})$ in negation normal form such that for all subformulas $[x \ W \ b]$ and $[x \ U \ b]$ the event b is propositional, then there is a SPF $\mathcal{P}(\vec{i},\vec{a})$ with new variables $\vec{a}$ such that for some $\Theta_j \in \{\forall, \exists\}$ the equation $\Phi(\vec{i}) = \Theta_1 a_1 \ldots \Theta_n a_{\|\vec{a}\|}.\mathcal{P}(\vec{i},\vec{a})$ holds.*

Proof. $\Theta_1 a_1 \ldots \Theta_n a_{\|\vec{a}\|}.\mathcal{P}(\vec{i},\vec{a})$ is computed by the function **LTL2QWHEN** of figure 5.1. The proof is done by structural induction and follows directly the implementation of **LTL2QWHEN**. In the induction steps, the closure theorems for the boolean operators, X, and W are used and the induction basis follows from the following theorems, where x and b have to be propositional:

$-\ [x \ W \ b] := \exists q. [(q = 0) \wedge G (Xq = q \vee b)] \wedge G (q \vee \neg b \vee x)$
$-\ [x \ U \ b] := \exists q. [(q = 0) \wedge G (Xq = q \vee b)] \wedge G (q \vee b \vee x)$
$-\ x = \exists p. [(p = 0) \wedge G (Xp = 1)] \wedge G (x \vee p)$

∎

Hence, the closure of SPFs under boolean operators, X and W allows together with theorem 5.6 to construct for certain LTL formulas an equivalent quantified SPF. In the following, it is shown how *arbitrary* LTL formulas can be translated to *universally quantified FPFs*. Thus the following drawbacks of the function LTL2QWHEN have to be circumvented:

1. LTL2QWHEN can only translate LTL formulas, whose events are propositional, i.e. for all subformulas $[x \, \mathsf{W} \, b]$ and $[x \, \mathsf{U} \, b]$ the event b has to be propositional.
2. LTL2QWHEN does not always produce universally quantified formulas. Unfortunately, there is no 'simple' decision procedure for arbitrarily quantified SPFs. Since $\models \forall v_1 \ldots v_n . \mathcal{P}(\vec{i}, \vec{v})$ holds iff $\models \mathcal{P}(\vec{i}, \vec{v})$ holds, the validity of universally quantified SPFs can be computed by a decision procedure for SPFs.

Both problems are circumvented by the following trick: whenever one of the above problems is detected, the corresponding subformula φ is replaced by a new variable ℓ. Of course, it has to be guaranteed that ℓ behaves always equivalent to φ. This is done simply by adding the assumption $\mathsf{G}(\ell = \varphi)$, according to the theorem: $\Phi(\varphi) = \forall \ell . [\mathsf{G}(\ell = \varphi)] \rightarrow \Phi(\ell)$. The detailed computation of the assumptions is given by the algorithm of figure 5.2 and stated in the following theorem:

Theorem 5.8. *Given an arbitrary LTL formula $\Phi(\vec{i})$ in negation normal form, the function LTL2FLTL in figure 5.2 generates a set of formulas $\mathcal{E} = \{\mathsf{G}(\ell_1 = \varphi_1(\vec{i}, \vec{\ell})), \ldots, \mathsf{G}(\ell_n = \varphi_n(\vec{i}, \vec{\ell}))\}$ with new variables ℓ_j (not occurring in $\Phi(\vec{i})$) and a simple prefix formula $\mathcal{P}(\vec{i}, \vec{\ell}, \vec{a})$ such that the following holds:*

$$\Phi(\vec{i}) = \forall \ell_1 \ldots \ell_n . \bigwedge_{j=1}^{n} [\mathsf{G}(\ell_j = \varphi_j(\vec{i}, \vec{\ell}))] \rightarrow \forall a_1 \ldots a_{\|\vec{a}\|} . \mathcal{P}(\vec{i}, \vec{\ell}, \vec{a})$$

Moreover, the formulas $\varphi_j(\vec{i}, \vec{\ell})$ are of one of the following forms: $\mathsf{G}x$, $\mathsf{F}x$, $[x \, \mathsf{U} \, b]$ or $[x \, \mathsf{W} \, b]$, where both b and x are propositional.

Proof. Given a LTL formula φ in negation normal form, the function top in figure 5.2 replaces all temporal subformulas of φ by new variables and adds corresponding assumptions to the set $\mathcal{E}$. For example, top($[[\mathsf{F}y] \, \mathsf{W} \, [\mathsf{G}x]] \wedge \mathsf{F}\mathsf{G}z$) adds the assumptions $\mathsf{G}(\ell_1 = \mathsf{G}x)$, $\mathsf{G}(\ell_2 = \mathsf{F}y)$, $\mathsf{G}(\ell_3 = [\ell_2 \, \mathsf{W} \, \ell_1])$, $\mathsf{G}(\ell_4 = \mathsf{G}z)$, $\mathsf{G}(\ell_5 = \mathsf{F}\ell_4)$ to $\mathcal{E}$ and finally returns $\ell_3 \wedge \ell_5$. Note that top(φ) is always propositional and note also that top(φ)=φ, if φ is propositional. top is used by the function GenFair to replace non-propositional events b of subformulas $[x \, \mathsf{U} \, b]$ and $[x \, \mathsf{W} \, b]$ by propositional ones. Thus the first of the two problems of LTL2QWHEN is circumvented. The introduction of $\exists$-quantifiers by LTL2QWHEN arises only when a subformula $\mathsf{F}\varphi_1$ is replaced by its equivalent WHEN expression. The call of the function top in the function GenFair (in case $\varphi = \mathsf{F}\varphi_1$) eliminates this case. Hence, the result of the function GenFair is – under the assumptions of the set $\mathcal{E}$ – an equivalent formula which can be translated by LTL2QWHEN to a universally quantified SPF. ∎

```
VAL ε := {};

FUNCTION add_eq(φ) = {ℓ = new_var; ε := ε ∪ {G(ℓ = φ)}; return ℓ; }

FUNCTION top(φ) =
  CASE φ of
    prop(φ)      : return φ;
    φ₁ ∧ φ₂      : return top(φ₁) ∧ top(φ₂);
    φ₁ ∨ φ₂      : return top(φ₁) ∨ top(φ₂);
    Xφ₁          : return add_eq(Xtop(φ₁));
    Gφ₁          : return add_eq(Gtop(φ₁));
    Fφ₁          : return add_eq(Ftop(φ₁));
    [φ₁ U b]     : return add_eq([top(φ₁) U top(b)]);
    [φ₁ W b]     : return add_eq([top(φ₁) W top(b)]);

FUNCTION GenFair(φ) =
  CASE φ of
    prop(φ)      : return φ;
    φ₁ ∧ φ₂      : return GenFair(φ₁) ∧ GenFair(φ₂);
    φ₁ ∨ φ₂      : return GenFair(φ₁) ∨ GenFair(φ₂);
    Xφ₁          : return X (GenFair(φ₁));
    Gφ₁          : return G (GenFair(φ₁));
    Fφ₁          : return F(top(φ₁));
    [φ₁ U b]     : return [(GenFair(φ₁)) U (top(b))];
    [φ₁ W b]     : return [(GenFair(φ₁)) W (top(b))];

FUNCTION LTL2FLTL(φ) =
  ε := {}; ψ:= GenFair(φ); P := LTL2QWHEN(ψ);
  return (ε,P);
```

Fig. 5.2. Generation of fairness constraints for arbitrary LTL formulas

It remains now to translate the assumptions $G(\ell_j = \varphi_j)$ into fairness constraints. As $Gx = [0 \text{ W } (\neg x)]$, $Fx = [1 \underline{\text{W}} x]$, $[x \text{ U } b] = [b \text{ W } (x \to b)]$ holds, it is sufficient to be able to translate $G(\ell = [x \text{ W } b])$ and $G(\ell = [x \underline{\text{W}} b])$ to fairness constraints. As $(\ell = [x \underline{\text{W}} b]) = (\neg \ell = [(\neg x) \text{ W } b])$ holds, it is even only necessary to be able to translate $G(\ell = [x \text{ W } b])$ into a fairness constraint. How this is done, is explained in the following lemma.

Lemma 5.1 (Replacing Definitions by Fairness Constraints).
$G(\ell = [x \text{ W } b])$ *is equivalent to:*

$$\left(\begin{array}{l} \exists p\, q. \\ \quad (p = 1) \wedge G(Xp = (\ell \wedge q \wedge [x \vee \neg b]) \vee (\neg \ell \wedge q \wedge b \wedge \neg x)) \wedge \\ \quad (q = 1) \wedge G(Xq = (\neg \ell \wedge q \wedge \neg [x \wedge b]) \vee (\ell \wedge p \wedge b \wedge x)) \wedge \\ \quad GFp \end{array} \right)$$

The above theorem has been constructed by computing the accepting Büchi automata according to [MaPn87b] and has been formally proved in the HOL system. A proof can be found in [Schn96a].

In order to reduce a LTL formula $\Phi(\vec{i})$ into a universally quantified FPF, simply apply the function LTL2FLTL to obtain $\forall \vec{\ell}. \bigwedge_{j=1}^{n} [G(\ell_j = \varphi_j(\vec{i}, \vec{\ell}))] \rightarrow \forall \vec{a}.\mathcal{P}(\vec{i}, \vec{\ell}, \vec{a})$. After that, shift the quantification $\forall \vec{a}$ outwards to $\forall \vec{\ell}$ and replace the assumptions $G(\ell_j = \varphi_j(\vec{i}, \vec{\ell}))$ according to the above lemma, where the new transitions are added to the transitions of $\mathcal{P}(\vec{i}, \vec{\ell}, \vec{a})$.

5.4 Strategies for Verifying Abstract Data Types

In this section, different strategies for handling the data expressions in verification goals are discussed. In order to process a higher order data type $\mathbb{C}$ with a digital system, a finite number of boolean values is usually collected and interpreted as single unit according to an interpretation function $\Phi_\mathbb{C}$. For example, if natural numbers are to be processed, then a tuple $[a_n, \ldots, a_0]$ of boolean values is often interpreted as the natural number $\Phi_\mathbb{N}([a_n, \ldots, a_0]) := \sum_{i=0}^{n} 2^i \times a_i$. In general, each specification that uses an abstract data type $\mathbb{C}$ can be described at three levels of abstractions as shown in figure 5.3:

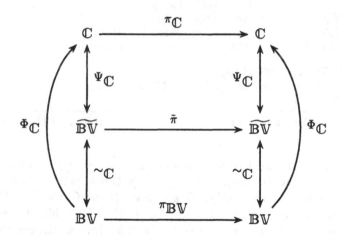

Fig. 5.3. Differnt levels of data abstraction

- The data operations can be formulated directly on the abstract data type $\mathbb{C}$ with operations $\pi_\mathbb{C}$ that are defined on that data type, as e.g. $\Phi_\mathbb{N}([b_2, b_1, b_0]) = \Phi_\mathbb{N}([a_1, a_0]) + 1$. This is the most readable version, but as the abstract data type $\mathbb{C}$ does not occur in hardware (structure) descriptions, the verification requires to lift bitvector operations π_{BV} used in the hardware description in SHDL by the data interpretation function $\Psi_\mathbb{C}$ to a corresponding operation $\pi_\mathbb{C}$ on the data type $\mathbb{C}$.
- Data operations may also be formulated at the 'bitvector level', where a bitvector is defined to be a boolean list with *an arbitrary length*. The set

of bitvectors is denoted by $\mathbb{BV}$ in the following. Using predefined bitvector operators $\pi_{\mathbb{BV}}$, the above example looks like $[b_2, b_1, b_0] = \mathsf{INC}\,([a_1, a_0])$, where INC is a bitvector operation that is defined as follows $((b \triangleright B)$ appends a boolean term b at the lefthand side of the bitvector B.):

$$(\mathsf{ALLONE}\,([]) = [1]) \wedge (\mathsf{ALLONE}\,((b \triangleright B)) = b \wedge \mathsf{ALLONE}\,(B))$$
$$(\mathsf{INC}\,([]) = [1]) \wedge (\mathsf{INC}\,((b \triangleright B)) = ((\mathsf{ALLONE}\,(B) \oplus b) \triangleright \mathsf{INC}\,(B)))$$

This level has many advantages: first its readability is comparable to descriptions at the abstract level $\mathbb{C}$. The only data type that is really used is 'bitvector' $\mathbb{BV}$, hence it is sufficient to have a decision procedure for bitvectors. This circumvents the problem of defining for each data type $\mathbb{C}$ an interpretation function $\Psi_{\mathbb{C}}$. For fixed lengths of bitvectors, it is also possible to translate these descriptions to propositional logic such that tautology checking can serve as a decision procedure.

- The data operations can directly be formulated as propositional formulas on the bits $a_n, \ldots, a_0$, where n has to be a concrete number in this case. Using two bits, the above example is represented as $(b_0 = \neg a_0) \wedge (b_1 = a_1 \oplus a_0) \wedge (b_2 = a_1 \wedge a_0)$. The advantage of this description is that simple tautology checking can be used as decision procedure for data types. However, these descriptions are not readable at all and do not allow n-bit specifications.

We usually prefer the second level for the verification, although the C@S logic would also allow us to use the most abstract level or the propositional level. For this reason, the following strategies can be found in C@S for verifying abstract data types:

Strategy Θ_B. This strategy transforms the equations $\Theta_1, \ldots, \Theta_n$ of a specification as given on page 273 into propositional formulas $\Theta_1^{prop}, \ldots, \Theta_n^{prop}$ that can be substituted in the temporal abstraction Φ. Of course, this strategy does only work with non-generic systems.

Strategy Θ_{BV}. This strategy translates operations on abstract data types such as addition on natural numbers in corresponding operations on bitvectors. If the specification is already given on bitvectors instead of (real) abstract data types, then this strategy is not required.

Strategy Induct. This strategy invokes an inductionless induction procedure to prove a specification that is given at the bitvector level. It is assumed that all operations are also given at this level or that strategy Θ_{BV} has been invoked before.

Strategy Decompose. If the system is recursively defined, then this strategy decomposes the verification goal into separate subgoals by applying a suitable induction rule which directly reflects the definition of the system. According to the induction hypothesis, the system structure is replaced by the system's specification in the induction step. Note that systems which are built up non-recursively but use recursively defined modules are not changed by this strategy.

Strategy Interact. This strategy allows to apply all interactive rules of the C@S. Hence, this strategy is not specialized for any system structure or any verification goal and allows arbitrary manipulations of the goal. It is however the intention to invoke this strategy only if a fully automated proof is not possible. In such cases, strategy Interact is used to prove additional lemmas or to change the design hierarchy such that one of the other rules can be applied.

5.5 Combining the Strategies

The above strategies have to be combined specifically for each verification problem. In general, the choice of the suitable strategy can be done according to the algorithm *Verify* below. *imp* is the implementation description, *sp* is the list of specifications for the currently considered module and *inv_list* is a possibly empty list of given invariants.

$$FUNCTION\ Verify(imp,sp,inv\_list)$$
$$IF\ Recursive\_Definition(imp)$$
$$THEN\ MAP\ Verify\ (Decompose(imp,sp,inv\_list))$$
$$ELSE$$
$$let\ sp1 = \Phi_{PF}(\Phi_{IV}(inv\_list, sp))$$
$$in\ IF\ fixed\_bv\_length(imp)$$
$$THEN\ HW\_Decide(HW\_IMP(imp,\Theta_B(sp1)))$$
$$ELSE$$
$$IF\ one\_safety(sp1)$$
$$THEN\ \mathsf{Induct}(imp,sp1)$$
$$ELSE\ \mathsf{Interact}(imp,sp,inv\_list)$$

Verify first tries to *decompose* recursively defined systems by applying induction rules. The induction principle follows directly the structure of the system, which is assumed to be well-defined as in the Boyer-Moore theorem prover [BoMo79]. According to the induction hypotheses, the implementation descriptions in the induction step(s) are immediately replaced by the specifications and the thereby obtained goals are also fed into *Verify*. If a non-recursive system is to be verified, *Verify* first translates the temporal abstraction of the specification into a prefix formula using strategy Φ_{PF}. If additional invariants are given in *inv_list*, these are used by applying strategy Φ_{IV} right before applying Φ_{PF}.

If only a concrete instantiation of the recursion scheme is to be proved (e.g. a 16-bit adder), then *Verify* translates the data equations into propositional formulas and calls the decision procedure *HW_Decide* based on symbolic model checking for prefix formulas. If this succeeds, the system has been verified without any interaction. Otherwise, it is checked, if the acceptance condition of the obtained prefix formula has only a safety property. If this is the case, then the prefix formula can be interpreted as a set of universally quantified equations and hence it can be proved by inductionless induction. The inductionless induction procedure requires some minor interaction for giving appropriate term orderings. In some cases, the inductive completion will run into infinite loops. In these cases, *Verify* will switch to interactive mode of C@S where further interactions of the user have to be applied to modify the proof goal such that another call to *Verify* will be more successful.

In general, the more information on the system and its specification (i.e. invariants and lemmas on the abstract data types) is given to the system, the more efficient its verification will be. In the next section, case studies of the common book examples and other systems will show in detail how C@S can be used for the verification of digital systems.

6. Experimental Results

In this section, different circuits are presented to show how different verification strategies are used for their verification. The first case study, a single pulser circuit is verified that has been investigated in [JoMC94] due to its complex temporal behavior. The circuit can be verified automatically by the temporal model checker. The second example, the Black Jack Dealer of the IFIP benchmarks, is also verified automatically by the same strategy though it contains some small parts for processing numerical data. The safe box example illustrates the generation of countermodels. The example is parameterised such that counterexamples of arbitrary length can be generated. Section 6.4 investigates the island tunnel controller presented in [FiJo95] that consists of three parallel processes controlling the restricted access of some agents (the cars) to a common resource (the island). This circuit could also be verified automatically by temporal logic model checking. The same holds also for the arbiter given in section 6.5, that has also been verified by symbolic CTL model checking in [McMi93a]. The generic von Neumann adder given in section 6.6, shows how circuits with abstract data types can be verified by the interactive application of invariant rules. It also shows that the complexity of the verification done by temporal logic model checking can be reduced by applying invariant rules. Note that the invariants express in a formal way the interaction between control and data path of the circuit. The case study given in 6.7 considers sequential n-bit multipliers in order to show that even the use of invariants does not always lead to such an improvement for model checking due to the structure of the invariant. In this case, inductive reasoning with invariant rules has to be used for the proof of additional lemmas in an interactive verification. The systolic arrays verified in section 6.8 do have a simple temporal behaviour, i.e. the specification is a simpy safety property. Hence, these examples are verified by term rewriting methods, and as these modules are defined in a generic manner, inductionless induction is used. The last example given in section 6.9, is a hardware realization of a part of the speech encoding algorithm of the GSM standard.

6.1 Single Pulser

The circuit has a boolean valued input *in* and a boolean valued output *out*. It is required that between two rising edges of *in* there is exactly one point of time, where *out* is high. The circuit is very small, however the specification is not trivial for some formalisms [JoMC94]. Using linear temporal logic, this can be expressed by the following items:

1. $G[\neg in \wedge Xin \rightarrow X([out \ B \ (\neg in \wedge Xin)] \vee [out \ W \ (\neg in \wedge Xin)])]$
2. $G[out \rightarrow [(X\neg out) \ U \ (\neg in \wedge Xin)]]$
3. $G[out \rightarrow X\neg out]$
4. $G[\neg in \wedge Xin \rightarrow XFout]$

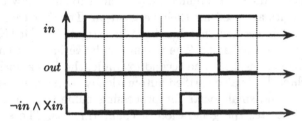

Fig. 6.1. Not a Single Pulser behavior.

There may be some confusion when the first and the second rising edge occurs. Hence, let us have a closer look at these points. We define a rising edge by the expression $\neg in \wedge Xin$. This evaluates to true if *in* is low at the current point of time and high at the following (see figure 6.1). Figure 6.1 indicates that there must be either a single pulse of *out* (properly) between two rising edges of *in* or at the time when the second rising edge occurs. In the latter case, it is however mandatory that the next output pulse does not follow immediately. Otherwise we would have the behavior shown in figure 6.1 which has certainly not a single pulse at *out*.

The first specification is satisfied iff after a rising edge at *in*, the output will be high before the next rising edge at *in* occurs. The second specification requires that each time *out* holds, it must stay low after that point of time until the next rising edge occurs. If we would only consider these two specifications, then it would be possible to have the behavior given in figure 6.1, since the second line trivially holds when *out* holds at a point of time when $\neg in \wedge Xin$ also holds. Hence the third specification explicitly states that *out* consists of single pulses. The fourth specification demands that *out* goes high at some time if there has been a rising edge at *in*.

The runtimes[4], the number of BDD nodes and the number of states for the verification of the specifications are given in the following table, where $\mathcal{C} \vdash i$ means the verification problem where specification i is to be verified:

Goal	User Time	BDD nodes	Reach. States	Poss. States
$\mathcal{C} \vdash 1$	0.08	5198	140	2^{15}
$\mathcal{C} \vdash 2$	0.08	1546	108	2^{11}
$\mathcal{C} \vdash 3$	0.02	112	14	2^5
$\mathcal{C} \vdash 4$	0.05	353	36	2^8

There are numerous other ways to describe the behavior of the single pulser, e.g. the following formulae are equivalent to specification 1:

5. $\mathsf{G}[\neg in \wedge \mathsf{X}in \rightarrow \mathsf{X}[out \vee [(\mathsf{X}out)\ \mathsf{B}\ (\neg in \wedge \mathsf{X}in)]]]$
6. $\mathsf{G}[\neg in \wedge \mathsf{X}in \rightarrow \mathsf{X}[out \vee [(\neg(\neg in \wedge \mathsf{X}in))\ \mathsf{U}\ out]]]$

The additional runtimes are given in the following table, where $\mathcal{C} \vdash i$ means the verification problem where specification i is to be verified and $i = j$ means to problem to verify that the specifications i and j are equivalent to each other.

Goal	User Time	BDD nodes	Reach. States	Poss. States
$\mathcal{C} \vdash 5$	0.08	3130	96	2^{14}
$\mathcal{C} \vdash 6$	0.09	3520	116	2^{13}
$1 = 5$	0.26	10128	312	2^{21}
$1 = 6$	0.15	10001	264	2^{21}

The example has no abstract data types at all, but requires to specify and to verify a quite complex behavior, that can not be done easily in temporal logics as e.g. in CTL. The specifications we gave are propositional linear temporal logic formulae that can be translated to prefix formulae as outlined in section 5.3. Note that a part from checking verification problems of the form $\mathcal{C} \vdash i$, this method allows us also to check the equivalence of specifications (of the form $i = j$).

6.2 Black Jack Dealer

Black Jack is one of the most popular Casino card games. There are at most seven players and a dealer who deals the cards from a card deck with 52 cards, collects the bets and pays the players. At the beginning, each player and also the dealer is dealt two cards, one face down and one face up. There are variations where both cards are face up. Aces may be counted freely either as 1 or 11. The objective of the game is to reach a score of 21 or to have a count under 21 but greater than the dealers. If the score is greater than 21, the player has lost the game. If the dealer goes over 21 the remaining

---

[4] All runtimes have been measured on a SUN SPARC10.

288   Klaus Schneider and Thomas Kropf

players are paid off. If the dealer's cards count sixteen or under the dealer must take a card. If the dealer has a score of seventeen or more he must not take additional cards.

The benchmark circuit (one of the common book examples, see Appendix) plays the dealer's hand in a Black Jack game, i.e. two players are getting cards from a staple of cards and sum up the corresponding values.

The benchmark circuit has a boolean valued input *card_rdy* and a 4-bit input *card*. *card_rdy* signals that the value of a card which is a binary number of $\{1,\ldots,10\}$ can be read from *card*. After that the card's value is added to an internal score. It is reasonable to count the first ace as 11 such that the score becomes greater than 16 as soon as possible. Note that it does not make sense to count further aces as 11 since the score would then exceed 21 and the game would be lost. For this reason, the circuit notes if an ace has been added with value 11 to the score. If the score exceeds 21 and an ace has been added with value 11 to the score, the game can still be won if this ace is counted as 1 instead of 11. Hence, in this case the circuit subtracts 10 from the score. If the score has a value greater than 16 but less than 22, the circuit indicates on the output *stand* that the game is over and that it has won the game. If the score exceeds 21 and no ace has been added with value 11 to the score, the circuit knows that it has lost the game and indicates that at output *broke*. Finally, if the score is less than 16 the circuit indicates at the output *hitme* that it demands a further card. The benchmark circuit has an additional *reset* input for resetting the circuit to its initial state and for clearing the internal score and registers.

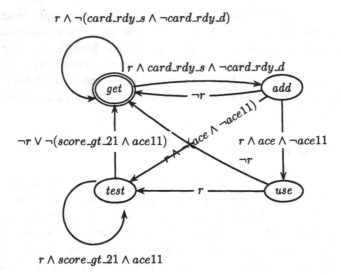

Fig. 6.2. State transition diagram of the controller of the Black Jack Dealer

Before presenting the detailed behavior of the benchmark circuit, we first verify that the controller of the Black Jack Dealer (see Appendix) has the state transition diagram shown in figure 6.2. The outputs of the controller are computed by the following equations:

1. $set\_broke = test \land score\_gt\_21 \land \neg ace11$
2. $clr\_broke = clr\_stand = get \land card\_rdy\_s \land \neg card\_rdy\_d$
3. $set\_stand = test \land score\_gt\_16 \land \neg score\_gt\_21$
4. $clr\_stand = clr\_broke = get \land card\_rdy\_s \land \neg card\_rdy\_d$
5. $set\_ace11 = use$
6. $clr\_ace11 = (get \land card\_rdy\_s \land \neg card\_rdy\_d \land (stand \lor broke)) \lor (test \land score\_gt\_21 \land ace11)$
7. $ld\_score = add \lor use \lor (test \land score\_gt\_21 \land ace11)$
8. $clr\_score = get \land card\_rdy\_s \land \neg card\_rdy\_d \land (stand \lor broke)$
9. $adder\_s1 = add$
10. $adder\_s0 = test \land score\_gt\_21 \land ace11$
11. $hitme = get \land \neg card\_rdy\_s$

Now we can give the detailed behavior of the circuit with the exact timing.

State get: This is the initial state and also the state which the circuit is brought to by *reset*. The circuit is in this state if either the game is over or if it demands a further card. In the first case, its internal score is greater than 16 and it has won the game if and only if the score is less than or equal to 21. This is indicated by the outputs *stand* and *broke*, respectively. Otherwise, the circuit indicates at the output *hitme* that it is ready to accept a new card value. The output functions in this state are as follows:

1. $set\_broke = set\_stand = set\_ace11 = ld\_score = 0$
2. $adder\_s1 = adder\_s0 = 0$
3. $clr\_broke = clr\_stand = card\_rdy\_s \land \neg card\_rdy\_d$
4. $clr\_ace11 = clr\_score = card\_rdy\_s \land \neg card\_rdy\_d \land (stand \lor broke))$
5. $hitme = \neg card\_rdy\_s$

The first line denotes that the values of the internal flipflops and the score remain unchanged in this state, hence, the values of *adder_s1* and *adder_s0* are irrelevant in this state since they are used to choose a value to be added to the score. The signal $card\_rdy\_s \land \neg card\_rdy\_d$ is high for one point of time exactly when there was a rising edge at input *card_rdy*. The circuit remains in state *get* until a new card is given. In this case, the circuit switches to state *add* and clears the flipflops of the data path which hold the values for *stand* and *broke*. Note that if the game was not already over, these values are low and hence this has no effect. However, if the game was over, i.e. either *stand* or *broke* hold, then the arrival of the new card value starts a new game by also resetting the internal score and the *ace11* flag.

For the understanding of the following facts that we have verified, note that $\neg card\_rdy \wedge Xcard\_rdy \wedge Xreset$ is equivalent to $card\_rdy_s$:

$$\text{Get1 } G\left[XX\left[\begin{array}{l}\neg card\_rdy \wedge Xcard\_rdy \wedge Xreset \rightarrow \\ \left[\begin{array}{l}get \wedge reset \rightarrow \\ \left(\begin{array}{l}(Xscore = ([stand \vee broke] \Rightarrow 0|\ score))\wedge \\ (Xstand = 0) \wedge (Xbroke = 0)\wedge \\ (Xace11 = ([stand \vee broke] \Rightarrow 0|\ ace11))\end{array}\right)\end{array}\right]\end{array}\right]\right]$$

Get2 $G[get \rightarrow X(get \vee add)]$

Get3 $G\left[\neg card\_rdy \wedge Xcard\_rdy \wedge Xreset \rightarrow XX\left[get \wedge reset \rightarrow Xadd\right]\right]$

Get4 $G[Xget \rightarrow (Xhitme = \neg(reset \wedge card\_rdy))]$

State add: In this state the circuits internal score is updated by adding the current card value to the internal score. All outputs are 0 in this state except for the outputs ld_score and $adder\_s1$ that command the operation unit to add the card value. If the card is not an ace, the next state will be *test*. Otherwise it is checked if this ace is the first one. If this is the case, the next state is state *use* which is used to set the *ace11* flag and to add 10 to the score, i.e. the first ace is counted as 11. On the other hand, if an ace has already been added to the score, the currently read ace must be counted as 1 and the next state is hence *test*.

$$\text{Add1 } G\left[add \wedge reset \rightarrow \left(\begin{array}{l}(Xscore = (score\dotplus card))\wedge \\ (Xstand = 0) \wedge (Xbroke = 0)\wedge \\ (Xace11 = ace11)\end{array}\right)\right]$$

Add2 $G[add \wedge reset \rightarrow X(use \vee test)]$

Add3 $G[add \wedge reset \wedge acecard \wedge \neg ace11 = Xuse]$

Add4 $G[add \wedge reset \wedge \neg(acecard \wedge \neg ace11) \rightarrow Xtest]$

Add5 $G[add \rightarrow \neg (16 \lesssim score)]$

State use: In this state all outputs of the controller are false except for set_ace11, ld_score. Hence, in this state the $ace11$ flag is set and 10 is added to the internal score as already mentioned.

$$\text{Use1 } G\left[use \wedge reset \rightarrow \left(\begin{array}{l}(Xscore = (score\dotplus 10))\wedge \\ (Xstand = 0) \wedge (Xbroke = 0)\wedge \\ (Xace11 = 1)\end{array}\right)\right]$$

Use2 $G[use \wedge reset \rightarrow Xtest]$

Use3 $G[use \rightarrow \neg (17 \lesssim score)]$

Use4 $G[use \wedge reset \rightarrow \neg ace11 \wedge \neg stand \wedge \neg broke \wedge \neg hitme]$

State test: In this state the internal score of the circuits data path is compared with 16 and 21 in order to find out whether the game is over or not. The output functions are as follows:

$set\_broke = score\_gt\_21 \wedge \neg ace11$

$clr\_broke = clr\_stand = set\_ace11 = clr\_score = hitme = adder\_s1 = 0$

$set\_stand = score\_gt\_16 \wedge \neg score\_gt\_21$

$clr\_ace11 = ld\_score = score\_gt\_21 \wedge ace11$

Note that $score\_gt\_21$ implies $score\_gt\_16$. There are three cases to distinguish in this state. In the first case, $score \leq 16$ holds though eventually a (the first) ace has been added with value 11. In this case, the game is neither won or lost for the circuit, the flipflops of the data path remain unchanged and the next state is *get*. In the second case, $16 < score \leq 21$ holds. In this case, the game is over for the dealer. This has the effect that the *stand* flag is set and the next state is also *get*. Finally, in the third case $21 < score$ holds. Then it is checked if an ace has been added with value 11 to the score. If this is the case then 10 is subtracted from the score such that this ace is counted only with 1 and the *ace11* flag is cleared. The next state will then also be *test* since it must then again be checked whether the game is over or not. Note however that the loop from state *test* to *test* can occur at most once since this transition clears the *ace11* flag which must be set for the transition.

$$
\text{Test1} \quad G \left[\begin{array}{l}
test \wedge reset \rightarrow \\
\left(\begin{array}{l}
([(21 \lesssim score) \wedge ace11] \rightarrow Xscore = (score - 10)) \wedge \\
(\neg[(21 \lesssim score) \wedge ace11] \rightarrow Xscore = score) \wedge \\
(Xstand = (16 \lesssim score) \wedge \neg(21 \lesssim score)) \wedge \\
(Xbroke = (21 \lesssim score) \wedge \neg ace11) \wedge \\
(Xace11 = \neg((21 \lesssim score) \wedge ace11) \wedge ace11)
\end{array} \right)
\end{array} \right]
$$

Test2 $G[test \wedge reset \rightarrow X(get \vee test)]$

Test3 $G[test \wedge reset \wedge (21 \lesssim score) \wedge ace11 \rightarrow Xtest]$

Test4 $G[test \wedge reset \wedge \neg((21 \lesssim score) \wedge ace11) \rightarrow Xget]$

Test5 $G[test \wedge reset \rightarrow X(test \rightarrow Xget)]$

Beneath the specifications that have to hold in the specific controller states, there are also global specification that have to hold for all states:

Global1 $(21 \lesssim score) \rightarrow (16 \lesssim score)$

Global2 $get \wedge Xget \wedge score = 0 \wedge \neg stand \wedge \neg broke \wedge \neg ace11 \wedge hitme$

Global3 $G[\neg reset \rightarrow X[get \wedge \neg stand \wedge \neg broke \wedge \neg ace11 \wedge score = 0]]$

Global4 $[GF card\_rdy] \rightarrow [GF get]$

Global5 $[G reset] \wedge [GF(\neg card\_rdy \wedge X card\_rdy)] \rightarrow [GF add]$

Global6 $[G reset] \wedge [GF(\neg card\_rdy \wedge X card\_rdy)] \rightarrow [GF use]$

Global7 $[G reset] \wedge [GF(\neg card\_rdy \wedge X card\_rdy)] \wedge [G(card \neq 0)] \rightarrow [GF(stand \vee broke)]$

The runtimes of all results are given table 6.1 where in each case the data path has been reduced to propositional logic with strategy Θ_B and the remaining specification has been translated to a prefix formula by strategy Φ_{PF} without any application of an invariant rule or other interactions. From the viewpoint of data handling, the example is a very simple one, but not all the specifications can be expressed in CTL (e.g. Global7). Hence, from a viewpoint of temporal reasoning, the example is not as trivial.

6.3 Safe Box

This example is due to William Keisler and is based on the Chinese Ring Puzzle. Assume there is a safe box that has n knobs $k_{n-1}, \ldots, k_0$, such that

Goal	Runtime [sec.]	BDD nodes	Trans. Relation	Reach. States/ Poss. States
Get1	0.83	11335	2207	$77632/2^{30}$
	3.36	36649	14064	$179008/2^{34}$
Get2	0.36	10083	1181	$41792/2^{20}$
Get3	0.66	10197	1761	$68032/2^{27}$
Get4	0.39	10578	1102	$31168/2^{21}$
Add1	2.00	21455	9540	$353920/2^{27}$
Add2	0.37	10561	1113	$37824/2^{21}$
Add3	0.46	11227	1270	$60224/2^{23}$
Add4	0.46	11227	1270	$60224/2^{23}$
Add5	0.37	10949	1126	$30976/2^{18}$
Use1	1.24	22376	7835	$115328/2^{26}$
Use2	0.37	10182	1117	$35264/2^{21}$
Use3	0.36	10949	1126	$30976/2^{18}$
Use4	0.36	10699	1093	$30976/2^{18}$
Test1	1.34	20188	9243	$129792/2^{28}$
	1.36	20210	9245	$129792/2^{28}$
	0.60	12040	1603	$49920/2^{27}$
Test2	0.37	10004	1153	$44480/2^{27}$
Test3	0.52	12615	1314	$49152/2^{23}$
Test4	0.49	12615	1314	$49152/2^{23}$
Test5	0.53	10077	1284	$61312/2^{24}$
Global1	0.35	10949	1126	$30976/2^{18}$
Global2	2.02	24900	10702	$132480/2^{30}$
Global3	0.41	10510	1097	$31104/2^{20}$
Global4	0.52	10907	1158	$930048/2^{24}$
Global5	0.71	10818	1392	$1262850/2^{28}$
Global6	0.69	10524	1437	$1482500/2^{28}$
Global7	9.10	15843	3130	$15231200/2^{33}$

Table 6.1. Runtime data for the verification of the Black Jack Dealer

each knob k_i has only two possible positions: open (0) or close (1). The knobs cannot be turned independently:

1. Initially, all knobs are in their closed position.
2. The leftmost knob k_{n-1} can always be turned.
3. If one doesn't choose to turn k_{n-1}, then the only knob that can be turned is the one directly to the right of the first closed knob from the left.
4. If the last knob k_0 is the only one in the closed position then the above rule does not apply, and the only choice left is to turn k_{n-1}.
5. At each point of time, only one knob can be turned.

The objective is to open the box, i.e. to set all knobs to their open position by a sequence of knob turns. The example can be used as benchmark especially for testing the generation of countermodels, if we state that the safe box can not be opened at all. The verification system must then disprove the statement by finding a knob pressing sequence that opens the safe. We will see that the minimal length $\ell(n)$ of the safe box problem with n knobs will have a length in $O(2^n)$, hence the length of the countermodel will grow rapidly. A knob k_i is pressed at a certain point of time iff $Xk_i = \neg k_i$ holds for that point of time. The following facts can be proved for the safe opener problem:

Lemma 6.1. *The n-knobs safe box problem is given by the following state transition equations:*

$$\left(\bigwedge_{i=0}^{n-1} (k_i = 1) \right) \wedge \left(\bigwedge_{i=0}^{n-2} G(Xk_i = \neg k_i = k_{i+1} \wedge \bigwedge_{j=i+2}^{n-1} \neg k_j) \right)$$

The following facts hold for the corresponding state decision diagram;

1. *Each state has at most two successor states: one of them is reached by pressing the leftmost knob k_{n-1} and the other one is reached by pressing the knob k_i for which $k_{i+1} \wedge \bigwedge_{j=i+2}^{n-1} \neg k_j$ holds if one such knob exists.*
2. *The only two states that have only one successor state are the states $(0,\dots,0)$ and $(0,\dots,0,1)$.*
3. *If a state s_1 is the successor of a state s_0 then s_0 is also a successor of s_1.*
4. *The minimal length of a knob pressing sequence to open the n-knob safe is:*

$$\ell(n) := \begin{cases} \frac{2}{3}(2^n - \frac{1}{2}) & : \text{ if } n \text{ is odd} \\[2mm] \frac{2}{3}(2^n - 1) & : \text{ if } n \text{ is even} \end{cases}$$

Proof. To prove 1. let us check that the second possibility for pressing a knob is uniquely defined if it exists. To see this, assume there are two such knobs k_i and k_p with that property. Without loss of generality assume $i < p$. This leads immediately to a contradiction as the property for knob k_i demands that $\neg k_{p+1}$ holds. To prove 2., note that if in a state only the leftmost knob k_{n-1} can be pressed, this means that for all $i \in \{0,\dots,n-2\}$ the knob k_i can not be pressed which in turn means that $k_{i+1} \wedge \bigwedge_{j=i+2}^{n-1} \neg k_j$ does not hold for all $i \in \{0,\dots,n-2\}$. Hence, all states who have only one successor are solutions of the following formula:

$$\neg k_1 \vee k_2 \vee k_3 \vee \dots \vee k_{n-1}$$
$$\wedge \neg k_2 \vee k_3 \vee \dots \vee k_{n-1}$$
$$\wedge \neg k_3 \vee \dots \vee k_{n-1}$$
$$\vdots$$
$$\wedge \neg k_{n-2} \vee k_{n-1}$$
$$\wedge \neg k_{n-1}$$

It can be easily seen that there are only two solutions: as all rows have to be fulfilled, the last row requires that $\neg k_{n-1}$ has to hold. This implies

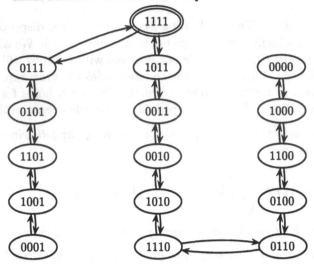

Fig. 6.3. State transition diagram for the 4-knob safe cracking problem

that $\neg k_{n-2}$ must hold, and in turn that $\neg k_{n-3}$ must hold and so on. Hence, the above formula is equivalent to $\bigwedge_{j=1}^{n-1} \neg k_j$, and hence the states which do only have one successor are the states $(0,\ldots,0)$ and $(0,\ldots,0,1)$. Hence, if s is the number of reachable states, then the number of edges is exactly $2(s-1)$. The third proposition is quite simple to prove. Assume s_1 is reached from s_0 by pressing the leftmost knob k_{n-1}. Then s_0 is reached by pressing the leftmost knob k_{n-1} again in state s_1. If on the other hand the second possibility is chosen, then it has to be noticed that the knob that has been pressed can be pressed also in the successor state because the decision whether a knob k_i may be pressed or not only depends on the state of the knobs $k_{i+1},\ldots,k_{n-1}$. Pressing the knob k_i does however not affect the state of the knobs $k_{i+1},\ldots,k_{n-1}$.

The propositions 1-3 show that the state transition diagram consists of two 'chains' of states, one ending in $(0,\ldots,0,1)$ and the other one ending in $(0,\ldots,0)$ (see figure 6.3 for an example). The latter one opens the box and is roughly 2 times longer than the other one. Let $\ell_0(n)$ and $\ell_1(n)$ be the lengths of the chains ending in $(0,\ldots,0)$ and $(0,\ldots,0,1)$, respectively. Then, a detailed analysis shows that

$$
\ell_0(n) := \begin{cases} 2\ell_0(n-1)+1 & : \text{if } n \text{ is odd} \\ 2\ell_0(n-1) & : \text{if } n \text{ is even} \end{cases}
$$

holds. By induction on n it follows that $\ell_0(n) + \ell_0(n-1) = 2^{n-1}$ holds. This in turn implies $\ell_0(n) - \ell_0(n-2) = (\ell_0(n) + \ell_0(n-1)) - (\ell_0(n-1) + \ell_0(n-2)) = (2^n - 1) - (2^{n-1} - 1) = 2^{n-1}$, hence $\ell_0(n) = 2^{n-1} + \ell_0(n-2)$. Now it is easy to see that $\ell_0(2n+1) = \sum_{i=0}^{n} 2^{2i}$ and $\ell_0(2n) = \sum_{i=0}^{n-1} 2^{2i+1}$. The final result follows from the general equation for geometrics sums:

$$\sum_{i=0}^{n} q^i = \frac{q^{n+1} - 1}{q - 1}$$

■

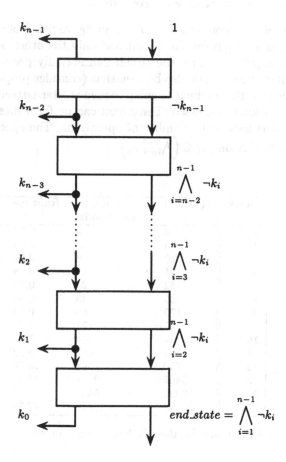

Fig. 6.4. Implementation of the safe opener circuit

Figure 6.3 shows the state transition diagram for the safe box problem of a safe with 4 knobs. The state transition diagram shown in figure 6.3 is typical for all $n \geq 3$: two chains of states are starting in the initial state. One chain 'ends' in the state $(0, \dots, 0)$ and the other one 'ends' in the state $(0, \dots, 0, 1)$. According to the previously proved facts, there are no more than two chains as the only possible end points are the states $(0,\ldots,0)$ and $(0,\ldots,0,1)$. The previous lemma also shows that for each number n there is a knob pressing sequence that opens the safe with n knobs in roughly $\frac{2}{3}2^n$ steps.

The transition equations given in the previous lemma can be implemented as a synchronous circuit that consists of n cells for the n-knob problem. Each cell has inputs l_node, l_conj_in and the outputs k_i and l_conj_out, which are defined as

- $k_i = 1 \land G[Xk_i = \neg k_i = l\_node \land l\_conj\_in]$
- $l\_conj\_out = \neg l\_node \land l\_conj\_in$

The cells are connected as shown in figure 6.4. Note that the leftmost knob k_{n-1} is an input for the circuit and only the other positions of the knobs k_i are computed. Hence, the circuit can actually press two knobs at a certain point of time. It can also be seen that (consider proposition 3 of the previous lemma) in the minimal opening sequence alternatively the leftmost knob and another knob is pressed. The circuit can perform these two steps at once and can thus halves the number of operations. The specification which is to be disproved is simply $G\left(\bigwedge_{i=0}^{n-1} k_i\right)$.

knobs	$\ell(knobs)$	states in countermodel	Runtime [sec]	BDD nodes	Trans-Rel
2	2	3	0.03	259	7
3	5	3	0.05	360	14
4	10	8	0.08	864	22
5	21	11	0.08	1395	32
6	42	24	0.20	2895	42
7	85	43	0.36	5426	52
8	170	88	0.72	10006	62
9	341	171	1.72	10004	72
10	682	344	3.97	13621	82
11	1365	683	8.98	26622	92
12	2730	1368	18.98	52762	102
13	5461	2731	44.28	104670	112
14	10922	5464	104.1	208650	122
15	21845	10923	277.36	416245	132

Table 6.2. Results for the safe box opening problem

The runtimes and the lengths of the countermodel we obtained are given in table 6.2. As can be seen, the lengths of the countermodels are almost optimal, i.e. half of $\ell(knobs)$ (as the circuit can perform these two steps at once). However, we have to admit that this is due to the very good implementation of SMV that is used as backend model checker of our LTL model checker. The runtimes and the storage requirements, however seem to grow exponentially such that about 39.6 MBytes of main memory are required to find the countermodel for the 15 knobs problem. We have only investigated for the automatic verification of the safe box opening problem. A machine checked proof of lemma 6.1 has not been done with C©S, but should be possible. However, we assume that it would require a considerable amount of user interactions.

6.4 The Island Tunnel Controller

This controller problem is an extension of the *traffic light control* problem and, as such it is a metaphor for coordinating access to a restricted resource [FiJo95]. Concurrent entry/exit is allowed under certain conditions, represented by a one-lane tunnel between the mainland and the island (see figure 6.5) that may at any time contain several vehicles, and an additional constraint on the number of vehicles that can be on the island. The circuit has the task to control the traffic lights at the each end of this tunnel. Of course, the specification requires that both traffic lights should never have a green light at the same time. Moreover, the system is always alive, i.e. if a car wants to enter the tunnel, it may do so after a finite amount of time. Furthermore, the traffic on the island is limited: at no point of time there should be more than 16 cars on the island.

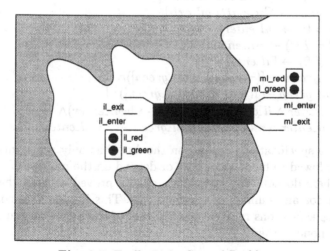

Fig. 6.5. Traffic Light Control Problem

There are four sensors for detecting the presence of vehicles: one at the tunnel entrance on the island side *il_enter*, one at the tunnel exit on the island side *il_exit*, one at the tunnel entrance on the mainland side *ml_enter*, and one at the tunnel exit on the mainland side *ml_exit*. The controller has two counters IC and TC, the latter for counting cars inside the tunnel, the other one for counting cars currently on the island or inside the tunnel traveling to the island. The design assumes the following assumptions on its environment, described in terms of the behavior of vehicles that entering the system:

A1: Cars are not produced in the tunnel, i.e. if there is no car in the tunnel, then no car can exit the tunnel (on none of its sides) at this point of time.

A2: Cars do not disappear in the tunnel.

A3: Cars are not produced in the island.

A4: The island counter counts the cars which are on the island or which are currently inside the tunnel traveling to the island. If all cars on the island are currently inside the tunnel then no car can enter the tunnel at the island side.

A5: Each car wants to leave the island after some time.

A6: If the island traffic light is green, then cars can only exit the tunnel at the mainland side until the mainland traffic light turns to green (analogously for the mainland).

A7: If a car wants to enter a side of the tunnel where the traffic light is green, it will actually enter the tunnel after some time.

The above assumptions can be expressed as follows in temporal logic:

A1: $G[(TC = 0) \rightarrow \neg il\_exit \wedge \neg ml\_exit]$

A2: $G[(TC \neq 0) \rightarrow F[il\_exit \vee ml\_exit]]$

A3: $G[(IC = 0) \rightarrow \neg il\_enter]$

A4: $G[(IC = TC) \rightarrow \neg il\_enter]$

A5: $G[(IC \neq 0) \rightarrow F il\_enter]$

A6: $G \left[\begin{array}{l} (il\_green \rightarrow [(\neg il\_exit)\ U\ ml\_green]) \wedge \\ (ml\_green \rightarrow [(\neg ml\_exit)\ U\ il\_green]) \end{array} \right]$

A7: $G \left[\begin{array}{l} (il\_enter \wedge il\_green \wedge X il\_green \rightarrow F \neg il\_enter) \wedge \\ (ml\_enter \wedge ml\_green \wedge X ml\_green \rightarrow F \neg ml\_enter) \end{array} \right]$

Some of the specifications depend on the fact that only a maximal number of cars is allowed to be on the island or depend on the number of cars in the tunnel, others do not. The latter ones can be proved without the counters and hence, for any number of maximal cars. This implies that some of the following specifications require only a subset of the above assumptions, and some even none of them.

S1: At no point of time, both lights are green.

S2: At no point of time, there are more than 16 cars on the island.

S3: If a car wants to enter the tunnel and persists to enter the tunnel, then it has the chance to do so after a finite time.

S4: Traffic lights can change only when the tunnel is empty.

The formalizations of the above specifications are as follows:

S1: $G \neg [ml\_green \wedge il\_green]$

S2: $G[IC \leq 16]$

S3: $G[\neg G[il\_enter \wedge il\_red] \wedge \neg G[ml\_enter \wedge ml\_red]]$

S4: $G \begin{bmatrix} (ml\_red \wedge Xml\_green \rightarrow (TC = 0)) \wedge \\ (il\_red \wedge Xil\_green \rightarrow (TC = 0)) \end{bmatrix}$

The specification S3 can also be expressed equivalently as follows: If a car wants to enter a side of the tunnel where the light is red and the car waits until the light changes, then the light will actually change after some time. Hence, the formula $G[\neg G[enter \wedge red]]$ is equivalent to

$$G[enter \wedge red \wedge [(enter \wedge red) \ U \ (enter \wedge green)] \rightarrow F(enter \wedge green)]$$

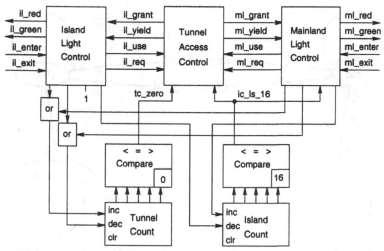

Fig. 6.6. Three-process architecture of the Island Tunnel Control problem.

An implementation of the tunnel controller was given in [FiJo95] and involves three processes implemented by three subcontrollers (figure 6.6) and two counters for counting the number of cars presently inside the tunnel (TC) and the number of cars presently on the island (IC). As subcontrollers there are two side controllers and a tunnel access controller. The latter is an arbiter which grants access to the tunnel either for the mainland or for the island controller. The side controllers behave equivalent, their inputs and outputs have the following meaning (each signal is prefixed later on by either $il\_$ or $ml\_$):

- *green* and *red* are the outputs for generating green and red lights for the particular side.
- *use* indicates that this side is currently allowed to use the tunnel.

- *req* indicates that the controller has currently no access to the tunnel but wants to access it.
- *yield* indicates that the controller is being instructed to release control of the tunnel
- *grant* indicates that the island has been granted control of the tunnel.
- *tc_inc* indicates that tunnel controller has to be incremented.
- *tc_dec* indicates that tunnel controller has to be decremented.
- *ic_change* indicates that island controller has to be modified (if the signal comes from the island side it has to be decremented, otherwise it has to be incremented).

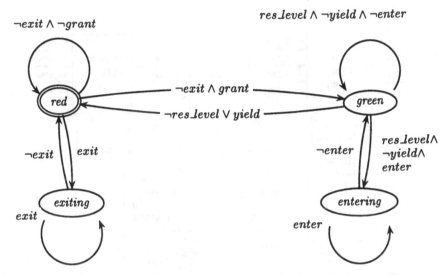

Fig. 6.7. State transition diagram for the island/mainland controller

The behaviour of the side controller is given in figure 6.7, where the reset behaviour is neglected for reasons of simplicity. The output equations of the side controllers are as follows:

- $green\_light := green \lor entering$
- $red\_light := red \lor exiting$
- $use := green \lor entering$
- $req := red \land enter$
- $tc\_inc := green \land \neg yield \land res\_level \land enter$
- $tc\_dec := red \land exit$
- $ic\_change := green \land \neg yield \land res\_level \land enter$

Initially, the side controllers are in state *red* and do not have access to the tunnel, both counters are zero, and there are no vehicles neither on the island nor in the tunnel. Cars are not allowed to enter the tunnel in this state as a red light is produced, but cars may exit the tunnel. Each time a car exits the tunnel at this side, this is detected via *exit* and the tunnel counter is decremented. While the car is exiting the tunnel, the controller is in the exiting state where also a red light is produced. After the car has exited the tunnel, the controller will be again in the *red* state. If a car wants to enter the tunnel in this state, the signal *enter* will indicate this and the controller signals via *req* to the tunnel access controller that it wants to access the tunnel. If the other side controller has the control over the tunnel, the tunnel access controller will instruct the other side controller to release it and will then wait until no more cars are inside the tunnel. After that, the tunnel access controller will allow access to the tunnel and the side controller switches to its *green* state. In this state, a green light is produced and cars may enter the tunnel. Each time the car enters the tunnel, the tunnel counter is incremented and the island counter is modified, i.e. incremented if we are on the mainland and decremented otherwise. However, if the side controller is on the mainland side, it has furthermore to check if more than 16 cars are on the island. This is checked via the input *res_level* which is always high on the island side controller and equals to $IC < 16$ on the mainland side. If the limit of 16 cars is reached, the controller switches automatically to its red state.

The behaviour of the tunnel access controller is given in figure 6.8. The initial state is *dispatch*, where no side controller has the control over the tunnel. Again, for reasons of simplicity, the reset behaviour is suppressed in figure 6.8, but it has to be noted that the controller always switches to state *dispatch* when *reset* becomes high. The output equations of the tunnel access controller are as follows:

$$- il\_grant := (dispatch \land il\_req \land \neg ml\_use \land (TC = 0))$$
$$\lor (mlclear \land (TC = 0))$$
$$- ml\_grant := (dispatch \land \neg il\_req \land ml\_req \land (IC < 16)\land$$
$$\neg il\_use \land (TC = 0))$$
$$\lor (ilclear \land (TC = 0))$$
$$- il\_yield := iluse$$
$$- ml\_yield := mluse$$

The states have the following meaning: the controller is in state *mlclear* when the island controller wants to access the tunnel and the mainland controller does not use the tunnel, but there are still cars inside the tunnel (from a previous use by the mainland controller). *mluse* is the state, where the mainland controller has control over the tunnel, but the island controller wants to access it. The states *ilclear* and *iluse* are used analogously.

The island controller has a higher priority than the mainland controller. If neither the island nor the mainland controller have control over the tunnel, and both request it at the same point of time, then there are two cases: if there are no cars in the tunnel, the island controller is granted immediately access to the tunnel. Otherwise, the tunnel access controller switches to state

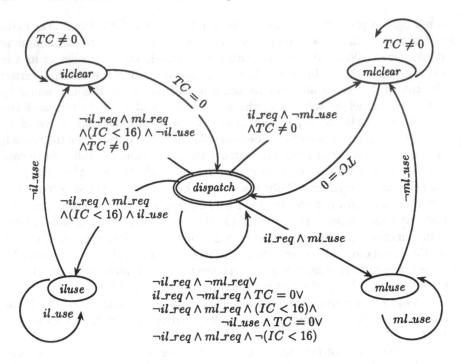

Fig. 6.8. State transition diagram for the tunnel access controller

mlclear where it remains until all cars have left the tunnel. Then there are no more cars in the tunnel and the island controller is granted access to it.

If, on the other hand, the island controller requests the control over the tunnel and the tunnel is currently used by the mainland controller, then the tunnel access controller switches to state *mluse* and sends the *yield* signal to the mainland controller, i.e. instructs it to release the control over the tunnel. The mainland controller will do so either immediately or, if currently a car is entering at the mainland side, after this car has entered. The mainland controller signals the release of the tunnel by the signal *ml_use*, which forces the tunnel access controller to switch to state *mlclear*, where it remains until all cars have left the tunnel. Then is switches again to state *dispatch* and grants access of the tunnel to the island controller. The island controller switches then to state *green* and outputs via *il_use* that it has now taken control over the tunnel.

Now, assume the mainland controller wants to access the tunnel, but the island controller does not. Then some more cases are considered:

1. If the island counter equals to 16, then the access is not granted.
2. If the island counter is less than 16 and the tunnel is currently not used by the island controller, then the tunnel access controller switches to state

ilclear to empty the tunnel and grants access to the mainland controller after the tunnel is emptied.

3. If the island counter is less than 16 and the tunnel is currently used by the island controller (though it does not request it at the moment), then the tunnel access controller switches to state *iluse*, sends the *yield* signal to the island controller. After the island controller has released the tunnel, the tunnel is emptied and the access is granted to the mainland controller.

The island tunnel controller can be implemented for arbitrary numbers of cars that may be on the island, hence, it is a generic problem where the number of cars is a parameter. In [FiJo95], the problem has been presented for 16 cars as in this chapter. Table 6.3 lists our verification results for some other numbers of cars. A generic verification of the problem has not been done, but we suppose that it should be possible without too much interactions.

6.5 Arbiter

The task of an arbitration circuit is to manage requests of components for a shared resource like a bus. The arbitration circuit obtains all requests and give at each point of time at most one component a grant for the common resource. The circuit that is considered in this section is due to David Dill and can also be found in [McMi93a]. The circuit can be implemented for an arbitrary number of components and has hence n inputs $req_0, \ldots, req_{n-1}$ where each component can request a grant for the shared resource. The circuit has also n outputs $ack_0, \ldots, ack_{n-1}$ for allowing access to one of the components. The implementation of the arbitration circuit of this section assumes that each access to the component does not take longer than one point of time. Hence, at each point of time a new component may access the resource. Furthermore, it is assumed that the component holds its access request until it either obtains the access or it does not need it any more. The implementation of the arbitration circuit is given in figure 6.9.

The implementation for n components consists of n cells, which are all implemented in the same manner, except for the first one. This is due to the fact, that the arbiter considers priorities as well as a round-robin scheme: there is a token which circulates through the cells of the arbiter. Hence, the initialisation of the cells requires that initially one cell obtains the token and the others do not have a token at that point of time. Beneath the token flag, each cell has another flag (the persistence flag) that is set when the corresponding component wants to access the resource, but does not achieve it though its cell has the token. This gives the cell dynamically a higher priority and it can definitely access the resource when the token arrives again in that cell. If the cell with the token has however not set the persistence flag, the access is computed by a static priority: the component with the lowest index will then receive the access.

spec	cars	Runtime [seconds]	BDD nodes	Trans- Relation	Reachable States
S1	2	1.30	13080	2420	$13472/2^{22}$
S2	2	2.51	15880	2600	$24224/2^{23}$
S3_il	2	2.85	16728	2836	$33376/2^{25}$
S3_ml	2	2.54	15532	2630	$35936/2^{25}$
S1	3	1.35	13838	2227	$15872/2^{22}$
S2	3	1.67	13141	2235	$15872/2^{23}$
S3_il	3	3.32	14925	2821	$40928/2^{25}$
S3_ml	3	2.79	16053	2470	$43040/2^{25}$
S1	4	3.57	18720	3362	$45216/2^{24}$
S2	4	6.84	19266	3576	$83104/2^{25}$
S3_il	4	8.21	17184	4154	$112736/2^{27}$
S3_ml	4	7.54	18060	3636	$121696/2^{27}$
S1	5	3.82	21547	3414	$52384/2^{24}$
S2	5	7.35	17479	3634	$98464/2^{25}$
S3_il	5	8.84	18154	4292	$132384/2^{27}$
S3_ml	5	8.13	19563	3704	$140640/2^{27}$
S1	6	3.94	19212	3315	$59552/2^{24}$
S2	6	7.75	19657	3582	$113824/2^{25}$
S3_il	6	9.68	18212	4301	$152160/2^{27}$
S3_ml	6	8.71	19865	3647	$159584/2^{27}$
S1	7	4.07	19631	2934	$64512/2^{24}$
S2	7	4.71	17965	2942	$64512/2^{25}$
S3_il	7	10.52	19149	4108	$167648/2^{27}$
S3_ml	7	9.16	20998	3285	$174112/2^{27}$
S1	16	28.54	30921	5300	$622752/2^{28}$
S2	16	62.90	49141	5568	$1.167520/2^{29}$
S3_il	16	73.21	41758	8092	$1.567330/2^{31}$
S3_ml	16	80.48	47646	5726	$1.689950/2^{31}$

Table 6.3. Results for the island traffic controller

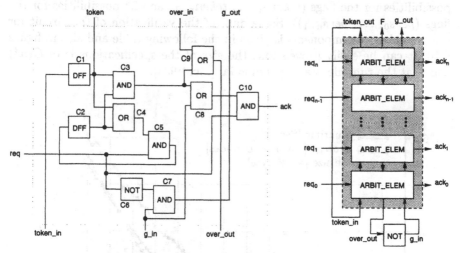

Fig. 6.9. Implementation of the arbiter

In general, an arbitration circuit must fulfill the following requirements:

– At each point of time, at most one component may access the common resource:

$$\forall t. \bigwedge_{k=0}^{n-1} \left(ack_k^{(t)} \to \bigwedge_{j=0,j\neq k}^{n-1} \neg ack_j^{(t)} \right)$$

– Only components have access to the resource, which request for it:

$$\forall t. \bigwedge_{k=0}^{n-1} \left(ack_k^{(t)} \to req_k^{(t)} \right)$$

– The arbitration is fair, i.e. after some time each component obtains access to the resource:

$$\bigwedge_{k=0}^{n-1} \neg \left(\forall t.req_k^{(t)} \wedge \neg ack_k^{(t)} \right)$$

– If none of the components wants to access the resource at a certain point of time, then at the next point of time, no persistence flag will be set and the access is given by the following static priorities:

$$\forall t. \left(\bigwedge_{k=0}^{n-1} \neg req_k^{(t)} \right) \to \left(\bigwedge_{k=0}^{n-1} ack_k^{(t+1)} = req_k^{(t+1)} \wedge \bigwedge_{j=0}^{k-1} \neg req_j^{(t+1)} \right)$$

As each cell has two flags, the circuit for n components has at most 2^{2n} reachable states. However, as always exactly one cell has the token, it can be easily seen that the circuit has only $n2^n$ reachable states (there are n

possibilities for the flags $(token_0, \ldots, token_{n-1})$ and 2^n possibilities for the flags $(pers_0, \ldots, pers_{n-1})$). Some data of the verification of the circuit for some number of components is given in the following table and also in figure 6.10. It can be easily proved that the size of the specifications is of $O(n^2)$ and that the runtime is also polynomial [McMi93a].

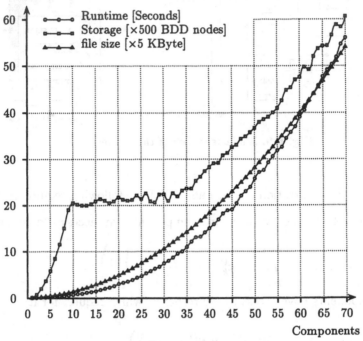

Fig. 6.10. Runtimes for the verification of the arbitration circuit

The automatic verification is in this case completely sufficient. Nevertheless, let us consider also a verification of the generic n-component case. The problem is here that the arbiter itself is not given as a recursive module, but as a generic one that is based on a recursive module. For this reason, we have to define a new subcircuit called ARBIT_CHAIN (gray shaded in figure 6.9), that can be recursively defined. ARBIT_CHAIN has the following specification:

$$\left(\begin{array}{l} \left(\forall t.g_{out}^{(t)} = g_{in}^{(t)} \wedge \bigwedge_{k=0}^{n-1} \neg req_k^{(t)}\right) \wedge \\[2ex] \left(\forall t.over_{out}^{(t)} = over_{in}^{(t)} \vee \bigvee_{k=0}^{n-1} \left[pers_k^{(t)} \wedge token_k^{(t)}\right]\right) \wedge \\[2ex] \left(\forall t. \left[\bigwedge_{k=0}^{n-1} pers_k^{(0)} = 0\right] \wedge \left[\bigwedge_{k=0}^{n-1} pers_k^{(t+1)} = req_k^{(t)} \wedge (pers_k^{(t)} \vee token_k^{(t)})\right]\right) \wedge \\[2ex] \left(\forall t. \left[\bigwedge_{k=1}^{n-1} token_k^{(0)} = 0\right] \wedge \left[\bigwedge_{k=1}^{n-1} token_k^{(t+1)} = token_{k-1}^{(t)}\right]\right) \wedge \\[2ex] \left(\forall t. \left[token_0^{(0)} = 1\right] \wedge \left[token_0^{(t+1)} = token_{n-1}^{(t)}\right]\right) \wedge \\[2ex] \left(\forall t. \bigwedge_{k=0}^{n-1} ack_k^{(t)} = \left[req_k^{(t)} \wedge pers_k^{(t)} \wedge token_k^{(t)}\right] \vee \left[req_k^{(t)} \wedge g_{in}^{(t)} \wedge \bigwedge_{j=0}^{k-1} \neg req_j^{(t)}\right]\right) \end{array}\right)$$

ARBIT_CHAIN of length $n+1$ can be defined by simply adding an arbitration cell to the chain of length n. The correctness proof of **ARBIT_CHAIN** can be done by a simple induction proof on the number of components. The resulting proof is the easily obtained by rewriting with the additional equations $token_{in} = token_{out}$ and $g_{in} = over_{out}$.

6.6 Von Neumann Adder

The sequential adder of this circuit goes back to John v. Neumann. A simple implementation of the algorithm is shown in figure 6.12. The behavior of the circuit is as follows: Initially, the circuit is ready for new computations. This is indicated by the output rdy. If the circuit is ready at any point of time, and a new computation is requested at req, then the inputs A and B are read into the registers R_A and R_B and the circuit switches into computation mode. In this mode, the circuit performs iterations and ignores further requests. In the computation mode, the output rdy is false. If the computation terminates, then the circuit leaves the computation mode $rdy = 1$ and stores the results until a new computation is requested. Note that the number of required iteration steps depends on the given numbers and that an upper bound is the length n of the given bitvectors. The implementation of the controller is given in Figure 6.11.

The correctness of the above algorithm is based on the following theorem, which guarantees that the sum of R_A and R_B remains unchanged during the iterations:

$$\left(\sum_{i=0}^{n} a_i 2^i\right) + \left(\sum_{i=0}^{n} b_i 2^i\right) = 2\left(\sum_{i=0}^{n}(a_i \wedge b_i)2^i\right) + \left(\sum_{i=0}^{n}(a_i \oplus b_i)2^i\right)$$

A specification of the circuit, whose verification is outlined in detail in the following, is therefore the following: $rdy \wedge req \rightarrow [(C = (A \vec{+} B))\ \mathbf{W}\ rdy]$. The

```
MODULE OneLoopCntrl(req, rdy1)
    C₁ : Not(rdy);
    C₂ : Or(req, C₁.o);
    C₃ : Dflipflop(C₂.o);
    C₄ : And(req, rdy);
    C₅ : Or(C₃.o2, rdy1);
OUTPUT
    rdy := C₅.o;
    ini := C₄.o;
    load := C₂.o;
END;
```

Fig. 6.11. Implementation of the Controller

verification of this specification can be done by various strategies, however, as many bitvector operations are involved, we suggest to verify at the bitvector level. Figure 6.13 shows the runtimes (in seconds) and figure 6.14 shows the storage requirements (in BDD nodes) on a Sun Sparc 10 for some strategies.

The simplest strategy that can be used is $\Phi_{PF} \circ \Theta_B$, i.e. translating the data equations into propositional logic and the temporal abstraction into a prefix formula. Finally, symbolic model checking is used as a decision procedure. It is well-known that the size of the BDDs is important for the efficiency of the verification and depends crucially on the variable ordering. For the verification of the adder, the following ordering has been used: $req \prec q \prec a_0 \prec b_0 \prec a_1 \prec b_1 \ldots a_{n-1} \prec b_{n-1}$.

If the number of iteration steps did not depend on the data, it is possible to replace the sequential circuit by a sequence of operation units, such that a combinational circuit is obtained which can then be verified by simple tautology checking. The number of iteration steps of the von Neumann adder however depends on the inputs. Nevertheless, this 'unrolling' can also be used for the verification of the adder, since an upper bound for the number of iterations is known and the values of the registers do not change if further iterations are performed after termination. The required results are given in figure 6.13 and 6.14.

In order to enhance the efficiency of the verification, invariants can be used to use the basic design ideas of the circuit. In this case, it is important to observe that the sum of the registers remains unchanged during the iteration, i.e. the formula $J_1 := q \wedge (A \vec{+} B) = (R_A \vec{+} (R_c \triangleright R_B))$ is an invariant of the loop $((R_c \triangleright R_B)$ means the bitvector which is obtained by concatenating R_c to the left hand side of R_B, and $(A \vec{+} B)$ means a bitvector addition). This invariant allows the elimination of all temporal operators. For reasons of readability, define the values R_A°, R_B° and R_c° of R_A, R_B and R_c at the next point of time by $R_A^\circ := (R_A \oplus R_B)$, $R_B^\circ := \mathsf{LSH}(R_A \barwedge R_B)$ and $R_c^\circ := R_c \vee \mathsf{HD}((R_A \barwedge R_B))$. Then the following subgoal remains to be proved:

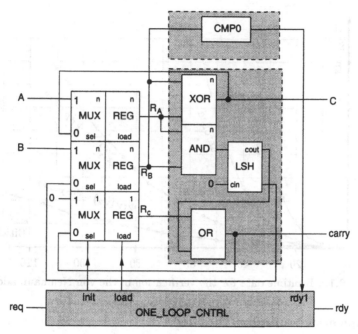

Fig. 6.12. Implementation of a sequential adder

$$(A \vec{+} B) = (R_A \vec{+} (R_c \triangleright R_B)) \rightarrow (A \vec{+} B) = (R_A^\circ \vec{+} (R_c^\circ \triangleright R_B^\circ))$$

Using then strategy Θ_B, the above goal can be translated into pure propositional logic and can be solved afterwards by simple tautology checking.

The number of variables can be reduced by using the more detailed invariant $J_2 := q \wedge ((A \vec{+} B) \sim_{\mathbb{N}} (R_A \vec{+} (R_c \triangleright R_B))) \wedge (R_c \rightarrow \Phi_{\mathbb{N}}((R_A \vec{+} R_B)) < 2^n)$ instead of J_1 (again, $(R_c \triangleright R_B)$ means the bitvector which is obtained by concatenating R_c to the left hand side of R_B, $(A \vec{+} B)$ means a bitvector addition, and $(A \sim_{\mathbb{N}} B)$ means that the bitvectors A and B represent the same number, i.e. differ only on leading zeros). J_2 contains explicitly the knowledge that there is at least one carry, i.e. if $R_c = 1$ holds, then the remaining sum of R_A and R_B has only n bits, otherwise the sum may have $n+1$ bits. Figure 6.13 and 6.14 show the detailed results of both invariant approaches. In this example it can be seen that the more information is given to the system, the more efficiently the circuit can be verified. However, this observation is not always valid as shown in the next section.

A generic proof has also been done for the circuit that makes use of the listed invariants. It is obvious that all used operators can be defined recursively on the length of the bitvectors. The correctness proof is then simply done by induction on the length of the arguments.

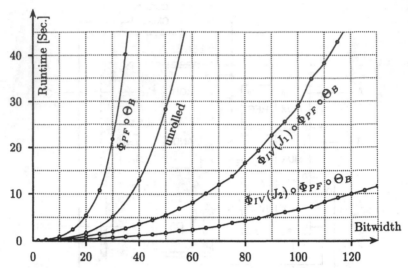

Fig. 6.13. Runtime data for the verification of the von Neumann adder

6.7 Sequential Multiplier Circuits

The circuit given in figure 6.15 implements a Russian multiplier with the same controller for the loop as in the previous section. The previously presented strategies can also be used for the verification of this circuit, the results are given in table 6.4 (runtime again in seconds; storage in number of BDD nodes).

Bits	direct		Invariant		unrolled	
	Time	Storage	Time	Storage	Time	Storage
2	0.22	3254	0.11	567	0.17	56
3	0.38	9725	0.26	3503	0.15	361
4	0.99	10923	0.74	10303	0.31	1820
5	3.32	31024	3.42	47990	0.53	7160
6	15.63	116755	20.13	221678	1.20	11049
7	107.94	467636	214.38	989823	3.50	39379
8	1203.04	1901547	3416.04	4256226	12.78	143568
9	—	—	—	—	62.86	449337
10	—	—	—	—	435.54	1518158
11	—	—	—	—	3468.37	4212190

Table 6.4. Results for the Russian multiplier

The first column shows the results of strategy $\Phi_{PF} \circ \Theta_B$. The corresponding prefix formula of the n-bit circuit has exactly $(n + 2)2^{2n} + 2^n$ reachable states and can not be verified for more than 8 bits (this required 31 MByte).

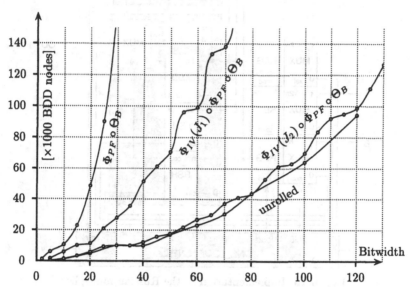

Fig. 6.14. Memory requirement for the verification of the von Neumann adder

Using the invariant $q \wedge ((A \vec{\times} B) \sim_{\mathbb{N}} (R_C \vec{+} \text{LASTN}_{2n} ((R_A \vec{\times} R_B))))$, it is also only possible to verify at least 8 bits. Unroling the circuit allows even to verify the 11 bit circuit, but more than 11 bits can not be verified with this approach. The strategies which were quite successful for the adder are not applicable for the verification of this circuit. In contrast to the adder, the complexity of this verification problem stems from the data equations and only minor complexity stems from the control part. It is however well-known that propositional formulae for multiplication have no variable ordering which gives the corresponding BDDs a non-exponential size. Hence, strategy Θ_B is not the best choice for this circuit. Instead, the n-bit circuit has to be verified by induction for handling the data equations and invariants for handling the control part. This strategy required a lot of user interaction and cannot be explained here in detail. For example. the following lemmas had to be proved:

$$(y \text{ MOD } 2) = 0 \vdash (2 \times x) \times ((y \text{ DIV } 2)) = x \times y$$
$$(y \text{ MOD } 2) = 1 \vdash x + (2 \times x) \times ((y \text{ DIV } 2)) = x \times y$$

The Russian multiplier given in figure 6.15 is usually not used, since it contains more flipflops than necessary. A more efficient (in terms of chip size) sequential multiplier is given in figure 6.16. However, this multiplier is slower, since it requires always n iteration steps in contrast to the multiplier of figure 6.15 that required at most n iteration steps. The savings of chip area lead

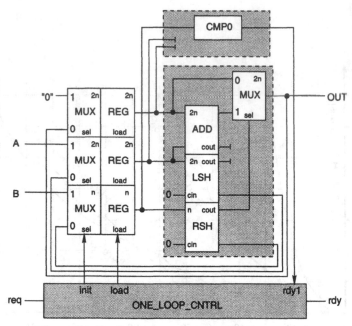

Fig. 6.15. Implementation of the Russian multiplier

however not to any savings for the complexity of the verification. The results for the multiplier of figure 6.16 are given in table 6.5.

6.8 Systolic Arrays

6.8.1 A Simple Version. This circuit has the task to compute the scalar product of a sequence of inputs in a sequential manner, i.e. the components of the vectors are given sequentially. More formally, the circuit has three m-bit inputs $\mathcal{A}_{in}$, $\mathcal{B}_{in}$ and $\mathcal{C}_{in}$ and one m-bit output $\mathcal{C}_{out}$. The specification of the circuit is as follows:

$$\forall t.\mathcal{C}_{out}^{t+n} = \mathcal{C}_{in}^t + \sum_{i=1}^n \mathcal{A}_{in}^t \times \mathcal{B}_{in}^t$$

The circuit is parameterized in two ways: first there is a parameter n which is the length of the input vectors and second there is a parameter m which is the bitwidth of the components of the vectors. The implementation of the circuit is given by a sequence of base cells, i.e. a one dimensional systolic array (figure 6.17). The inputs $\mathcal{A}_j$ and $\mathcal{B}_j$ of cell j are just passed to the outputs $\mathcal{A}_{j+1}$ and $\mathcal{B}_{j+1}$ of the cell, while the output $\mathcal{C}_{n-j}$ of cell j is the sum of the input $\mathcal{C}_{n-j}$ and the product of $\mathcal{A}_j$ and $\mathcal{B}_j$ of the previous point of time.

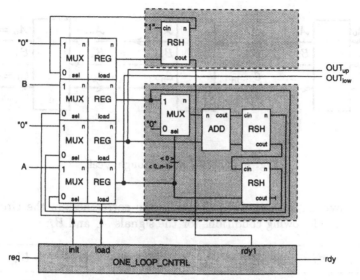

Fig. 6.16. Implementation of an ADD/SHIFT multiplier

Bits	direct		unrolled	
	Time	Storage	Time	Storage
2	0.24	2805	0.10	41
3	0.39	7513	0.13	268
4	1.04	10429	0.27	1553
5	4.02	23286	0.39	6669
6	22.87	87532	1.02	11587
7	138.09	357527	3.13	39768
8	1599.76	1484104	11.94	153791
9	—	—	64.56	430916
10	—	—	469.30	1567273
11	—	—	4546.16	4812990

Table 6.5. Results for the ADD/SHIFT multiplier

The behavior of a single cell determines the relations between the signals and we obtain the following description of the signal flow for $j \in \{0, \ldots, n-1\}$:

$$
\begin{array}{lll}
(1)\ \mathcal{A}_{j+1}^{t+1} := \mathcal{A}_j^t & (4)\ \mathcal{B}_{j+1}^{t+1} := \mathcal{B}_j^t & (7)\ \mathcal{C}_{j+1}^{t+1} := \mathcal{C}_j^t + \mathcal{A}_{n-(j+1)}^t \times \mathcal{B}_{n-(j+1)}^t \\
(2)\ \mathcal{A}_{j+1}^0 := 0 & (5)\ \mathcal{B}_{j+1}^0 := 0 & (8)\ \mathcal{C}_{j+1}^0 := 0 \\
(3)\ \mathcal{A}_0^t := \mathcal{A}_{in}^t & (6)\ \mathcal{B}_0^t := \mathcal{B}_{in}^t & (9)\ \mathcal{C}_0^t := \mathcal{C}_{in}^t
\end{array}
$$

Equation (7) is not given immediately from the implementation. From figure 6.17, it can be derived that $\mathcal{C}_{n-j+1}^{t+1} := \mathcal{C}_{n-j}^t + \mathcal{A}_j^t \times \mathcal{B}_j^t$ holds for $j \in \{1, \ldots, n\}$. If we substitute $j := n - i$, we obtain: $\mathcal{C}_{n-(n-i)+1}^{t+1} := \mathcal{C}_{n-(n-i)}^t + \mathcal{A}_{n-i}^t \times \mathcal{B}_{n-i}^t$

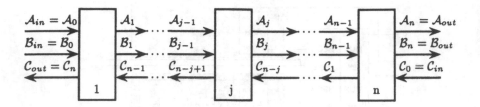

Fig. 6.17. Data flow through a cell

which is obviously equation (7). To show the correctness of the circuit, we first prove the following conditions for the signals A_j and B_j:

$$(10)\ \forall j \le n.\forall t.A_j^{t+j} = A_{in}^t \qquad (11)\ \forall j \le n.\forall t.B_j^{t+j} = B_{in}^t$$

Both propositions can be proved via induction on j. In the base case, we have $\forall t.A_0^{t+0} = A_{in}^t$, which does hold according to equation (3). In the induction step, we have $A_{j+1}^{t+j+1} \stackrel{(1)}{=} A_j^{t+j} \stackrel{(10)}{=} A_{in}^t$. The proof of (11) is done analogously. Now, we can prove

$$(12)\ \forall j \le n.\forall t.C_j^{t+j} = C_{in}^t + \sum_{i=1}^{j} A_{n-i}^{t+i-1} \times B_{n-i}^{t+i-1}$$

Again, we perform induction on j: the base case is trivial, since we have $\forall t.C_0^{t+0} = C_{in}^t + \sum_{i=1}^{0} \ldots = C_{in}^t$ which holds by equation (9). The induction step is proved as follows:

$$C_{j+1}^{t+j+1} \stackrel{(7)}{=} C_j^{t+j} + A_{n-(j+1)}^{t+j} \times B_{n-(j+1)}^{t+j}$$
$$\stackrel{(12)}{=} C_{in}^t + \left(\sum_{i=0}^{j} A_{n-i}^{t+i-1} \times B_{n-i}^{t+i-1} \right) + A_{n-(j+1)}^{t+j} \times B_{n-(j+1)}^{t+j}$$
$$= C_{in}^t + \sum_{i=0}^{j+1} A_{n-i}^{t+i-1} \times B_{n-i}^{t+i-1}$$

Our final aim is to prove the flow of the output C_{out}. The combination of the equations (10), (11) and (12) yields in:

$$(13)\ \forall t.C_n^{t+2n} = C_{in}^{t+n} + \sum_{i=1}^{n} A_{in}^{t+2i-1} \times B_{in}^{t+2i-1}$$

In order to do the proofs by RRL, the equations (1-9) are introduced by the translation of the circuit structure into equation systems. Furthermore, the following equations are added by the translation of the abstract data type definition with its recursive definition of the operators below):

$$
\begin{array}{ll}
\text{(14)} \; 0 \leq S(x) & \text{(18)} \; \mathcal{M}_i^t := \mathcal{A}_i^t \times \mathcal{B}_i^t \\
\text{(15)} \; S(x) \leq S(y) := x \leq y & \text{(19)} \; \mathcal{S}_{0,n}^t := 0 \\
\text{(16)} \; x - 0 := x & \text{(20)} \; \mathcal{S}_{j+1,n}^t := \mathcal{S}_{j,n}^t + \mathcal{M}_{n-(j+1)}^{t+1} \\
\text{(17)} \; S(x) - S(y) := x - y &
\end{array}
$$

Equations (14) and (15) are the axioms for $\leq$, (16) and (17) are the axioms for $-$. Equation (18) is not necessary, but leads to shorter terms in that the product of $\mathcal{A}_i^t$ and $\mathcal{B}_i^t$ is abbreviated by $\mathcal{M}_i^t$. Finally, equations (19) and (20) define the scalar product of $\mathcal{A}_i^t$ and $\mathcal{B}_i^t$. Note that the scalar product is normally a function in the input vector $\mathcal{A}_i^t$ and $\mathcal{B}_i^t$. Such a view on the scalar product makes it however a higher-order function that can not be handled by first order term rewriting. Hence, we define the scalar product only for our given vectors. Obviously, we have $\mathcal{S}_{j,n}^t = \sum_{i=1}^{j} \mathcal{M}_{n-i}^{t+i-1} = \sum_{i=1}^{n} \mathcal{A}_{n-i}^{t+i-1} \times \mathcal{B}_{n-i}^{t+i-1}$.

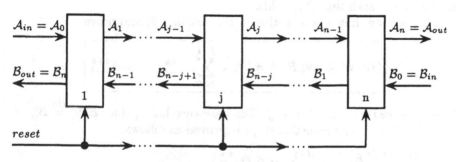

Fig. 6.18. Data flow through the array

6.8.2 A more complicated version. This array is essentially the array as given in the last section. However, there is no input B anymore. Instead the scalar product is done by fixed weights that are stored in the cells. In an initial phase, these weights are loaded from the input $\mathcal{A}$. From figure 6.18, we can read the following equations for $j \in \{0, \dots, n-1\}$:

$$
\begin{array}{ll}
\text{(1)} \; \mathcal{A}_{j+1}^{t+1} := \mathcal{A}_j^t & \text{(4)} \; \mathcal{B}_{j+1}^{t+1} := \mathcal{B}_j^t + \mathcal{A}_{n-(j+1)}^t \times \mathcal{W}_{n-j}^t \\
\text{(2)} \; \mathcal{A}_{j+1}^0 := 0 & \text{(5)} \; \mathcal{B}_{j+1}^0 := 0 \\
\text{(3)} \; \mathcal{A}_0^t := \mathcal{A}_{in}^t & \text{(6)} \; \mathcal{B}_0^t := \mathcal{B}_{in}^t \\
& \text{(7)} \; \mathcal{W}_{j+1}^{t+1} := \left(reset^{(t)} \Rightarrow \mathcal{A}_j^t \mid \mathcal{W}_{j+1}^t \right) \\
& \text{(8)} \; \mathcal{W}_{j+1}^0 := 0
\end{array}
$$

$\mathcal{W}_j^t$ is the value of the weight that is stored by cell j at time t. We want to prove a relationship between the signals $\mathcal{A}_{in}$ and $\mathcal{B}_{out}$. For this reason, as in the last example we establish some lemmata which are helpful for this purpose. Furthermore, we give an input restriction for the input r: we assume that

$$(9)\ reset^{(n)} := 1 \qquad\qquad (10)\ \forall t.reset^{(t+n+1)} := 0$$

holds. This means we initialize the weights of the array at time $n-1$ with the weights $\mathcal{A}_{in}^0, \ldots, \mathcal{A}_{in}^n$. In order to prove our result, we prove:

$$(11)\ \forall j \le n.\forall t.\mathcal{A}_j^{t+j} = \mathcal{A}_{in}^t \qquad\qquad (12)\ \forall j \le n.\forall t.\mathcal{W}_{j+1}^{t+n+1} = \mathcal{A}_j^n$$

(11) can be proved via induction on j. In the base case, we have $\forall t.\mathcal{A}_0^{t+0} = \mathcal{A}_{in}^t$, which does hold according to equation (3). In the induction step, we have $\mathcal{A}_{j+1}^{t+j+1} \stackrel{(1)}{=} \mathcal{A}_j^{t+j} \stackrel{(11)}{=} \mathcal{A}_{in}^t$. (12) is proved by induction on t: in the base case, we have $\mathcal{W}_{j+1}^{n+1} \stackrel{(7)}{=} \left(reset^{(n)} \Rightarrow \mathcal{A}_j^n \,|\, \mathcal{W}_{j+1}^n\right) \stackrel{(9)}{=} \mathcal{A}_j^n$ and in the induction step $\mathcal{W}_{j+1}^{t+n+2} \stackrel{(7)}{=} \left(reset^{(t+n+1)} \Rightarrow \mathcal{A}_j^{t+n+1} \,|\, \mathcal{W}_{j+1}^{t+n+1}\right) \stackrel{(10)}{=} \mathcal{W}_{j+1}^{t+n+1} \stackrel{(12)}{=} \mathcal{A}_j^n$. Hence, we know that the array has been initialized correctly, i.e. at time n all weights are initialised such that (12) holds.

Now we can draw our attention to the streams $\mathcal{B}_j$ and prove

$$(13)\ \forall j \le n.\forall t.\mathcal{B}_j^{t+j} = \mathcal{B}_{in}^t + \sum_{k=0}^{j-1} \mathcal{A}_{n-(k+1)}^{t+k} \times \mathcal{W}_{n-k}^{t+k}$$

This is proved by induction on j. The base case holds, since $\mathcal{B}_0^{t+0} \stackrel{(6)}{=} \mathcal{B}_{in}^t = \mathcal{B}_{in}^t + \sum_{k=0}^{-1} \ldots$. The induction steps is proved as follows:

$$\mathcal{B}_{j+1}^{t+j+1} \stackrel{(4)}{=} \mathcal{B}_j^{t+j} + \mathcal{A}_{n-(j+1)}^{t+j} \times \mathcal{W}_{n-j}^{t+j}$$
$$\stackrel{(13)}{=} \mathcal{B}_{in}^t + \left(\sum_{k=0}^{j-1} \mathcal{A}_{n-(k+1)}^{t+k} \times \mathcal{W}_{n-k}^{t+k}\right) + \mathcal{A}_{n-(j+1)}^{t+j} \times \mathcal{W}_{n-j}^{t+j}$$
$$= \mathcal{B}_{in}^t + \left(\sum_{k=0}^{j} \mathcal{A}_{n-(k+1)}^{t+k} \times \mathcal{W}_{n-k}^{t+k}\right)$$

In particular $(j := n)$, we find after a simple index transformation:

$$(13a)\ \forall t.\mathcal{B}_{out}^{t+n} = \mathcal{B}_{in}^t + \sum_{j=1}^{n-1} \mathcal{A}_{j-1}^{t+n-j} \times \mathcal{W}_j^{t+n-j}$$

If we instantiate $t+n$ for t in equation (13a) and use equation (11), we obtain finally the following behavior of the circuit:

$$(13b)\ \forall t.\mathcal{B}_{out}^{t+2n} = \mathcal{B}_{in}^{t+n} + \sum_{j=1}^{n-1} \mathcal{A}_{in}^{t+2(n-j)+1} \times \mathcal{W}_j^{t+2n-j}$$

Using again an index transformation this is the same as

$$(13c) \ \forall t. \mathcal{B}_{out}^{t+2n} = \mathcal{B}_{in}^{t+n} + \sum_{k=1}^{n-1} \mathcal{A}_{in}^{t+2k+1} \times \mathcal{W}_{j}^{t+n+k}$$

Hence, we see that the circuit computes two scalar products, one at the even points of time and another one at the odd points of time.

6.9 GSM Speech Encoding

6.9.1 The Algorithm. The verification example given in this section is part of the European digital cellular telecommunication system, in particular it belongs to a 'full rate speech transcoder'. It's implementation is based on the standardisation document [GSM94] of the European Institute for Telecommunication Norms.

A detailed evaluation of the entire description of [GSM94] shows that in the contextof a HW/SW codesign the hardware implementation of the module for the *short term analysis filtering* (STAF for short) leads to a significant enhancement of the performance of the full rate speech transcoder. In this section, a hardware implementation generated by a high-level synthesis tool and the verification of the module is shown.

The short term analysis filtering due to [GSM94] requires special arithmetic functions add and mult_r which are defined in the standard document [GSM94] for signed integers. Bitvectors are thereby interpreted as signed integers by the following function Φ_s that interprets the bitvectors as two-complement encodings:

$$\Phi_s([b_n, \dots, b_0]) := -b_n \cdot 2^n + \sum_{i=0}^{n-1} b_i \cdot 2^i$$

Given a bitwidth $n + 1$, i.e. bitvectors of the form $[a_n, \dots, a_0]$, the largest representable number is $\mathsf{Max\_Int} := 2^n - 1$ and the smallest (negative) number $\mathsf{Min\_Int} := -2^n$. The definition of the functions add and mult_r used in the definition of STAF are defined as follows: $\mathsf{add}(x, y)$ performs the addition of x and y with overflow control and saturation; the result is set at $\mathsf{Max\_Int}$ when overflow occurs or at $\mathsf{Min\_Int}$ when underflow occurs. $\mathsf{mult\_r}(x, y)$ performs multiplications of x by y and scales the result (with rounding). The formal definition of the two operations in terms for the bitwidth $n + 1$ is as follows:

$$\mathsf{add}(x, y) := \begin{cases} 2^n - 1 & : overflow(x + y) \\ -2^n & : underflow(x + y) \\ x + y & : otherwise \end{cases} = \begin{cases} \mathsf{Max\_Int} & : overflow(x + y) \\ \mathsf{Min\_Int} & : underflow(x + y) \\ x + y & : otherwise \end{cases}$$

$$\mathsf{mult\_r}(x, y) := \begin{cases} 2^n - 1 & : for \ x = y = -2^n \\ (x \cdot y + 2^{n-1}) \gg_s n & : otherwise \end{cases}$$
$$= \begin{cases} \mathsf{Max\_Int} & : for \ x = y = \mathsf{Min\_Int} \\ (x \cdot y + 2^{n-1} \ \mathrm{DIV} \ 2^n) & : otherwise \end{cases}$$

Division of x by 2^m means in the bit representation of the number x a right shift by m bits, where the sign bit (the most significant and also the leftmost one) is replicated. Given two signed $n + 1$-bit numbers x and y in 2-complement, the bitwidths of $\text{add}(x, y)$ is also $n + 1$, since the cases where either an underflow or an overflow occurred are defined as Min_Int and Max_Int, respectively. $\text{mult\_r}(x, y)$ also requires only $n + 1$ bits since $2n - 1$ bits are sufficient to store the intermediate result $(x \cdot y) + 2^{n-1}$:

$$\text{mult\_r}(\Phi_s(A), \Phi_s(B)) \leq \big((-2^n)(-(2^n - 1)) + 2^{n-1} \text{ DIV } 2^n\big)$$
$$= \lfloor 2^n - 1 + \tfrac{1}{2} \rfloor = 2^n - 1 = \text{Max\_Int}$$
$$\text{mult\_r}(\Phi_s(A), \Phi_s(B)) \geq \big((-2^n)(2^n - 1) + 2^{n-1} \text{ DIV } 2^n\big)$$
$$= \lfloor -(2^n - 1) + \tfrac{1}{2} \rfloor = -(2^n - 1) > \text{Min\_Int}$$

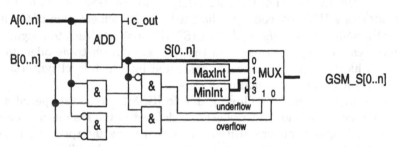

Fig. 6.19. Implementation of the GSM addition

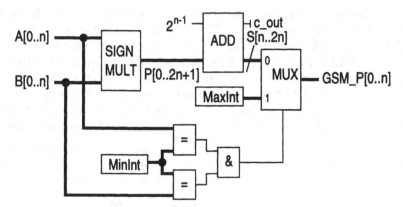

Fig. 6.20. Implementation of the GSM multiplication

Note that $\lfloor x + \frac{1}{2} \rfloor = x$ for $x \in \mathbb{Z}$. Combinational hardware implementations for the computation of these arithmetic operations are given in figures 6.19 and 6.20. Note that in figure 6.20 the result of the adder has $2n+2$ bits, i.e. the array $S[0 \ldots 2n + 1]$, and that the leftmost bit as well as the lowest $n - 1$ bits are neglected. This is legal due to the inequalities above that assert that the result can be stored with $n+1$ bits. Note that the mult_r(Min_Int,Min_Int) has to be declared in a special case as the usual term would lead to an overflow, as can be seen as follows:

$$((-2^n)(-2^n) + 2^{n-1} \text{ DIV } 2^n) = (2^{n+1} + 1 \text{ DIV } 2) = 2^n = \text{Max\_Int} + 1$$

The implementation of mult_r as given in figure 6.20 has however the disadvantage that intermediate computation steps have a bitwidth larger than $n + 1$. This has serious drawbacks when the multiplication, the addition of 2^{n-1} and the shift-operation are done at different points of time, such that the intermediate results have to be stored. The implementation given in figure 6.21 removes this drawback. The implementations given in figures 6.20 and 6.21 can be shown to be equivalent to each other. The correctness can be shown as follows:

$$\left(\Phi_s(A) \cdot \Phi_s(b) + 2^{n-1} \text{ DIV } 2^n \right)$$
$$= \left(\left(\Phi_s(A) \cdot \Phi_s(b) \text{ DIV } 2^{n-1} \right) + 1 \text{ DIV } 2 \right)$$
$$= \begin{cases} (\Phi_s(A) \cdot \Phi_s(b) \text{ DIV } 2^n) & : \text{ for } \Phi_s(A) \cdot \Phi_s(b)[n-1] = 0 \\ (\Phi_s(A) \cdot \Phi_s(b) \text{ DIV } 2^n) + 1 & : \text{ for } \Phi_s(A) \cdot \Phi_s(b)[n-1] = 1 \end{cases}$$

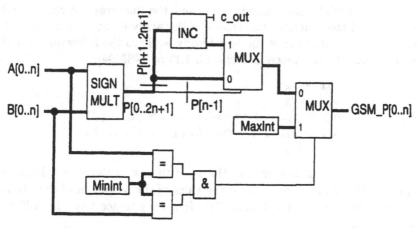

Fig. 6.21. Implementation of the GSM multiplication

A description of the short term analysis filtering algorithm of section 4.2.10 in [GSM94] in the programming language C is given in figure 6.22. The algorithm needs to keep the array $u[0 \ldots 7]$ in memory for each call, and the initialisation of this array is 0.

```
static void STAF(int u[8], rp[8], k_n, s[160])
{int d,z,ui,sav,rpi,i,j;

  for(k=0;k < k_n;k++)
    {
    d = s[k];
    sav = d;
    for(i=0;i¡8;i++)
      {
      ui = u[i];
      rpi = rp[i];
      u[i] = sav;
      z = mult_r(rpi,d);
      sav = add(ui,z);
      z = mult_r(rpi,ui);
      d = add(d,z);
      }
    s[k] = d;
    }
}
```

Fig. 6.22. The short term analysis filtering algorithm in C

Note that the given function **STAF** overwrites the array s with the computed values. The standard document does not allow this, instead a new array $d[0\ldots159]$ is to be assigned. However, the given C function does only model the hardware module and before it is invoked, the the array s is copied. After termination of the function, the values of the array d can be read from the local array s. An unroling of the loops body leads to the following recursive equations that are also given in section 3.1.11 of [GSM94]:

$$d_{0,k} := s_k$$
$$u_{0,k} := s_k$$
$$d_{i,k} := d_{i-1,k} + rp_i \cdot u_{i-1,k-1} \quad i \in \{1,\ldots,8\}$$
$$u_{i,k} := u_{i-1,k-1} + rp_i \cdot d_{i-1,k} \quad i \in \{1,\ldots,8\}$$
$$d_k := d_{8,k}$$

$d_{i,k}$ is the value of the output d of the $i-1$-th row in the k-th column (for $i \in \{1,\ldots,8\}$) and $d_{0,k} = s_k$. $u_{i,k}$ is the value of the output ui of the i-th row in the k-th column (for $i \in \{0,\ldots,7\}$). It is easy to see that the following equations hold:

$$d_{i,k} = s_k + \sum_{j=1}^{i} rp_j u_{j-1,k_1} \qquad u_{i,k} = s_k + \sum_{j=1}^{i} rp_j d_{j-1,k}$$

It it is possible to implement the entire function as a combinatorial circuit according to the above equations. Figure 6.23 shows an example implementation where $8 * k_n$ modules are required that are themselves implemented by two GSM multipliers and two GSM adders (see figures 6.19 and 6.20). The implementation of figure 6.23 is a matrix with 8 columns and k_n rows. The initial values $u[i]$ in row $k = 0$ are 0 according to the standard document.

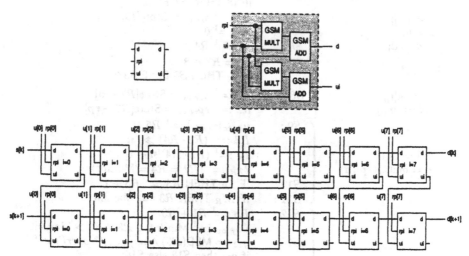

Fig. 6.23. Implementation of STAF as combinatorial circuit.

Figure 6.23 should be read as rows of columns, since each row corresponds to the execution of the inner loop. The figure illustrates the data flow through the computation in the module. While the array rp is not changed, the array u is updated every time the inner loop is performed. After termination of the inner loop, the value $d[k]$ is available (and stored in $s[k]$ in the C version).

For obvious reasons, some parts of the circuit have to be implemented in a sequential manner such that the number of GSM multipliers and GSM adders can be reduced, i.e. the same GSM multipliers and GSM adders can be used for different computations. This requires the implementation of a controller that manages the different computations in the data path. In the next section, a sequential implementation of STAF is presented that requires only one (GSM) multiplier and one (GSM) adder.

6.9.2 Register-Transfer Implementation of STAF. A sequential hardware implementation of STAF must first schedule the arithmetical operations to certain points of time. A simple way to do this without further optimizations is to map each instruction of the C code to one clock cycle. A result of such a simple mapping leads then to the algorithm of figure 6.24, where some statements are labeled with corresponding control states $S0, \ldots$.

The next step is to implement the control flow of the program and the corresponding data path. The sequential implementation of STAF based on figure 6.24 contains two data paths and a controller. The first data path

$k := 0$ $\left\{ \begin{array}{l} S0: \\ S3: \end{array} \right.$ $R6 :=^1 0$ falls $Start$
$l_8 := R6 < k_n$

$k < k_n$? $Ready := \neg l_8$
IF $l8$ THEN $S4$ ELSE $S0$;

$d = s[k]$; $\left\{ \begin{array}{l} S4: \end{array} \right.$ $R4 :=^1 s_{read} = $ Store$[R6 + s]$
$i := 0$; $R7 :=^1 0$

$sav := d$; $S5:$ $R5 :=^1 R4$

       $S7:$ $l8 := R7 < 8$;
$k < k_n$?   IF $l8$ THEN $S8$ ELSE $S37$;

$ui := u[i]$; $S8:$ $R1 :=^1 u_{read} = $ Store$[R7 + u]$
$rpi := rp[i]$; $R2 :=^1 rp_{read} = $ Store$[R7 + rp]$

       $S9:$ $u_{data}[u_{addr}] :=^1 R5$
$R3 :=^1 (R2 \cdot R4) \gg_s n$

$u[i] := sav$ $I_R :=^1 (R2 \cdot R4)[n-1]$
$z := $ mult_r(rpi, d) $M_R :=^1 (R2 = R4 = $ Min_Int$)$

$S10:$ IF I_R THEN $R3 :=^1 R3 + 1$
$S11:$ IF M_R THEN $R3 :=^1 $ Max_Int

$S12:$ $R3 :=^1 R1 + R3$
$uo_{flow} :=^1 $ Min_Int $< R3 \vee R3 > $ Max_Int

$S13:$ $g_{32} := $ MSB$(R3) \wedge uo_{flow}$; /* overflow */
if g_{32} then $S18$ else $S15$;

$S15:$ $g_{32} := \neg(\neg$MSB$(R3) \wedge uo_{flow})$;
$sav := $ add_r(ui, z) /* $g_{32} = $ no underflow */
if g_{32} then $S16$ else $S20$;

$S16:$ $R5 :=^1 R3$; goto $S22$;
$S18:$ $R5 :=^1 $ Max_Int; goto $S22$;
$S20:$ $R5 :=^1 $ Min_Int; goto $S22$;

$S22:$ $R1 :=^1 (R2 \cdot R1) \gg n$
$I_R :=^1 (R2 \cdot R1)[n-1]$
$z := $ mult_r(rpi, ui) $M_R :=^1 (R2 = R1 = $ Min_Int$)$

$S23:$ IF I_R THEN $R1 :=^1 R1 + 1$
$S24:$ IF M_R THEN $R1 :=^1 $ Max_Int

$S25:$ $R3 :=^1 R4 + R1$
$uo_{flow} :=^1 $ Min_Int $< R3 \vee R3 > $ Max_Int

$S26:$ $g_{32} := $ MSB$(R3) \wedge uo_{flow}$; /* overflow */
if g_{32} then $S31$ else $S28$;

$S28:$ $g_{32} := \neg(\neg$MSB$(R3) \wedge uo_{flow})$;
$d := $ add_r(d, z) /* $g_{32} = $ no underflow */
if g_{32} then $S29$ else $S33$;

$S29:$ $R4 :=^1 R3$; goto $S35$;
$S31:$ $R4 :=^1 $ Max_Int; goto $S35$;
$S33:$ $R4 :=^1 $ Min_Int; goto $S35$;

$i := i + 1$; $S35:$ $R7 :=^1 R7 + 1$
$s[k] := d$ $S37:$ Store$[R6 + s] := R4$
$k := k + 1$ $S38:$ $R6 :=^1 R6 + 1$

Fig. 6.24. A finite state algorithm for STAF with $n + 1$ bits in the data path.

is used to perform the computations according to the last section, and the second data path is used for updating the loop variables k and i.

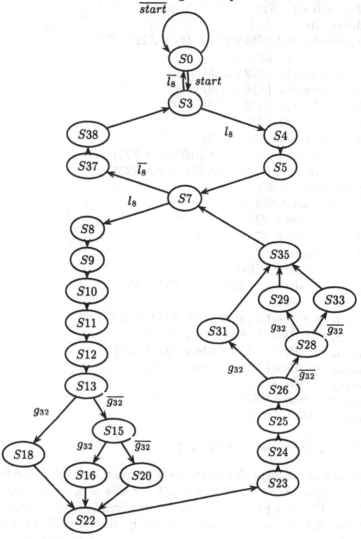

Fig. 6.25. State transition diagram of the correct version of the STAF controller

The state transition diagram of the controller is given in figure 6.25. There is also a transition from each of the states to the initial state $S0$ that is performed when a *reset* signal occurs. For reasons of readability, these transitions are not drawn in figure 6.25. The output equations of the controller are given as follows:

1. $Ready := \neg Reset \wedge S3 \wedge \neg l8$

2. $sel1_1 := \neg Reset \wedge (S23 \vee S24)$
3. $sel1_0 := \neg Reset \wedge (S22 \vee S24)$
4. $sel2 := \neg Reset \wedge S22$
5. $sel3_1 := \neg Reset \wedge S11$
6. $sel3_0 := \neg Reset \wedge (S10 \vee S11 \wedge M_R \vee S12 \vee S25)$
7. $sel4_1 := \neg Reset \wedge (S31 \vee S33)$
8. $sel4_0 := \neg Reset \wedge (S29 \vee S33)$
9. $sel5_1 := \neg Reset \wedge (S18 \vee S20)$
10. $sel5_0 := \neg Reset \wedge (S16 \vee S20)$
11. $sel6_1 := \neg Reset \wedge S25$
12. $sel6_0 := \neg Reset \wedge (S12 \vee S23)$
13. $sel7_1 := \neg Reset \wedge (S25 \vee ir \wedge S10 \vee ir \wedge S23)$
14. $sel7_0 := \neg Reset \wedge (S12 \vee ir \wedge S10 \vee ir \wedge S23)$
15. $sel8 := \neg Reset \wedge S38$
16. $sel9 := \neg Reset \wedge S38$
17. $sel10 := \neg Reset \wedge S35$
18. $sel11 := \neg Reset \wedge S7$
19. $sel12 := \neg Reset \wedge S7$
20. $sel13 := \neg Reset \wedge (S15 \vee S28)$
21. $R1_{enable} := \neg Reset \wedge (S8 \vee S22 \vee S23 \vee S24)$
22. $R2_{enable} := \neg Reset \wedge S8$
23. $R3_{enable} := \neg Reset \wedge (S9 \vee S10 \vee S11 \vee S12 \vee S25)$
24. $R4_{enable} := \neg Reset \wedge (S4 \vee S29 \vee S31 \vee S33)$
25. $R5_{enable} := \neg Reset \wedge (S5 \vee S16 \vee S18 \vee S20)$
26. $R6_{enable} := \neg Reset \wedge (S0 \wedge Start \vee S38)$
27. $R7_{enable} := \neg Reset \wedge (S4 \vee S35)$
28. $s_{enable} := \neg Reset \wedge S37$
29. $u_{enable} := \neg Reset \wedge S9$
30. $rp_{enable} := 0$
31. $mult_{enable} := \neg Reset \wedge (S9 \vee S22)$

The mapping of the algorithm to the circuit structure involves the following register mapping: $R1 \sim ui$, $R2 \sim rpi$, $R3 \sim z$, $R4 \sim d$, $R5 \sim sav$, $R6 \sim k$, and $R7 \sim i$. The data paths of the circuit are given in figures 6.26 and 6.27, where a multiplex-based implementation has been chosen. The data path of figure 6.27 is used for controlling the loops and the data path of figure 6.26 performs the additions and multiplications for the filtering function.

The standard requires that the bitwidths of the upper half are 16 bits, but the verification can also be done with smaller bitwidths, if the constants Max_Int and Min_Int are adapted properly in the data path of figure 6.26. In contrast to the data path of figure 6.26, the values in the registers R6 and R7 can be interpreted as *unsigned* integers in figure 6.27. It has to be noted, that register R7 is used for storing the loop variable i and hence, only contains the values $0, \ldots, 7$. Bitwidths wider than 3 bits for R7 do not make sense at all. Register R6 corresponds to the loop variable k and runs through the

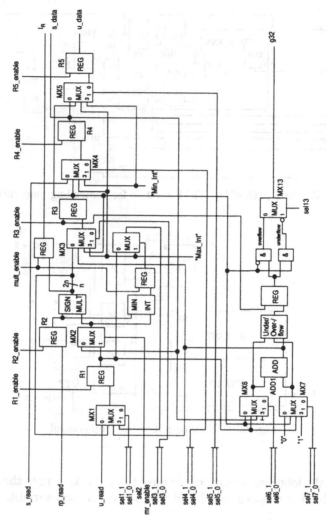

Fig. 6.26. Data path of the STAF module for performing the filter function.

values $0, \ldots, k_n$. An upper bound for k_n is 159, and hence a bitwidth greater than 8 for R7 does also not make sense.

6.9.3 Specifications and Verification Results. In order to verify the correctness of the circuit, the circuits 6.19 and 6.20 are used to implement verification circuits according to figures 6.29 and 6.29.

The circuit given in figure 6.29 is used to check the correctness of the loop control of both loops. There are two states of the controller that are important for the verification of the control behaviour: the initial state $S0$ and the state $S7$, where it is checked whether the inner loops is to be terminated or not. In

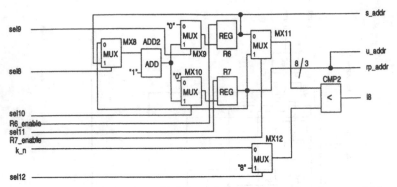

Fig. 6.27. Data path of the STAF module for controlling loop variables.

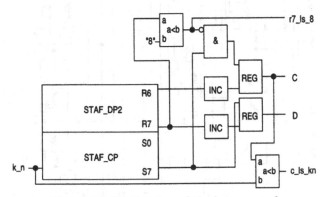

Fig. 6.28. Verification of the loop control

particular the following specifications have to be checked (note that the inner loop variable i is mapped to $R7$ and that the outer loop variable is mapped to $R6$):

Initial Control State. Initially, the controller is in state $S0$ and the next state is also $S0$ when a reset occurs:

C1a: $S0$

C1b: $G[reset \rightarrow X S0]$

Initialisation of the Loop Variables. If a *start* signal occurs, then the controller moves to state $S7$ and sets the registers $R6 = R7 = 0$:

C2a: $G(\neg reset) \rightarrow G[S0 \wedge start \rightarrow X[R6 = 0]]$

C2b: $G[S0 \rightarrow [([R6 = 0] \wedge [R7 = 0]) \ W \ S7]]$

C2c: $G(\neg reset) \rightarrow G[S0 \wedge start \wedge X[k_n \neq 0] \rightarrow$
$XXXX[S7 \wedge [R6 = 0] \wedge [R7 = 0]]]$

Updating Loop Variables during inner Loop. If the inner loop's body is executed then the inner loop variable i $(R7)$ is incremented (C3a) and the outer loop variable k $(R6)$ is not changed (C3b):

C3a: $G(\neg reset) \to G[S7 \wedge (R7 < 8) \to X[(R7 = D) \underline{W} S7]]$

C3b: $G(\neg reset) \to G[S7 \wedge (R7 < 8) \to [(R6 = XR6) \underline{U} S7]]$

Terminating the inner loop. If the outer loop's body is executed (after inner loop has been executed) and another iteration of the outer loop is necessary, i.e. $k < k_n$ $(R6 + 1 < k_n$ or $C + 1 < k_n)$, then the inner loop variable i $(R7)$ is reinitialized to 0 (C4a) and the outer loop variable k $(R6)$ is incremented (C4b):

C4a: $G(\neg reset) \to G[S7 \wedge \neg(R7 < 8) \to X[(R7 = 0) \underline{W} S7]]$

C4b: $G(\neg reset) \to G[S7 \wedge \neg(R7 < 8) \wedge X(C < k_n) \to [(\neg S0) \underline{U} S7]]$

C4c: $G(\neg reset) \to G[S7 \wedge \neg(R7 < 8) \wedge X(C < k_n) \to$
   $X[(R6 = C \wedge R7 = 0) \underline{W} S7]]$

Terminating the outer Loop. If the outer loop's body is executed (after inner loop has been executed) and another no further iteration of the outer loop is necessary, then the controller moves to the initial state $S0$:

C5: $G(\neg reset) \to G[S7 \wedge \neg(R7 < 8) \wedge \neg X(C < k_n) \to XXXXS0]$

The above specifications describe the entire behaviour of the loop control. Therefore, only the states $S0$ and $S7$ have to be considered. The runtimes for the verification of the above specifications with the LTL model checker of C@S are given in table 6.6.

Spec	Runtime [sec]	BDD nodes total	BDD nodes for Trans. Rel.
C1a	22.74	10091	393
C1b	23.11	10011	386
C2a	32.11	10362	421
C2b	24.45	10028	427
C2c	31.69	10092	800
C3a	46.94	10692	1174
C3b	48.68	10469	1397
C4a	45.58	10472	1153
C4b	60.56	10664	1328
C4c	58.36	11584	1378
C5	87.59	22563	3731

Table 6.6. Runtimes for the STAF control behavior.

The verification of the data path that computes the filtering (figure 6.26), similar hardware extensions have been added to the circuit (see figure 6.29). The circuit given in figure 6.29 is used to check the correctness of the filter function. The additional hardware contains the module that has been used in section 6.9.2 to implement a combinational circuit for STAF. It is activated,

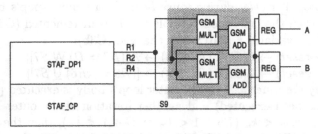

Fig. 6.29. Verification of the filter function

i.e. the results of the computation is stored into the registers A and B exactly when the controller is in state $S9$ as this is the first state, where the inputs for the computations have been read and are available in the registers $R1$, $R2$, and $R4$. The specifications that have been checked are that the values $A := \mathsf{add}(\mathsf{mult\_r}(rpi, ui)), d)$ and $B := \mathsf{add}(\mathsf{mult\_r}(rpi, d)), ui)$ are computed correctly. Note that rpi, ui, and d are mapped to the registers $R2$, $R1$, $R4$, respectively. The values A and B can be computed in state $S9$, as this is the first state where the current values of $R1$ and $R2$, i.e. rpi and ui are available. They are compared to the values of the registers $R4$ and $R5$ when the controller is again in state $S7$:

$$D1 : \quad \mathsf{G}[S9 \to [(A = R4) \ \underline{\mathsf{W}} \ S7]]$$
$$D2 : \quad \mathsf{G}[S9 \to [(B = R5) \ \underline{\mathsf{W}} \ S7]]$$
$$D3 : \quad \mathsf{G}[S9 \to [(C = R7) \ \underline{\mathsf{W}} \ S7]]$$

The runtimes for the verification of the above specifications with the LTL model checker of C@S are given in table 6.7.

Spec	Runtime [sec]	BDD nodes total	BDD nodes for Trans. Rel.
D1	282.55	327467	93321
D2	197.04	289418	74329

Table 6.7. Runtimes for the STAF data path.

7. Conclusions

We have presented a new approach for the verification of digital circuits. Different decision procedures are coupled by transforming specification and implementation descriptions into a common representation based on a restricted class of higher order logic. Dependent on the type of the verification problem, the most suited solution is selected and the necessary decision procedures are invoked. In contrast to previous approaches, several solution strategies may be combined to verify circuits for which a specification includes complex timing as well as abstract data types. Using different examples, different verification strategies have been demonstrated and compared with regard to the circuit size and necessary amount of interaction. The C@S system is a research prototype and currently extended by other decision procedures and verification strategies, which have not been presented in this book section. For an actual and up-to-date overview of the current implementation status and availability see the online documentation at `http://goethe.ira.uka.de/hvg/cats/`.

Appendix: The Common Book Examples

Thomas Kropf

1. Introduction

One of the distinguishing aspects of this book is the use of common circuit examples to exemplify the different approaches and systems. They have been taken from a hardware verification benchmark suite, originally proposed by IFIP WG 10.5 [Krop94b]. The set of benchmark examples comprises different circuits, ranging from combinational circuits to a simple microprocessor. More details on these circuits, including detailed documentation with VDHL descriptions etc. as well as general information about the benchmark suite can be found online[1].

From all available circuits the following four have been chosen:

- *The Single Pulser* as this example is well suited to demonstrate the ability to specify complex timing behavior.
- *An Arbiter* as an example for a scalable controller type circuit.
- *The Black-Jack Dealer* as an example for a circuit, consisting of a controller and a data path.
- *A one-dimensional Systolic Array* as an example of a complex scalable data path.

In the following these four circuits will be described briefly. In some book sections, additional examples of the benchmark suite have been treated. For detailed information about those see the online documentation. There, for all circuits including the common book examples a detailed documentation with VDHL descriptions etc. as well as information about the benchmark suite in general can be found.

2. Single Pulser

Despite its small size, this circuit has shown to be hard to specify in many formalisms like temporal logics and is well-suited as an introductory example.

---

[1] More can be found at http://goethe.ira.uka.de/benchmarks/ and ftp://goethe.ira.uka.de/pub/benchmarks/.

2.1 Introduction

A Single Pulser is a clocked-sequential device with a one-bit input, I, and a one-bit output O. The purpose of the circuit is described as follows [WiPr80]:

> We have a debounced push-button, on (true) in the down position, off (false) in the up position. Devise a circuit to sense the depression the button and assert an output signal for one clock pulse. The system should not allow additional assertions of the output until after the operator has released the button.

A design specification of the Single Pulser can be found in the text book *The Art of Digital Design* by D. Winkel and F. Prosser [WiPr80]. A detailed treatment of this example for comparing different specification formalisms can be found in [JoMC94].

2.2 Specification

Assuming that the input is synchronous and debounced, the specification may be stated as:

For each input pulse on I, the Single Pulser issues exactly one pulse of unit duration on O regardless of the duration of I.

The specification may be also stated by the following three properties [JoMC94]:

1. Whenever there is a rising edge at I, O becomes true some time later.
2. Whenever O is true it becomes false in the next time instance and it remains false at least until the next rising edge on I.
3. Whenever there is a rising edge, and assuming that the output pulse doesn't happen immediately, there are no more rising edges until that pulse happens (There can't be two rising edges on I without a pulse on O between them).

In [JoMC94] the specification is given in different formalisms like PVS and CTL [ORSS94, ClEm81].

2.3 Implementation

The implementation is taken from [WiPr80]. The incoming, not yet debounced asynchronous signal $Pulse\_In$ is fed to a D-flipflop and thus becomes the synchronized signal $Pulse\_sync$, which is then delayed for one clock cycle by using another D-flipflop. Its output is negated, and the AND-connection of the synchronous pulse with its own delay generates the resulting signal, lasting one clock cycle (Fig. 2.1).

In Fig. 2.2 waveforms are given to illustrate the behavior of the circuit.

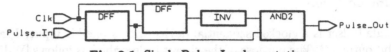

Fig. 2.1. Single Pulser Implementation

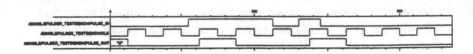

Fig. 2.2. Example Waveform

2.4 Acknowledgements

A design specification of the Single Pulser can be found in the text book "The Art of Digital Design" by D. Winkel and F. Prosser [WiPr80]. A nice overview about the verification of this circuit in different reasoning systems can be found in [JoMC94].

3. Arbiter

This arbiter is a good example for a synchronous scalable state machine.

3.1 Introduction

The purpose of the bus arbiter is to grant access on each clock cycle to a single client among a number of clients contending for the use of a bus (or another resource). The inputs to the circuit are a set of request signals $req_0, \dots, req_{k-1}$ and the outputs are a set of acknowledge signals $ack_0, \dots, ack_{k-1}$ (Fig. 3.1). Normally the arbiter asserts the acknowledge signal of the requesting client with the lowest index. However, as requests become more frequent, the arbiter is designed to fall back on a round robin scheme, so that every requester is eventually acknowledged. This is done by circulating a token in a ring of arbiter cells, with one cell per client. The token moves once every clock cycle. If a given client's request persists for the time it takes for the token to make a complete circuit, that client is granted immediate access to the bus.

3.2 Specification

The desired properties to be verified are:

1. No two acknowledge outputs are asserted simultaneously.

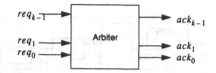

Fig. 3.1. Black box view of the arbiter

2. Every persistent request is eventually acknowledged.
3. Acknowledge is not asserted without request.

By restricting k to be two (or some other small constant), the problem becomes simple yet illustrative. Ultimately, it should be possible to verify a design where k is a parameter, i.e. do the verification for an arbitrary k.

3.2.1 Formal specification example. Data path free control circuits may be described easily using propositional temporal logics like CTL [ClEm81]. The above properties result in the following CTL expressions:

1. $\bigwedge_{i \neq j} \mathsf{AG}\neg(ack_i \wedge ack_j)$
2. $\bigwedge_i \mathsf{AGAF}(req_i \rightarrow ack_i)$
3. $\bigwedge_i \mathsf{AG}(ack_i \rightarrow req_i)$

3.3 Implementation

The basic cell of the arbiter is shown in Fig. 3.2. This cell is repeated k times, as shown in Fig. 3.3 for $k = 4$. Each cell has a request input and an acknowledge output. The grant of cell i is passed to cell $i + 1$, and indicates that no client of index less than or equal to i are requesting. Hence a cell may assert its acknowledge output if its grant input is asserted. Each cell has a register T which stores a one when the token is present. The T registers form a circular shift register which shifts up one place each clock cycle. Each cell also has a register W (for *waiting*) which is set to one when the request input is asserted and the token is present. The register remains set while the request persists, until the token returns. At this time, the cell's override and acknowledge outputs are asserted. The override signal propagates through the cells below, negating the grant input of cell 0, and thus preventing any other cells from acknowledging at the same time.

The circuit is initialized so that all of the W registers are reset and exactly one T register is set. To achieve this, one cell has a different implementation as shown in Fig. 3.4.

Fig. 3.5 and Fig. 3.6 show different simulation runs with and without request collisions.

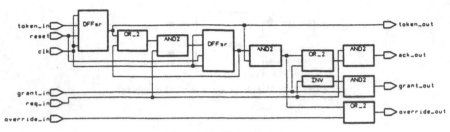

Fig. 3.2. Cell_2_plus

3.4 Acknowledgments

The implementation and formal specification has been taken from [McMi93a], p. 40ff. Parts of the description are directly quoted from there.

4. Black-Jack Dealer

4.1 Specification

A Black-Jack dealer is a device which plays the dealer's hand of a card game. Its inputs are *go* (true/false) and *card* (Ace of Spades,...,2 of clubs). Its outputs are *hitme*, *stand*, and *broke* (all truth-valued).

The *go/hitme* signals are used for a 4-cycle handshake with the operator. Cards are valued from 2 to 10, and aces may be valued as either 1 or 11 by choice of the player. The Black-Jack dealer is repeatedly presented with cards. It must assert *stand* when its accumulated score reaches 16; and it must assert *broke* when its score exceeds 21. In either case the next card starts a new game.

A design specification of the Black-Jack dealer can be found in the text book *The Art of Digital Design* by D. Winkel and F. Prosser [WiPr80].

4.2 Implementation

The implementation of the black jack dealer has been taken from [WiPr80]. From the abstract description of an FSM controlling the datapath, the following realization has been derived (Fig. 4.1). The FSM has been realized by encoding the states with two flipflops A and B and generating the next state by choosing the right signal with two 4:1 multiplexers (table 4.1).

The flipflop output signals are fed to a combinational logic to generate the controller output signals as described in the equations from table 4.2 and table 4.3.

The second part of the dealer is BlackJack_DataPath (Fig. 4.2), including several flipflops to hold the status information for the outside world, *hit*,

State	Next State	B	A	Condition
0 Get	Get	0	0	nGet_2
	Add	0	1	Get_2
1 Add	Use	1	0	Acecard ∧ nAcellflag
	Test	1	1	¬(Acecard ∧ nAcellflag)
2 Use	Test	1	1	true
3 Test	Get	0	0	nTest_3
	Test	1	1	Test_3

Table 4.1. State transition table

Get_1	=	S_Get ∧ nCard_r_s
Get_2	=	S_Get ∧ Card_r_s ∧ nCard_r_d
Get_3	=	Get_2 ∧ (Stand ∨ Broke)
Test_1	=	S_Test ∧ ScoreGT16 ∧ nScoreGT21
Test_2	=	S_Test ∧ ScoreGT16 ∧ ScoreGT21 ∧ nAcellflag = S_Test ∧ ScoreGT21 ∧ nAcellflag
Test_3	=	S_Test ∧ ScoreGT16 ∧ ScoreGT21 ∧ Acellflag = S_Test ∧ ScoreGT21 ∧ Acellflag

Table 4.2. Internal signals

Hit	=	Get_1
Set_Stand	=	Test_1
Clr_Stand	=	Get_2
Set_Broke	=	Test_2
Clr_Broke	=	Get_2
Set_Acellflag	=	S_Use
Clr_Acellflag	=	Get_3 ∨ Test_3
Ld_Score	=	S_Add ∨ S_Use ∨ Test_3
Clr_Score	=	Get_3
AdderS0	=	S_Add
AdderS1	=	Test_3

Table 4.3. Output signals

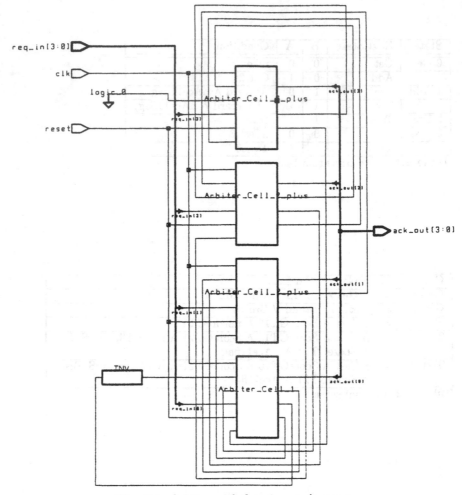

Fig. 3.3. Arbiter with four inputs/outputs

stand, broke, to debounce the input signal from a card ready button and to hold the *ace11flag*. The circuit uses the *card_value* input signal to discriminate the value of an ace card from others, indicating this by setting the signal *acecard = true*. *card_value* is then expanded by one bit and fed to a 4:1 multiplexer, which has to choose between D'+10', D'-10' and the actual card value according to table 4.4.

The multiplexer is followed by a 5-bit carry ripple adder, adding the actual score *internal_score* and the output of the multiplexer. The sum *internal_sum* is then fed to a 5-bit register with load and clear, which is controlled by the signals *Ld_Score* and *Clr_Score*, to produce the score. The score is then compared with D'16' and D'21' to generate the signals

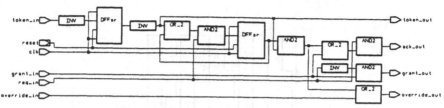

Fig. 3.4. Cell_1

Mux input	AdderS1	AdderS0	action
0	0	0	+10
1	0	1	-10
2	1	0	value
3	1	1	-

Table 4.4. Data path function

ScoreGT 16 and *ScoreGT* 21. Both units exchange control signals as viewed in Fig. 4.3.

The waveform diagram in Fig. 4.4 shows the exchange of control signals between the units. The signals $test(0)$ to $test(3)$ contain the state signals $S\_Get, S\_Use, S\_Add, S\_Test$. The signal $test\_dp(4:0)$ contains the internal sum before it is taken over by the register.

The waveform diagram in Fig. 4.5 shows the behavior of the black jack dealer as a black box. Moreover, the behavior of an algorithmical VHDL description has been added.

4.3 Acknowledgments

A design specification of the Black-Jack dealer can be found in the text book "The Art of Digital Design" by D. Winkel and F. Prosser [WiPr80].

5. 1Syst (Filter)

5.1 Introduction

The filter circuit has been chosen as a witness of a one-dimensional systolic architecture. The notion of *systolic arrays* has first been introduced by Kung and Leierson [KuLe78]. Basics on these specialized regular architectures can be found in [Kung82].

Systolic arrays require additional efforts for specifying the intended behavior, since here data have to be applied at different times in a special order to achieve a correct functioning of the circuit. Moreover, more-dimensional systolic arrays are a good illustration of more-dimensional generic circuits.

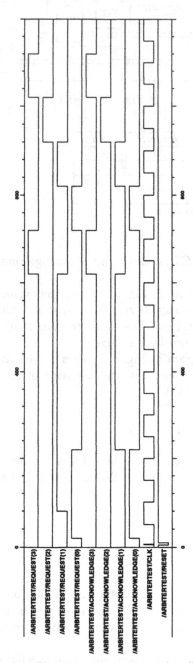

Fig. 3.5. Example simulation with collision

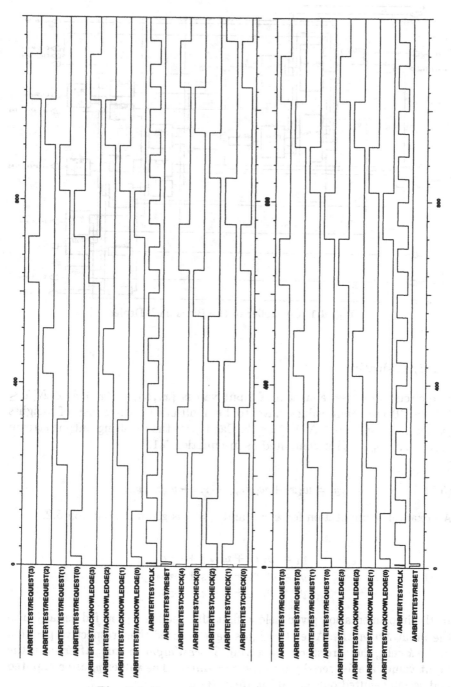

Fig. 3.6. Example simulation without collision

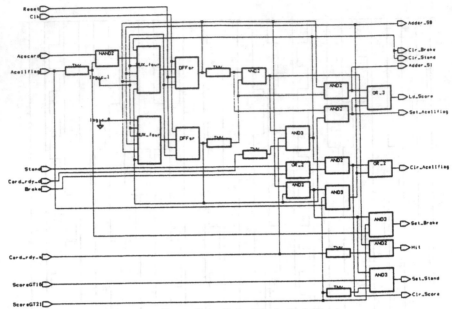

Fig. 4.1. Controller of the Black-Jack Dealer

5.2 Specification

The circuit processes a stream of input values $\{x_1, x_2, \ldots, x_n\}, x_i \in N, 1 \le i \le n$. These input values have to be multiplied with a list of weights $\{w_1, w_2, \ldots, w_k\}, w_j \in N, 1 \le i \le k$. Each y_i of the resulting output stream $\{y_1, y_2, \ldots, y_{n+1-k}\}$ is computed as in equation 5.1.

$$(5.1) \qquad y_i = w_1 x_i + w_2 x_{i+1} + \ldots + w_k x_{i+k-1}$$

An example computation for $k = 3$ and $n = 5$ is given in equation 5.2.

$$(5.2) \qquad \begin{aligned} y_1 &= w_1 x_1 + w_2 x_2 + w_3 x_3 \\ y_2 &= w_1 x_2 + w_2 x_3 + w_3 x_4 \\ y_3 &= w_1 x_3 + w_2 x_4 + w_3 x_5 \end{aligned}$$

In the implementation given below, first the weights w_i are fed serially into the input *StreamIn*. After k weights have reached their respective position in the k cells, they are stored by setting *StoreWeight* to true. Afterwards the input values are fed serially into the *StreamIn*. The values y_i are computed and serially shifted out on output *ResultOut*.

This specification can be verified for concrete values of k weights or as a generic circuit.

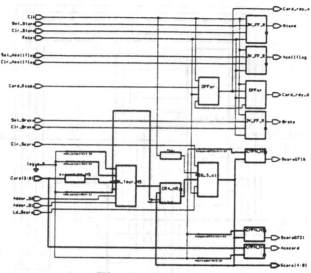

Fig. 4.2. Datapath

5.3 Implementation

5.3.1 General architecture. The implementation is realized using overlapping additions and multiplications. In the first cycle t_1 the input value is multiplied with the weight stored in the respective stage. In the second cycle t_2 this product is to be added to the now available intermediate result y_i. In order to achieve this, the input values x_i must be separated by two clock ticks (Fig. 5.1).

During an initialization phase the weights w_j are loaded into the stages by using the first k input values of the input stream as weights. In the following, inputs values are treated as n bit integers and output values as m bit integers (Fig. 5.2).

Let w be a weight and z be an internal variable. If there is a rising edge at t_1, then $x^{t+1} = x^t$ and $z = x^t * w$ is computed. If there is a rising edge at t_2, then $y^{t+1} = z + y^t$ is computed.

5.3.2 Implementation of one stage. The stage has two global inputs $StreamIn\langle n-1:0\rangle$ and $ResultIn\langle m-1:0\rangle$. Input $StreamIn\langle n-1:0\rangle$ is fed via a 1:2 n bit demultiplexer into two n bit registers. Using the control signal $SelectWgtStr$ of the demultiplexer either the register for storing the weight ($SelectWgtStr = 0$) or the register for the input value ($SelectWgtStr = 1$) is selected. Both register store their respective inputs, if $StoreWgt = 1$ ($StoreStr = 1$) and there is a rising edge at t_1.

The content of the input register is available at the output $EStreamOut\langle n-1:0\rangle$ and the input of an n bit multiplier. The multiplier also gets the weight.

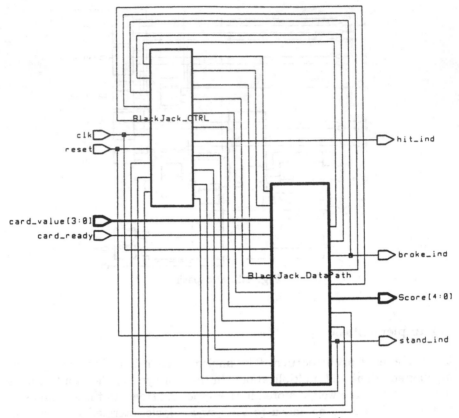

Fig. 4.3. Control - data path communication

The $2n$ bit output of the multiplexer is stored in a register if $StoreRes = 1$ and there is a rising edge at clock t_2 (Fig. 5.3).

Intermediate results, available at input $ResultIn\langle m - 1 : 0\rangle$ are stored in a m bit register if $StoreRes = 1$ and a rising edge at t_2. The content of this register as well as the multiplication results are added via an adder to get the result at $ResultOut\langle x : 1\rangle$ with $x \geq max(n, m) + 1$.

The general clocking scheme must be such that the rising edge of t_1 occurs before the rising edge of t_2.

The realization of a stage in a commercial design system is given in Fig. 5.3 and Fig. 5.4.

Fig. 5.5 shows an example computation with $w_1 = 4, w_2 = 2, w_3 = 1$ and $x_1 = 9$, $x_2 = 8$, $x_3 = 7$, $x_4 = 6$ and $x_5 = 5$, respectively. The output of a behavioral VHDL description produces the output stream $59, 52, 45$, the structural VHDL description (the implementation) produces $3B, 34, 2D$ (hexadecimal).

5.4 Acknowledgements

The circuit has been taken from [Kung82]. It has first been proposed as a benchmark circuit for the Second Conference on Theorem Provers in Circuit Design 1994 (TPCD94, [KuKr94]).

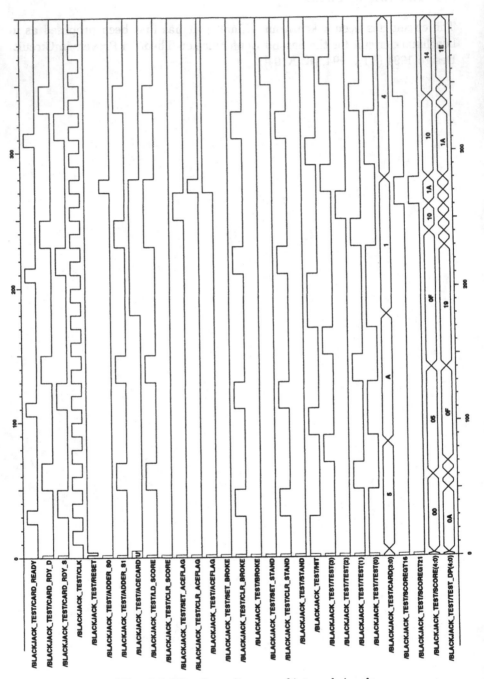

Fig. 4.4. Waveform diagram of internal signals

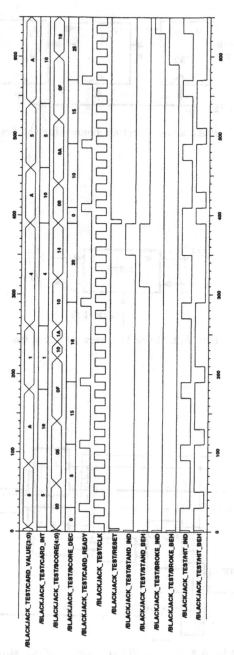

Fig. 4.5. Blackbox simulation of the circuit

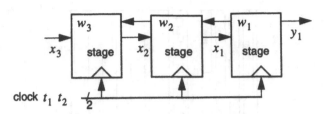

Fig. 5.1. Basic array structure computing the first result

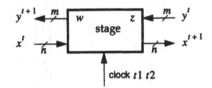

Fig. 5.2. Black box view of one stage

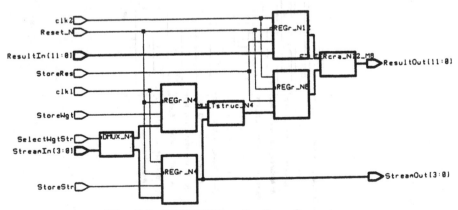

Fig. 5.3. SYNOPSIS realization of one stage

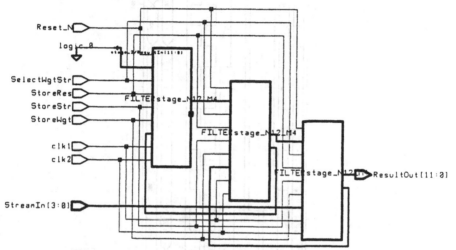

Fig. 5.4. SYNOPSIS realization of a complete structure

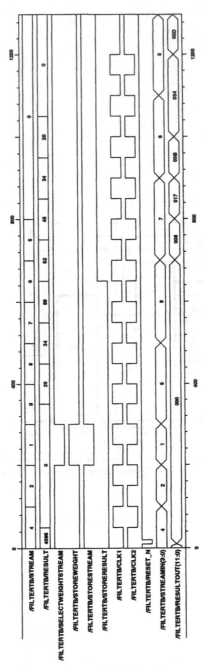

Fig. 5.5. Detailed timing diagram

References

Bibliography

[AaSe95]   M. Aagaard and C.-J.H. Seger. The formal verification of a pipelined double-precision IEEE floating-point multiplier. In *ACM/IEEE International Conference on Computer-Aided Design*, pages 7–10, November 1995.

[ABCH94]   A. Aziz, F. Balarin, S.-T. Cheng, R. Hojati T. Kam, S.C. Krishnan, R.K. Ranjan, T.R. Shiple, V. Singhal, S. Tasiran, H.-Y. Wang, R.K. Brayton, and A.L. Sangiovanni-Vincentelli. HSIS: A BDD-Based Environment for Formal Verification. In *31st ACM/IEEE Design Automation Conference (DAC)*, San Diego, CA, June 1994. San Diego Convention Center.

[AbLa88]   M. Abadi and L. Lamport. The Existence of Refinement Mappings. In *Proc. of the 3rd Symposium on Logic in Computer Science*, pages 165–175, Edinburgh, July 1988. IEEE.

[AbLa95]   M. Abadi and L. Lamport. Conjoining specifications. *ACM Transactions on programming languages and systems*, 17(3):507–534, may 1995.

[AnSW94]   H.R. Andersen, C. Stirling, and G. Winskel. A compositional proof system for the modal μ-calculus. In *Proceedings of the 9th Annual Symposium on Logic in Computer Science*, June 1994.

[ASSS94]   V. Singhal A. Aziz, T. R. Shiple and A. L. Sangiovanni-vincentelli. Formula-dependent equivalence for compositional CTL model checking. In David L. Dill, editor, *Proceedings of the sixth International Conference on Computer-Aided Verification CAV*, volume 818 of *Lecture Notes in Computer Science*, pages 324–337, Standford, California, USA, June 1994. Springer-Verlag.

[Atki68]   D.E. Atkins. Higher-radix Division Using Estimates of the Divisor and Partial Remainders. *IEEE Transactions on Computers*, C-17(10):925–934, October 1968.

[Aubi79]   R.Aubin. Mechanizing structural induction. *Theoretical Computer Science*, 9:329–362, 1979.

[Barr95]   H. Barringer. Symbolic verification of hardware systems. In *In Proceedings of the IFIP Conference on Hardware Description Languages and their Applications (CHDL'95)*. Chiba, Japan, August 1995.

[BCDM86]   M.C. Browne, E.M. Clarke, D.L. Dill, and B. Mishra. Automatic Verification of Sequential Circuits Using Temporal Logic. *IEEE Transactions on Computers*, C-35(12):1034–1044, December 1986.

[BCLM94]   J.R. Burch, E.M. Clarke, D.E. Long, K.L. MacMillan, and D.L. Dill. Symbolic model checking for sequential circuit verification. *IEEE Transactions on Computer-Aided Design of Integrated Circuits and Systems*, 13(4):401–424, April 1994.

[BCMD90a] J.R. Burch, E.M. Clarke, K.L. McMillan, and D.L. Dill. Sequential Circuit Verification Using Symbolic Model Checking. In *Proceedings of the 27th ACM/IEEE Design Automation Conference*, pages 46–51, Los Alamitos, CA, June 1990. ACM/IEEE, IEEE Society Press.

[BCMD92] J.R. Burch, E.M. Clarke, K.L. McMillan, D.L. Dill, and L.J. Hwang. Symbolic Model Checking: 10^{20} States and Beyond. *Information and Computing*, 98(2):142–170, June 1992.

[Beat93] D.L. Beatty. *A Methodology for Formal Hardware Verification with Application to Microprocessors*. PhD thesis, Carnegie-Mellon University, School of Computer Science, 1993.

[BeBr94] D.L. Beatty and R.E. Bryant. Formally Verifying a Microprocessor Using a Simulation Methodology. In *31st ACM/IEEE Design Automation Conference (DAC)*, San Diego, CA, June 1994. San Diego Convention Center. ch. 38.1.

[BeBR96] N. Berregeb, A. Bouhoula, and M. Rusinowitch. Automated verification by induction with associative-commutative operators. In Rajeev Alur and Thomas A. Henzinger, editors, *Proceedings of the Eighth International Conference on Computer Aided Verification CAV*, volume 1102 of *Lecture Notes in Computer Science*, pages 220–231, New Brunswick, NJ, USA, July/August 1996. Springer Verlag.

[BeBS91] D.L. Beatty, R.E. Bryant, and C.-J.H. Seger. Formal Hardware Verification by Symbolic Ternary Trajectory Evaluation. In *Proceedings of the 28th ACM/IEEE Design Automation Conference*. IEEE, IEE Computer Society Press, June 1991.

[Beln77] N.D. Belnap. A useful four-valued logic. In J.M. Dunn and G. Epstein, editors, *Modern Uses of Multiple Valued Logic*. D. Reidel, Dordrecht, 1977.

[BHMS84] R. Brayton, G. Hachtel, C. McMullen, and A. Sangiovanni-Vincentelli, editors. *Logic Minimization Algorithms for VLSI Synthesis*. The Kluwer International Series in Engineering and Computer Science. Kluwer Academic Publishers, 1984.

[BiMa88] J.P. Billon and J.C. Madre. Original concepts of PRIAM, an industrial tool for efficient formal verification of combinational circuits. In G.J. Milne, editor, *The Fusion of Hardware Design and Verification*, pages 487–501, Glasgow, Scotland, 1988. IFIP WG 10.2, North-Holland. IFIP Transactions.

[Binm77] K.G. Binmore. *Mathematical Analysis*. Cambridge University Press, Cambridge, 1977.

[Boch82] G.V. Bochmann. Hardware Specification with Temporal Logic: An Example. *IEEE Transactions on Computers*, C-31(3):223–231, March 1982.

[BoFi89a] S. Bose and A.L. Fisher. Automatic Verification of Synchronous Circuits Using Symbolic Simulation and Temporal Logic. In L.J.M. Claesen, editor, *Proceedings of the IMEC-IFIP International Workshop on Applied Formal Methods for Correct VLSI Design*, November 1989.

[BoMo79] R.S. Boyer and J.S. Moore. *A Computational Logic Handbook*. Academic Press, 1979.

[BoMo88] R.S. Boyer and J.S. Moore. *A Computational Logic Handbook*. Academic Press, 1988.

[BoRu93] A. Bouhoula and M. Rusinowitch. Automatic case analysis in proof by induction. In *International Joint Conference on Artificial Intelligence, IJCAI93*, Chambery, France, 1993.

[Bouh94] A.Bouhoula. Spike: a system for sufficient completeness and parameterized induction proof. In A. Bundy, editor, *Proceedings of 12th International Conference on Automated Deduction*, pages 836–840, Nancy, France, 1994. Springer Verlag, LNCS No. 814.

[BoWo94]   B. Boigelot and P. Wolper. Symbolic verification with periodic sets. In David L. Dill, editor, *Proceedings of the sixth International Conference on Computer-Aided Verification CAV*, volume 818 of *Lecture Notes in Computer Science*, pages 55–67, Standford, California, USA, June 1994. Springer-Verlag.

[BrBS91]   R.E. Bryant, D.L. Beatty, and C.-J.H. Seger. Formal hardware verification by symbolic ternary trajectory evaluation. In *28th ACM/IEEE Design Automation Conference*, pages 397–402, 1991.

[BrCh94]   R.E. Bryant and Y.-A. Chen. Verification or Arithmetic Functions with Binary Moment Diagrams. Technical Report CMU-CS-94-160, Carnegie Mellon University, Pittsburgh, 1994.

[BrCh95]   R.E. Bryant and Y.-A. Chen. Verification or Arithmetic Circuits with Binary Moment Diagrams. In *32nd ACM/IEEE Design Automation Conference*, Pittsburgh, June 1995. Carnegie Mellon University.

[BrSe91]   R.E. Bryant and C.-J.H. Seger. Formal verification of digital circuits using symbolic ternary system models. In E.M. Clarke and R.P. Kurshan, editors, *Proceedings of the Workshop on Computer-Aided Verification (CAV90)*, volume 3 of *DIMACS Series in Discrete Mathematics and Theoretical Computer Science*, New York, 1991. American Mathematical Society, Springer-Verlag.

[Brya85]   R.E. Bryant. Can a simulator verify a circuit? In G. Milne and P.A. Subrahmanyam, editors, *Formal Aspects of VLSI Design*, pages 125–136. North-Holland, 1985.

[Brya85a]   R.E. Bryant. Symbolic verification of MOS circuits. In H. Fuchs, editor, *Chapel Hill Conference on VLSI*, pages 419–438, Rockville, MD., 1985. Computer Science Press.

[Brya86]   R.E. Bryant. Graph-Based Algorithms for Boolean Function Manipulation. *IEEE Transactions on Computers*, C-35(8):677–691, August 1986.

[Brya91a]   R.E. Bryant. On the Complexity of VLSI Implemetations and Graph Representations of Boolean Functions with Application to Integer Multiplication. *IEEE Transactions on Computers*, 40(2):205–213, February 1991.

[Brya91c]   R.E. Bryant. Formal verification of memory-circuits by symbolic-logic simulation. *IEEE Transactions on Computer-Aided Design of Integrated Circuits and Systems*, 10(1):94–102, 1991.

[Brya91d]   R.E. Bryant. A method for hardware verification based on logic simulation. *Journal of the ACM*, 38(2):299–328, 1991.

[Brya92]   R.E. Bryant. Symbolic Boolean Manipulation with Ordered Binary-Decision Diagrams. *ACM Computing Surveys*, 24(3):293–318, September 1992.

[Brya95]   R. E. Bryant. Bit-level analysis of an SRT divider circuit. Technical Report CMU-CS-95-140, Carnegie-Mellon University, April 1995. ftp://reports.adm.cs.cmu.edu/usr/anon/1995/CMU-CS-95-140.ps.

[BuDi94]   J. R. Burch and D. L. Dill. Automatic verification of pipelined microprocessors control. In David L. Dill, editor, *Proceedings of the sixth International Conference on Computer-Aided Verification CAV*, volume 818 of *Lecture Notes in Computer Science*, pages 68–80, Standford, California, USA, June 1994. Springer-Verlag.

[Burs69]   R.M. Burstall. Proving properties of programs by structural induction. *Computer Journal*, 12:41–48, 1969.

[CaJB78] W.C. Carter, W.H. Joyner Jr., and D. Brand. Microprogram Verification Considered Necessary. In *National Computer Conference*, volume 48, pages 657–664. AFIPS Conference Proceedings, 1978.

[CaJB79] W.C. Carter, W.H. Joyner Jr., and D. Brand. Symbolic simulation for correct machine design. In *16th ACM/IEEE Design Automation Conference*, pages 280–286, 1979.

[Camp85] R. Camposano. Synthesis techniques for digital system design. In *22nd Design Automation Conference*, pages 475–481. IEEE, 1985.

[CGHJ93a] E.M. Clarke, O. Grumberg, H. Hiraishi, S. Jha, D.E. Long, K.L. McMillan, and L.A. Ness. Verification of the Futurebus+ Cache Coherence Protocol. In D. Agnew, L. Claesen, and R. Camposano, editors, *The Eleventh International Symposium on Computer Hardware Description Languages and their Applications*, pages 5–20, Ottawa, Canada, April 1993. IFIP WG10.2, CHDL'93, IEEE COMPSOC, Elsevier Science Publishers B.V., Amsterdam, Netherland.

[CHJP90] H. Cho, G. Hachtel, S.-W. Jeong, B. Plessier, E. Schwarz, and F. Somenzi. ATPG aspects of FSM verification. In *International Conference on Computer-Aided Design*, 1990.

[ChMi89] K.M. Chandy and J. Misra. *Parallel Program Design*. Addison-Wesley, Austin, Texas, May 1989.

[ChPC88] S. Chandrasekhar, J.P. Privitera, and K.W. Conradt. Application of term rewriting techniques to hardware design verification. In *24th Design Automation Conference*, 1988.

[Chri90] J. Christensen. Efficient synthesis of hierarchical circuit descriptions. Master's thesis, Department of Computer Science, Technical University of Denmark, 1990.

[ClEm81] E.M. Clarke and E.A. Emerson. Design and Synthesis of Synchronization Skeletons using Branching Time Temporal Logic. In D. Kozen, editor, *Proceedings of the Workshop on Logics of Programs*, volume 131 of *Lecture Notes in Computer Science*, pages 52–71, Yorktown Heights, New York, May 1981. Springer-Verlag.

[ClES86] E.M. Clarke, E.A. Emerson, and A.P. Sistla. Automatic Verification of Finite-State Concurrent Systems Using Temporal Logic Specifications. *ACM Transactions on Programming Languages and Systems*, 8(2):244–263, April 1986.

[ClFJ93] E. M. Clarke, T. Filkorn, and S. Jha. Exploiting symmetry in temporal logic model checking. In Courcoubetis, editor, *Proceedings of The Fifth Workshop on Computer-Aided Verification*, pages 450–462, June/July 1993.

[ClFZ95] M. Fujita E. Clarke and X. Zhao. Applications of multi-terminal binary decision diagrams. Technical Report CMU-CS-95-160, School of Computer Science Carnegie Mellon University, Pittsburgh, PA 15213, April 1995.

[ClFZ95a] E. Clarke, M. Fujita, and X. Zhao. Hybrid decision diagrams: Overcoming the limitations of MTBDDs and BMDs. Technical Report CMU-CS-95-159, School of Computer Science, Carnegie Mellon University, April 1995. ftp://reports.adm.cs.cmu.edu/usr/anon/1995/CMU-CS-95-159.ps.

[ClGL92] E.Clarke, O. Grumberg, and D. Long. Model Checking and Abstraction. In *Proceedings of the Nineteenth Annual ACM Symposium on Principles of Programming Languages*, New York, January 1992. ACM.

[ClGL94]   E.Clarke, O. Grumberg, and D. Long. Model checking and abstraction. *ACM Transactions on Programming Languages and systems*, 16(5):1512–1542, September 1994.

[ClGr87a]   E.M. Clarke and O. Grumberg. Research on Automatic Verificatipon of Finite-State Concurrent Systems. In *Annual Review of Computer Science*, pages 269–290, Carnegie Mellon University, Pittsburgh, 1987.

[ClGZ96]   E. M. Clarke, S. M. German, and X.Zhao. Verifying the srt division algorithm using theorem proving techniques. In Rajeev Alur and Thomas A. Henzinger, editors, *Proceedings of the Eighth International Conference on Computer Aided Verification CAV*, volume 1102 of *Lecture Notes in Computer Science*, pages 111–122, New Brunswick, NJ, USA, July/August 1996. Springer Verlag.

[ClKu96]   E. M. Clarke and R. P. Kurshan. Computer-aided verification. *IEEE Spectrum*, 33(6):61–67, 1996.

[ClKZ96]   E. M. Clarke, M. Khaira, and X. Zhao. Word level symbolic model checking – a new approach for verifying arithmetic circuits. In *Proceedings of the 33rd ACM/IEEE Design Automation Conference*. IEEE Computer Society Press, June 1996.

[ClLM89a]   E.M. Clarke, D.E. Long, and K.L. McMillan. Compositional Model Checking. In *Proceedings of Fourth Annual Symposium on Logic in Computer Science*, pages 353–361, Washington D.C., June 1989. IEEE Computer Society Press.

[ClLM89b]   E.M. Clarke, D.E. Long, and K.L. McMillan. A Language for Compositional Specification and Verification of Finite State Hardware Controllers. In J.A. Darringer and F.J. Rammig, editors, *International Symposioum on Computer Hardware Description Languages and their Applications*, pages 281–295, Amsterdam, 1989. IFIP North-Holland.

[ClZh95]   E. Clarke and X. Zhao. A new approach for verifying arithmetic circuits. Technical Report CMU-CS-95-161, School of Computer Science, Carnegie Mellon University, Pittsburgh, PA 15213, May 1995.

[CMCH96]   E. M. Clarke, K. L. McMillan, S. Campos, and V. Hartonas-Garmhausen. Symbolic model checking. In Rajeev Alur and Thomas A. Henzinger, editors, *Proceedings of the Eighth International Conference on Computer Aided Verification CAV*, volume 1102 of *Lecture Notes in Computer Science*, pages 419–422, New Brunswick, NJ, USA, July/August 1996. Springer Verlag.

[CoBM89a]   O. Coudert, C. Berthet, and J.C. Madre. Verification of synchronous sequential machines using symbolic execution. In *Proceedings of the International Workshop on Automatic Verification Methods for Finite State Systems*, volume 407 of *Lecture Notes in Computer Science*, pages 365–373, Grenoble, France, June 1989. Springer-Verlag.

[CoBM89b]   O. Coudert, C. Berthet, and J.C. Madre. Verification of sequential machines using boolean functional vectors. In L.J.M. Claesen, editor, *Proceedings of the IFIP International Workshop Applied Formal Methods for Correct VLSI Design*, pages 111–128, Leuven, Belgium, November 1989. North-Holland.

[CoHe94]   B. Cousin and J. Helary. Performance improvement of state space exploration by regular and differential hashing functions. In David L. Dill, editor, *Proceedings of the sixth International Conference on Computer-Aided Verification CAV*, volume 818 of *Lecture Notes in Computer Science*, pages 364–376, Standford, California, USA, June 1994. Springer-Verlag.

[CoMa90] O. Coudert and J.C. Madre. A unified framework for the formal verification of sequential circuits. In *Proceedings of the 1990 IEEE International Conference on Computer-Aided Design*. IEEE Computer Society Press, November 1990.

[Cons86] R. Constable. *Implementing Mathematics with the Nuprl Proof Development System*. Prentice–Hall, Englewood Cliffs, New Jersey, 1986.

[Cook86] J.V. Cook. Final Report for the C/30 Microcode Verification Project. Technical Report ATR-86(6771)-3, Computer Science Laboratory, The Aerospace Corporation, El Segundo, CA, September 1986. Export-controlled.

[Core93] F. Corella. Automated high-level verification against clocked algorithmic specification. In D. Agnew, L. Claesen, and R. Camposano, editors, *Computer Hardware Description Languages and their Applications*, pages 135–142, Ottawa, Canada, April 1993. IFIP WG10.2, CHDL'93, IEEE COMPSOC, Elsevier Science Publishers B.V., Amsterdam, Netherland.

[Core94] F. Corella. Automated verification of behavioral equivalence for microprocessors. *IEEE Transactions on Computers*, 43(1):115–117, January 1994.

[CORS95] J. Crow, S. Owre, J. Rushby, N. Shankar, and M. Srivas. A Tutorial Introduction to PVS. Presented at WIFT '95: Workshop on Industrial-Strength Formal Specification Techniques, Boca Raton , Florida, April 1995.

[CPVM91] L. Claesen, F. Proesmans, E. Verlind, and H. De Man. A Methodology for the Automation Verification of MOS Transistor Level Implementation from High Level Behavioral Specifications. In *Proceedings of the 1991 International Workshop on Formal Methods in VLSI Design*, January 1991.

[CRSS94] D. Cyrluk, S. Rajan, N. Shankar, and M.K.Srivas. Effective theorem proving for hardware verification. In T. Kropf and R. Kumar, editors, *Proc. 2nd International Conference on Theorem Provers in Circuit Design (TPCD94)*, volume 901 of *Lecture Notes in Computer Science*, pages 203–222, Bad Herrenalb, Germany, September 1994. Springer-Verlag. published 1995.

[Curz94] P. Curzon. Tracking design changes with formal verification. In T.F. Melham and J. Camilleri, editors, *International Workshop on Higher Order Logic Theorem Proving and its Applications*, volume 859 of *Lecture Notes in Computer Science*, pages 177–192, Malta, September 1994. Springer-Verlag.

[CyLS96] D. Cyrluk, P. Lincoln, and N. Shankar. On Shostak's Decision Procedure for Combinations of Theories. In M.A. McRobbie and J.K. Slaney, editors, *Automated Deduction–CADE-13*, volume 1104 of *Lecture Notes in Artificial Intelligence*, pages 463–477, New Brunswick, NJ, July/August 1996. Springer Verlag.

[Cyrl93] D. Cyrluk. Microprocessor Verification in PVS: A Methodology and Simple Example. Technical Report SRI-CSL-93-12, Computer Science Laboratory, SRI International, Menlo Park, CA, December 1993.

[Cyrl96] D. Cyrluk. Inverting the abstraction mapping: A methodology for hardware verification. In M. Srivas and A. Camilleri, editors, *First international conference on formal methods in computer-aided design*, volume 1166 of *Lecture Notes in Computer Science*, pages 172–186, Palo Alto, CA, USA, November 1996. Springer Verlag.

[CZSL94] F. Corella, Z. Zhou, X. Song, M. Langevin, and E. Cerny. Multiway decision graphs for automated hardware verification. In *Formal Methods in System Design*, volume 10(1), 1994.

[Darw94] M. Darwish. Formal verification of a 32-bit pipelined RISC processor. Master's thesis, University of British Columbia, Department of Electrical Engineering, 1994. MASc Thesis.

[Ders83] N. Dershowitz. Applications of the knuth-bendix completion procedure. Technical Report ATR-83(8478)-2, Aerospace Corporation, 1983.

[Ders87] N. Dershowitz. Termination of rewriting. *Journal of Symbolic Computation*, 3(1):69–115, 1987.

[Ders90] N. Dershowitz and J.-P. Jouannaud. *Handbook of Theoretical Computer Science*, chapter Rewrite Systems. North-Holland, 1990.

[DrBR95] R. Drechsler, B. Becker, and S. Ruppertz. K*BMDs: a new data structure for verification. In *IFI WG 10.5 Advanced Research Working Conference on Correct Hardware Design and Verification Methods*, Frankfurt, Germany, October 1995.

[Emer90] E.A. Emerson. Temporal and Modal Logic. In J. van Leeuwen, editor, *Handbook of Theoretical Computer Science*, volume B, pages 996–1072, Amsterdam, 1990. Elsevier Science Publishers.

[EmLe85a] E.A. Emerson and C.-L. Lei. Temporal model checking under generalized fairness constraints. In *Proceedings of the Eighteenth Hawaii International Conference on System Sciences*, volume 1, pages 277–288, North-Holland, CA, 1985. Western Periodicals Company.

[EmLe86a] E.A. Emerson and C.-L. Lei. Efficient model checking in fragments of the propositional mucalculus. In *Proccedings of the First Annual Symposium on Logic in Computer Science*, pages 267–278, Washington, D.C., 1986. IEEE Computer Society Press.

[EmLe87] E.A. Emerson and C.-L. Lei. Modalities for model checking: Branching time strikes back. *Science of Computer Programming*, 8:275–306, 1987.

[FiJo95] Kathryn Fisler and Steven D. Johnson. Integrating design and verification envrionments through a logic supporting hardware diagrams. In *Procedings of the 1995 IFIP International Conference on Computer Hardware Description Languages and Their Applications*, pages 669–674. IEEE Cat. No. 95TH8102, September 1995. CHDL proceedings pp. 493-696 of the "ACV'95" held August 29 to September 1, 1995, Chiba, Japan.

[Fitt89] M. Fitting. Bilattices and the theory of truth. *Journal of Philosophical Logic*, 18(3):225–256, August 1989.

[Fitt90] M. Fitting. *First-Order Logic and Automated Theorem Proving*. Texts and Monographs in Computer Science. Springer-Verlag, 1990.

[Fitt91] M. Fitting. Bilattices and the semantics of logic programming. *The Journal of Logic Programming*, 11(2):91–116, August 1991.

[Floy67] R.W. Floyd. Assigning meaning to programs. In *Proceedings of Symposia in Applied Mathematics: Mathematical Aspects of Computer Science*, volume 19, pages 19–31, 1967.

[FoHa89] M. P. Fourman and R. L. Harris. Lambda – logic and mathematics behind design automation. In *Proceedings of the 26th ACM/IEEE design Automation Conference*. ACM, 1989.

[Frib86] L. Fribourg. A strong restriction of the inductive completion procedure. In *13th International Colloqium on Automata, Languages and Programming*, pages 105–115, 1986.

[FSSS94]   T. Filkorn, H. A. Schneider, A. Scholz, A. Strasser, and P. Warkentin. *SVE User's Guide.* Siemens AG, TR ZFE BT SE 1-SVE-1, Munich, 1994.

[Fuji92]   M. Fujita. RTL design verification by making use of datapath information. In *Proceedings of the 1992 IEEE International Conference on Computer Design.* IEEE Computer Society Press, October 1992.

[GaGS88]   S.J. Garland, J. V. Guttag, and J. Staunstrup. Verification of VLSI circuits using LP. In G.J. Milne, editor, *The Fusion of Hardware Design and Verification*, pages 329–346, Glasgow, Scotland, July 1988. IFIP WG 10.2, North-Holland.

[GaGu89]   S.J. Garland and J. v. Guttag. An overview over LP, the Larch Prover. In N. Dershowitz, editor, *Proc. 3rd Conference on Rewriting Techniques and Applications*, Chapel Hill, North Carolina, USA, 1989. Springer-Verlag.

[Galt94]   D. Galter. Symbolic verification of instruction-set processors. Master's thesis, Dept of Computer Science, University of Waterloo, 1994.

[GoMe93]   M.J.C. Gordon and T.F. Melham. *Introduction to HOL: A Theorem Proving Environment for Higher Order Logic.* Cambridge University Press, 1993.

[GoMW79]   M.J.C. Gordon, R. Milner, and C.P. Wadsworth. *A Mechanized Logic of Computation*, volume 78 of *Lecture Notes in Computer Science.* Springer-Verlag, New York, 1979.

[Gord83b]   M.J.C. Gordon. Proving a Computer Correct. Technical Report TR 42, Computer Laboratory, University of Cambridge, 1983.

[Gord86]   M.J.C. Gordon. Why higher-order logic is a good formalism for specifying and verifying hardware. In G.J. Milne and P.A. Subrahmanyam, editors, *Formal Aspects of VLSI Design*, pages 153–177. North-Holland, Computer Laboratory, University of Cambridge, 1986.

[GrLo91]   O. Grumberg and D. E. Long. Model checking and modular verification. In J. C. M. Baeten and J. F. Groote, editors, *Proceedings of CONCUR'91: Second International Conference on Concurrency Theory*, volume 527 of *Lecture Notes in Computer Science*, New York, August 1991. Springer-Verlag.

[GrLo93]   S. Graf and C. Loiseaux. A Tool for Symbolic Program Verification and Abstraction. In T. Kropf, R. Kumar, and D. Schmid, editors, *GI/ITG Workshop Formale Methoden zum Entwurf korrekter Systeme*, pages 122–138, Bad Herrenalb, March 1993. GI/ITG, Universität Karlsruhe, Interner Bericht Nr. 10/93.

[GrLo94]   O. Grumberg and D.E. Long. Model checking and modular verification. *ACM Transactions on Programming Languages and systems*, 16(3):843–871, may 1994.

[GSM94]   European digital cellular telecommunications systems, phase 2: full reate speech transcoding, 1994.

[Gupt92]   A. Gupta. Formal hardware verification methods: A survey. *Journal of Formal Methods in System Design*, 1:151–238, 1992.

[HaDa86]   F.K. Hanna and N. Daeche. Specification and verification of digital systems using higher-order predicate logic. *IEE Proceedings*, 133 Part E(5):242–254, September 1986.

[HaDa92a]   F. Hanna and N. Daeche. The Veritas Design Logic: A User's View. In C. Koomen and T. Moto-oka, editors, *Hardware Description Languages and Their Applications (CHDL85)*, pages 418–433. Elsevier Science Publishers B.V. (North Holand), 1985.

[HaHK96] R. H. Hardin, Z. Har'El, and R. P. Kurshan. COSPAN. In Rajeev Alur and Thomas A. Henzinger, editors, *Proceedings of the Eighth International Conference on Computer Aided Verification CAV*, volume 1102 of *Lecture Notes in Computer Science*, pages 423–427, New Brunswick, NJ, USA, July/August 1996. Springer Verlag.

[HaKu90b] Z. Har'El and R.P. Kurshan. Software for Analytical Developement of Communications Protocols. *AT&T Technical journal*, 69(1):45–59, Jan-Feb 1990.

[HaMM83] J. Halpern, Z. Manna, and B. Moszkowski. A hardware semantics based on temporal intervals. In *Proceeding of the Tenth International Colloquium on Automata, Languages, and Programming*, volume 154 of *Lecture Notes in Computer Science*, pages 278–291, New York, 1983. Springer-Verlag.

[HaSe95] S. Hazelhurst and C.-J.H. Seger. A Simple Theorem Prover Based on Symbolic Trajectory Evaluation and BDD's. *IEEE Transactions on Computer-Aided Design of Integrated Circuits and Systems*, 14(4):413–422, April 1995.

[HaSh96] K. Havelund and N. Shankar. Experiments in Theorem Proving and Model Checking for Protocol Verification. In *Formal Methods Europe FME '96*, volume 1051 of *Lecture Notes in Computer Science*, pages 662–681, Oxford, UK, March 1996. Springer-Verlag.

[HaTu93] N.A. Harman and J.V. Tucker. Algebraic models and the correctness of microprocessors. In G.J. Milne and L. Pierre, editors, *Correct Hardware Design and Verification Methods*, volume 683 of *Lecture Notes in Computer Science*, pages 92–108, Arles, France, May 1993. IFIP WG10.2, Springer-Verlag.

[Haze96] S. Hazelhurst. *Compositional Model Checking of Partially-Ordered State Spaces*. PhD thesis, University of British Columbia, Department of Computer Science, 1996.

[HoBr95] R. Hojati and R. K. Brayton. Automatic datapath abstraction in hardware systems. In P. Wolper, editor, *Proceedings of the 7th International Conference On Computer Aided Verification*, volume 939 of *Lecture Notes in Computer Science*, pages 98–113, Liege, Belgium, July 1995. Springer Verlag.

[HoDB97] R. Hojati, D. Dill, and R.K. Brayton. Verifying linear temporal properties of data insensitive controllers using finite instantitations. In C. Delgado Kloos and E. Cerny, editors, *Proceedings of 13th International Conference on Computer Hardware Description Languages and Their Applications*, pages 60–73, Toledo, Spain, April 1997. Chapman & Hall.

[Hoom94] J. Hooman. Correctness of Real Time Systems by Construction. In H. Langmaack, W.-P. de Roever, and J. Vytopil, editors, *Formal Techniques in Real-Time and Fault-Tolerant Systems*, volume 863 of *Lecture Notes in Computer Science*, pages 19–40, Lübeck, Germany, September 1994. Third International Symposium Organized Jointly with the Working Group Provably Correct Systems-ProCoS, Springer-Verlag.

[Hoom95] J. Hooman. Verifying Part of the ACCESS.bus Protocol Using PVS. In P.S. Thiagarajan, editor, *15th Conference on the Foundations of Software Technology and Theoretical Computer Science*, volume 1026 of *Lecture Notes in Computer Science*, pages 96–110, Bangalore, India, December 1995. Springer-Verlag.

[HuBr92] W.A. Hunt, Jr. and B.C. Brock. A Formal HDL and its Use in the FM9001 Verification. In C.A.R. Hoare and M.J.C. Gordon, editors, *Mechanized Reasoning and Hardware Design*, pages 35–47, Hemel

Hempstead, UK, 1992. Prentice Hall International Series in Computer Science.

[HuGD95] H. Hungar, O. Grumberg, and W. Damm. What if model checking must be truly symbolic. In P.E. Camurati and H. Eveking, editors, *Correct Hardware Design and Verification Methods*, volume 987 of *Lecture Notes in Computer Science*, pages 1–20, OFFIS Oldenburg (Germany), The Technion (Haifa, Israel) Oldenburg University (Germany), October 1995. IFIP WG10.5 Advanced Research Working Conference CHARME'95, Springer-Verlag.

[HuHu82] G. Huet and J.-M. Hullot. Proofs by induction in equational theories with constructors. *Journal of Computer and System Sciences*, 25:239–266, 1982.

[Hunt85] W.A. Hunt. *FM8501: A Verified Microprocessor*. PhD thesis, University of Texas, Austin, 1985.

[Hunt86] W.A. Hunt. The mechanical verification of a microprocessor design. In D. Borrione, editor, *From HDL Descriptions to Guaranteed Correct Circuit Designs*, pages 89–132, Amsterdam, 1986. North-Holland.

[Hunt89] W.A. Hunt. Microprocessor design verification. *Journal of Automated Reasoning*, 5(4):429–460, 1989.

[Hunt94] W.A. Hunt. *FM8501: A Verified Microprocessor*, volume 795 of *Lecture Notes in Artificial Intelligence*. Springer Verlag, Berlin, 1994.

[INTE93] Intel Corporation. *Pentium Processor User's Manual, Volume I: Pentium Processor Data Book*, order number 241428 edition, 1993.

[JaGo94] P. Jain and G. Gopalakrishnan. Efficient symbolic simulation-based verification using the parametric form of boolean expressions. *IEEE Transactions on Computer-Aided Design of Integrated Circuits and Systems*, 13(8):1005–1015, August 1994.

[Jans93] G. L. J. M. Janssen. *ROBDD Software*. Department of Electrical Engineering, Eindhoven University of Technology, October 1993.

[JoDi95] R. B. Jones and D. L. Dill. Efficient validity checking for processor verification. In *IEEE International Conference on Computer-Aided Design*, San Jose, California, USA, November 1995.

[JoKo86] J.-P. Jouannaud and E. Kounalis. Automatic proofs by induction in equational theories without constructors. In *1st Conference on Logic in Computer Science*, pages 358–366, Cambrigde, Massachusetts, USA, 1986.

[JoKo89] J.-P. Jouannaud and E. Kounalis. Proofs by induction in equational theories without constructors. *Information and Computation*, 82:1–33, 1989.

[JoMC94] S.D. Johnson, P.S. Miner, and A. Camilleri. Studies of the single pulser in various reasoning systems. In T. Kropf and R. Kumar, editors, *Proc. 2nd International Conference on Theorem Provers in Circuit Design (TPCD94)*, volume 901 of *Lecture Notes in Computer Science*, pages 126–145, Bad Herrenalb, Germany, September 1994. Springer-Verlag. published 1995.

[JoSe93] J.J. Joyce and C.-J.H. Seger. Linking BDD-Based Symbolic Evalutation to Interactive Theorem-Proving. In *Proceedings of the 30th Design Automation Conference*, Dallas, Texas, 1993.

[Joyc88a] J.J. Joyce. Generic structures in the formal specification and verification of digital circuits. In G.J. Milne, editor, *The Fusion of Hardware Design and Verification*, pages 51–76, Glasgow, Scotland, July 1988. IFIP WG 10.2, North-Holland.

[Joyc88b] J.J. Joyce. Formal verification and implementation of a microprocessor. In G. Birtwistle and P.A. Subrahmanyam, editors, *VLSI Specification, Verification and Synthesis*, pages 129–158. Kluwer Academic Publishers, 1988.

[Joyc90]   J.J. Joyce. *Formal Specification and Verification of Synthesized MOS Structures*, pages 113–122. IFIP. Elsevier Science Publishers, 1990.

[KaKu86]   J. Katzenelson and R.P. Kurshan. S/R: A language for specifying protocols and other communicating processes. In *Proceedings of the Fifth IEEE International Conference on Computer Communications*, pages 286–292, New York, 1986. IEEE.

[KaSu96]   D. Kapur and M. Subramaniam. Mechanically verifying a family of multiplier circuits. In Rajeev Alur and Thomas A. Henzinger, editors, *Proceedings of the Eighth International Conference on Computer Aided Verification CAV*, volume 1102 of *Lecture Notes in Computer Science*, pages 135–146, New Brunswick, NJ, USA, July/August 1996. Springer Verlag.

[KaZh88]   D. Kapur and H. Zhang. RRL: a rewrite rule laboratory. In Lusk and Overbeek, editors, *9th International Conference on Automated Deduction*, pages 768–769. Springer-Verlag, 1988.

[KMOS95]   R.P. Kurshan, M. Meritt, A. Orda, and S.R. Sachs. Modelling asynchrony with a synchronus model. In P. Wolper, editor, *Proceedings of the 7th International Conference On Computer Aided Verification*, volume 939 of *Lecture Notes in Computer Science*, pages 339–352, Liege, Belgium, July 1995. Springer Verlag.

[KnBe70]   D.E. Knuth and P.B. Bendix. Simple word problems in universal algebra. In J.Leech, editor, *Computational problems in Abstract Algebra*, pages 263–297. Pergamon Press, 1970.

[Koho77]   T. Kohonen. *Associative Memory*. Communication and Cybernetics. Springer-Verlag, 1977.

[Kore93]   I. Koren. *Computer Arithmetic Algorithms*. Prentice-Hall, 1993.

[Krip75]   S.A. Kripke. Outline of a theory of truth. *Journal of Philosophy*, pages 690–716, 1975.

[Krop94a] T. Kropf. Benchmark-circuits for hardware-verification. Technical Report SFB358-C2-4/94, Universität Karlsruhe, Institut für Rechnerentwurf und Fehlertoleranz, 1994. ftp://goethe.ira.uka.de/pub/techreports/SFB358-C2-4-94.ps.gz.

[Krop94b] T. Kropf. Benchmark-Circuits for Hardware-Verification. In T. Kropf and R. Kumar, editors, *Proc. 2nd International Conference on Theorem Provers in Circuit Design (TPCD94)*, volume 901 of *Lecture Notes in Computer Science*, pages 1–12, Bad Herrenalb, Germany, September 1994. Springer-Verlag. published 1995.

[KuKr94]   R. Kumar and T. Kropf, editors. *2nd International Conference on Theorem Proving in Circuit Design*, volume 901 of *Lecture Notes in Computer Science*, Berlin, Heidelberg, September 1994. IFIP WG10.2, Springer-Verlag.

[KuLe78]   H.T. Kung and C.E. Leierson. Systolic arrays (for VLSI). In *Sparse Matrics Proceedings*, pages 256–282. Society for Industrial and Applied Mathematics 1979, 1978.

[KuMc91]   R.P. Kurshan and K.L. McMillan. Analysis of digital circuits through symbolic reduction. *IEEE Transactions on Computer-Aided Design Integrated Circuits and Systems*, 10(11):1356–1371, November 1991.

[Kung82]   H.T. Kung. Why systolic architectures. *IEEE Computer*, pages 37–46, January 1982.

[Kurs89]   R.P. Kurshan. Analysis of discrete event coordination. In J.W. de Bakker, W.-P. de Roever, and G. Rozenberg, editors, *Proceedings of REX Workshop on Stepwise Refinement of Distributed Systems: Models, Formalisms, Correctness*, volume 430 of *Lecture Notes in Computer Science*, New York, May 1989. Springer-Verlag.

[Kurs94a]   R. P. Kurshan. *Computer-aided Verification of Coordinating Processes – The Automata-Theoretic Approach*. Princeton Univ. Press, 1994.

[KuSK93a]   R. Kumar, K. Schneider, and T. Kropf. Structuring and Automating Hardware Proofs in a Higher-Order Theorem-Proving Environment. *International Journal of Formal System Design*, pages 165–230, 1993.

[LaCe91]   M. Langewin and E. Cerny. Verification of processor-like circuits. In P. Prinetto and E. Camurati, editors, *Workshop on Correct Hardware Design Methodologies*, North-Holland, June 1991.

[LaCe91a]   M. Langewin and E. Cerny. Comparing generic state machines. In K. G. Larsen and A. Skou, editors, *Workshop on Computer-Aided Verification*. Springer Verlag, July 1991.

[LaCe94]   M. Langevin and E. Cerny. An extended OBDD representation for extended FSMs. In *The European Design and Test Conference*, pages 208–303, Paris, February 1994. IEEE Computer Society Press. EDAC94.

[Lamp80]   L. Lamport. "sometime" is sometimes "not never"-on the temporal logic of programs. In *Proceedings of the Seventh ACM Symposium on Principles of Programming Languages*, pages 174–185, New York, 1980. ACM.

[Lamp83]   L. Lamport. Specifying concurrent program modules. *ACM Transactions on Programming Languages and Systems*, 5(2):190–222, April 1983.

[LeCB74]   G.B. Leeman, W.C. Carter, and A. Birman. Some Techniques for Microprogram Validation. In *Information Processing 74 (Proc. IFIP Congress 1974)*, pages 76–80. North-Holland Publishing Co, 1974.

[Lees88a]   M. Leeser. *Reasoning about the Function and Timing of Integrated Circuits with PROLOG and Temporal Logic*. PhD thesis, University of Cambridge Computer Laboratory, February 1988.

[LeGS93]   T.W.S. Lee, M.R. Greenstreet, and C.-J.H. Seger. Automatic verification of asynchronous circuits. Technical Report 93–140, University of British Columbia Department of Computer Science, Vancouver, Canada, November 1993.

[LeLe95]   M. Leeser and J. O'Leary. Verification of a Subtractive Radix-2 Square Root Algorithm and Implementation. In *Proceedings of the IEEE International Conference on Computer Design (ICCD '95)*, Austin, Texas, October 1995. Cornell University.

[LiPn85]   O. Lichtenstein and A. Pnueli. Checking that finite state concurrent programs satisfy their linear specification. In *Proceedings of the Twelfth Annual ACM Symposium on Principles of Programming Languages*, pages 97–107, New York, January 1985. ACM.

[Long93]   D.E. Long. *Model Checking, Abstraction, and Compositional Verification*. PhD thesis, Carnegie Mellon University, 1993.

[LoSt92]   H. H. Løvengreen and J. Staunstrup. Synchronous realization of asynchronous computations. In V. Stavridou and T. Melham, editors, *Proceedings of the IFIP WG 10.2 International Conference on Theorem Provers in Circuit Design: Theory, Practice and Experience*, pages 95–110. Elsevier, June 1992.

[LTZS96] M. Langevin, S. Tahar, Z. Zhou, X. Song, and E. Cerny. Behavioral verification of an atm switch fabric using implicit abstract state enumeration. In *International Conference on Computer Design*, pages 20–26, Austin, TX, USA, October 1996.

[MaOw81] Y. Malachi and S.S. Owicki. Temporal specifications of self-timed systems. In B. Spoull H.T. Kung and G. Steele, editors, *VLSI Systems and Computations*, pages 203–212, Rockville, MD, 1981. Computer Science Press.

[MaPn81] Z. Manna and A. Pnueli. Verification of concurrent programs: Temporal proof principles. In *Proceedings of the Workshop on Logics of Programs*, volume 131 of *Lecture Notes in Computer Science*, pages 200–252, New York, 1981. Springer-Verlag.

[MaPn82] Z. Manna and A. Pnueli. Verification of concurrent programs: The temporal framework. In R.S. Boyer and J.S. Moore, editors, *Correctness Problems in Computer Science*, pages 215–273, London, 1982. Academic Press.

[MaPn83] Z. Manna and A. Pnueli. How to cook a temporal proof system for your pet language. In *Proceedings of the Tenth Annual ACM Symposium on Principles of Programming Languages*, pages 141–154, New York, 1983. ACM.

[MaPn87b] Z. Manna and A. Pnueli. On the relation of programs and computations to models of temporal logic. In B.Banieqbal, H.Barringer, and A.Pnueli, editors, *Temporal Logic in Specification*, pages 124–164, Altrincham,UK, 1987. Springer-Verlag.

[Mare94] N. Maretti. Mechanized verification of refinement. In T. Kropf and R. Kumar, editors, *Proc. 2nd International Conference on Theorem Provers in Circuit Design (TPCD94)*, volume 901 of *Lecture Notes in Computer Science*, pages 185–202, Bad Herrenalb, Germany, September 1994. Springer-Verlag. published 1995.

[Mare95] N. Maretti. *Mechanized Implementation Verification*. PhD thesis, Department of Computer Science,Technical University of Denmark, DK 2800, Lyngby, Denmark, June 1995.

[MaWo84] Z. Manna and P. Wolper. Synthesis of communicating processes form temporal logic specifications. *ACM Transactions on Programming Languages and Systems*, 6:68–93, 1984.

[MaYo78] M. Machtey and P.Young. *An Introduction to the General Theory of Algorithms*. North-Holland, 1978.

[McMi92a] K.L. McMillan. *Symbolic Model Checking: An Approach to the State Explosion Problem*. PhD thesis, School of Computer Science, Carnegie Mellon University, Pittsburgh, PA, May 1992. CMU-CS-92-131.

[McMi93a] K.L. McMillan. *Symbolic Model Checking*. Kluwer Academic Publishers, Norwell Massachusetts, 1993.

[MeCo80] C. Mead and L. Conway. *Introduction to VLSI Design*. Addison-Wesley, 2. edition, October 1980.

[Melh88d] T.F. Melham. Automating recursive type definitions in higher order logic. Technical Report 146, University of Cambridge Computer Laboratory, Cambridge CB2 3QG, England, September 1988.

[Mell94] N. Mellergaard. *Mechanized Design Verification*. PhD thesis, Department of Computer Science, Technical University of Denmark, 1994.

[MeST94] N. Mellergaard and J. Staunstrup. Tutorial on design verification with synchronized transitions. In T. Kropf and R. Kumar, editors, *Proc. 2nd International Conference on Theorem Provers in Circuit Design (TPCD94)*, volume 901 of *Lecture Notes in Computer Science*, pages

239–257, Bad Herrenalb, Germany, September 1994. Springer-Verlag. published 1995.

[MiJo96]   P.S. Miner and S.D. Johnson. Verification of an Optimized Fault-Tolerant Clock Synchronization Circuit: A Case Study exploring the Boundary between Formal Reasoning Systems. In *Designing Correct Circuits*, Bastad, Sweden, September 1996.

[MiLe96]   P. S. Miner and J. F. Leathrum. Verification of IEEE compliant subtractive division algorithms. In M. Srivas and A. Camilleri, editors, *First international conference on formal methods in computer-aided design*, volume 1166 of *Lecture Notes in Computer Science*, pages 64–78, Palo Alto, CA, USA, November 1996. Springer Verlag.

[Miln89a] R. Milner. *Communication and Concurrency*. Prentice-Hall International, London, 1989.

[Mine93]   P.S. Miner. Verification of Fault-Tolerant Clock Synchronization Systems. NASA Technical Paper 3349, NASA Langley Research Center, Hampton, VA, November 1993.

[MiSr95]   S.P. Miller and M.K. Srivas. Formal Verification of the AAMP5 Microprocessor: A Case Study in the Industrial Use of Formal Methods. In *WIFT '95: Workshop on Industrial-Strength Formal Specification Techniques*, pages 2–16, Boca Raton, FL, 1995. IEEE Computer Society.

[Mosz85]   B. Moszkowski. A temporal logic for multilevel reasoning about hardware. *IEEE Computer Magazine*, pages 10–19, February 1985.

[Mosz85a] B. Moszkowski. Executing temporal logic programs. Technical Report 71, University of Cambridge Computer Laboratory, August 1985.

[Muss80]   D.R. Musser. On proving inductive properties of abstract data types. In *Proceedings 7th ACM Symposium on Principles of Programming Languages*, pages 154–162, 1980.

[NaSt88]   P. Narendran and J. Stillman. Formal verification of the Sobel image processing chip. In *Proceedings of Design Automation Conference*, 1988.

[NiSt95]   L.S. Nielsen and J. Staunstrup. Design and verification of a self-timed RAM. In *Proceedings from VLSI*, June 1995.

[ObF194]   S.F. Oberman and M.J. Flynn. Design Issues in Floating-Point Division. Technical Report CSL-TR-94-647, Dept. of Computer Science, Stanford University, Stanford, CA 94305-2140, December 1994.

[ORRS96]   S. Owre, S. Rajan, J. M. Rushby, N. Shankar, and M. K. Srivas. PVS: Combining specification, proof checking, and model checking. In Rajeev Alur and Thomas A. Henzinger, editors, *Proceedings of the Eighth International Conference on Computer Aided Verification CAV*, volume 1102 of *Lecture Notes in Computer Science*, pages 411–414, New Brunswick, NJ, USA, July/August 1996. Springer Verlag.

[ORSH95]   S. Owre, J. Rushby, N. Shankar, and F.W. von Henke. Formal Verification for Fault-Tolerant Architectures: Prolegomena to the Design of PVS. *IEEE Transactions on Software Engineering*, 21(2):107–125, February 1995.

[ORSS94]   S. Owre, J.M. Rushby, N. Shankar, and M.K. Srivas. A tutorial on using PVS for hardware verification. In T. Kropf and R. Kumar, editors, *Proc. 2nd International Conference on Theorem Provers in Circuit Design (TPCD94)*, volume 901 of *Lecture Notes in Computer Science*, pages 258–279, Bad Herrenalb, Germany, September 1994. Springer-Verlag. published 1995.

[OwRS92]   S. Owre, J.M. Rushby, and N. Shankar. PVS: A Prototype Verification System. In D. Kapur, editor, *11th International Conference on Auto-*

mated Deduction (CADE), volume 607 of *Lecture Notes in Artificial Intelligence*, pages 748–752, Saratoga, NY, June 1992. Springer-Verlag.

[OwRS95] S. Owre, J. Rushby, and N. Shankar. Analyzing Tabular and State-Transition Specifications in PVS. Technical Report SRI-CSL-95-12, Computer Science Laboratory, SRI International, Menlo Park, CA, July 1995.

[PaDi96a] S. Park and D.L. Dill. Verification of the FLASH Cache Coherence Protocol by Aggregation of Distributed Transactions. In *8th ACM Symposium on Parallel Algorithms and Architectures*, pages 288–296, Padua, Italy, June 1996.

[Parn95] D.L. Parnas. Using Mathematical Models in the Inspection of Critical Software. In Michael G. Hinchey and Jonathan P. Bowen, editors, *Applications of Formal Methods*, International Series in Computer Science, chapter 2, pages 17–31. Prentice Hall, 1995.

[Paul94] L. Paulson. *Isabelle: A Generic Theorem Prover*, volume 828 of *Lecture Notes in Computer Science*. Springer, 1994.

[Pnue77] A. Pnueli. The temporal logic of programs. In *Proceedings of the Eighth Annual Symposium on Foundations of Computer Science*, volume 18, pages 46–57, New York, 1977. IEEE.

[Pnue84] A. Pnueli. In transition from global to modular temporal reasoning about programs. In K. Apt, editor, *Logic and Models of Concurrent Systems*, volume 13 of *NATO ASI series in Computer and System Sciences*, pages 123–144, New York, 1984. Springer-Verlag.

[Prat95] V. Pratt. Anatomy of the Pentium Bug. In P.D. Mosses, M. Nielsen, and M.I. Schwartzbach, editors, *TAPSOFT'95: Theory and Practice of Software Development*, number 915 in Lecture Notes in Computer Science, pages 97–107. Springer Verlag, May 1995.

[RaSS95] S. Rajan, N. Shankar, and M. K. Srivas. An integration of model checking with automated proof checking. In P. Wolper, editor, *Proceedings of the 7th International Conference On Computer Aided Verification*, volume 939 of *Lecture Notes in Computer Science*, pages 84–97, Liege, Belgium, July 1995. Springer Verlag.

[Robe58] J.E. Robertson. A new Class of Digital Division Methods. In *IRE Trans. on Electron. Computers*, volume EC-7, pages 218–222, 1958.

[Rues96] H. Ruess. Hierarchical verification of two-dimensional high-speed multiplication in PVS: A case study. In M. Srivas and A. Camilleri, editors, *First international conference on formal methods in computer-aided design*, volume 1166 of *Lecture Notes in Computer Science*, pages 79–93, Palo Alto, CA, USA, November 1996. Springer Verlag.

[Rush93] J. Rushby. A Fault-Masking and Transient-Recovery Model for Digital Flight-Control Systems. In J. Vytopil, editor, *Formal Techniques in Real-Time and Fault-Tolerant Systems*, Kluwer International Series in Engineering and Computer Science, chapter 5, pages 109–136. Kluwer, Boston, Dordecht, London, 1993.

[Rush94] J. Rushby. A Formally Verified Algorithm for Clock Synchronization Under a Hybrid Fault Model. In *13th ACM Symposium on Principles of Distributed Computing*, pages 304–313, Los Angeles, CA, August 1994. Association for Computing Machinery.

[RuSS96] H. Rueß, N. Shankar, and M. K. Srivas. Modular verification of srt division. In Rajeev Alur and Thomas A. Henzinger, editors, *Proceedings of the Eighth International Conference on Computer Aided Verification CAV*, volume 1102 of *Lecture Notes in Computer Science*, pages 123–134, New Brunswick, NJ, USA, July/August 1996. Springer Verlag.

[RuSt95]   J. Rushby and D.W.J. Stringer-Calvert. A Less Elementary Tutorial for the PVS Specification and Verification System. Technical Report SRI-CSL-95-10, Computer Science Laboratory, SRI International, Menlo Park, CA, June 1995.

[RySa92]   M. Ryan and M. Sadler. Valuation systems and consequence relations. In *Handbook of Logic in Computer Science*, volume 1 (Background: Mathematical Structures), chapter 1, pages 1–78. Clarendon Press, Oxford, 1992.

[Schn96a]   K. Schneider. *Ein einheitlicher Ansatz zur Unterstützung von Abstraktionsmechanismen der Hardwareverifikation*, volume 116 of *DISKI (Dissertationen zur Künstlichen Intelligenz)*. Infix Verlag, Sankt Augustin, 1996. ISBN 3-89601-116-2.

[Schn96b]   K. Schneider. Translating LTL Model Checking to CTL Model Checking. Technical Report SFB358-C2-3/96, Universität Karlsruhe, Institut für Rechnerentwurf und Fehlertoleranz, January 1996. ftp://goethe.ira.uka.de/pub/techreports/SFB358-C2-3-96.ps.gz.

[ScKK91a]   K. Schneider, R. Kumar, and T. Kropf. Structuring Hardware Proofs: First steps towards Automation in a Higher-Order Environment. In A. Halaas and P.B. Denyer, editors, *International Conference on Very Large Scale Integration*, pages 81–90, Edinburgh, Scotland, August 1991. IFIP Transactions, North-Holland.

[ScKK91b]   K. Schneider, R. Kumar, and T. Kropf. Automating most Parts of Hardware Proofs in HOL. In K.G. Larsen and A. Skou, editors, *Workshop on Computer Aided Verification*, volume 575 of *Lecture Notes in Computer Science*, pages 365–375, Aalborg, July 1991. Springer-Verlag.

[ScKK92b]   K. Schneider, R. Kumar, and T. Kropf. Efficient Representation and Computation of Tableau Proofs. In L.J.M. Claesen and M.J.C. Gordon, editors, *International Workshop on Higher Order Logic Theorem Proving and its Applications*, pages 39–58, Leuven, Belgium, September 1992. North-Holland.

[ScKK93a]   K. Schneider, R. Kumar, and T. Kropf. Hardware-Verification using First Order BDDs. In D. Agnew, L. Claesen, and R. Camposano, editors, *Computer Hardware Description Languages and their Applications*, pages 35–52, Ottawa, Canada, April 1993. IFIP WG10.2, CHDL'93, IEEE COMPSOC, Elsevier Science Publishers B.V., Amsterdam, Netherland.

[ScKK94a]   K. Schneider, T. Kropf, and R. Kumar. Control-Path Oriented Verification of Sequential Generic Circuits with Control and Data Path. In *Proceeding of the European Design and Test Conference*, pages 648–652, Paris, France, March 1994. IEEE Computer Society Press.

[ScKK94b]   K. Schneider, T. Kropf, and R. Kumar. Accelerating Tableaux Proofs using Compact Representations. *Journal of Formal Methods in System Design*, 5:145–176, 1994.

[SeBr95]   C.-J.H. Seger and R.E. Bryant. Formal verification by symbolic evaluation of partially-ordered trajectories. *Formal Methods in Systems Design*, 6:147–189, Mars 1995.

[Sege93]   C.-J.H. Seger. Voss — a formal hardware verification system user's guide. Technical Report 93-45, Department of Computer Science, University of British Columbia, November 1993. ftp://ftp.cs.ubc.ca/pub/local/techreports/1993/TR-93-45.ps.gz.

[SGGH94]   J.B. Saxe, S.J. Garland, J.V. Guttag, and J.J. Horning. Using Transformations and Verification in Circuit Design. *Formal Methods in System Design*, 4(1):181–210, 1994.

[Shan92]   N. Shankar. Mechanical Verification of a Generalized Protocol for Byzantine Fault-Tolerant Clock Sync hronization. In J. Vytopil, editor, *Formal Techniques in Real-Time and Fault-Tolerant Systems*, volume 571 of *Lecture Notes in Computer Science*, pages 217–236, Nijmegen, The Netherlands, January 1992. Springer-Verlag.

[Shan93]   N. Shankar. Verification of Real-Time Systems Using PVS. In Costas Courcoubetis, editor, *Computer-Aided Verification, CAV '93*, volume 697 of *Lecture Notes in Computer Science*, pages 280–291, Elounda, Greece, June/July 1993. Springer-Verlag.

[Shan96]   N. Shankar. PVS: Combining specification, proof checking, and model checking. In M. Srivas and A. Camilleri, editors, *First international conference on formal methods in computer-aided design*, volume 1166 of *Lecture Notes in Computer Science*, pages 257–264, Palo Alto, CA, USA, November 1996. Springer Verlag.

[Shos84]   R.E. Shostak. Deciding Combinations of Theories. *Journal of the ACM*, 31(1):1–12, January 1984.

[Sief70]   D. Siefkes. *Decidable Theories I: Büchi's Monadic Second Order Successor Arithmetic*. Lecture Notes in Mathematics. Springer-Verlag, 1970.

[SKMB90] A. Srinivasan, T. Kam, S. Malik, and R.K. Brayton. Algorithms for discrete function manipulation. In *Proceedings of the IEEE International Conference on Computer Design*, pages 92–95. IEEE, 1990.

[Sorl61]   O.L. McSorley. High-speed Arithmetic in Binary Computers. In *Proc. of IRE*, pages 67–91, 1961.

[SpSt93]   J. Sparsø and J. Staunstrup. Delay-insensitive multi-ring structures. *INTEGRATION, the VLSI Journal, North-Holland*, 15(3), 1993.

[SrBi90]   M.K. Srivas and M. Bickford. Formal Verification of a Pipelined Microprocessor. *IEEE Software*, 7(5):52–64, September 1990.

[SrMi95]   M. K. Srivas and S. P. Miller. Applying formal verification to a commercial microprocessor. In *IFIP Conference on Hardware Description Languages and their Applications*, Chiba, Japan, August 1995.

[SrMi95a] M.K. Srivas and S.P. Miller. Formal verification of the AAMP5 microprocessor. In Michael G. Hinchey and Jonathan P. Bowen, editors, *Applications of Formal Methods*, Prentice Hall International Series in Computer Science, chapter 7, pages 125–180. Prentice Hall, Hemel Hempstead, UK, 1995.

[SrMi96]   M.K. Srivas and S.P. Miller. Applying Formal Verification to the AAMP5 Microprocessor: A Case Study in the Industrial Use of Formal Methods. *Formal Methods in Systems Design*, 8(2):153–188, March 1996.

[Stau94a] Jørgen Staunstrup. *A Formal Approach to Hardware Design*. Kluwer Academic Publishers, 1994.

[StGG89]   S. J. Garland J. Staunstrup and J. V. Guttag. Localized verification of circuit descriptions. In *Proceedings of the Workshop on Automatic Verification Methods for Finite State Systems, LNCS 407*. Springer Verlag, 1989.

[Stir92]   C. Stirling. Modal and temporal logics. In S. Abramsky, D.M. Gabbay, and T.S. Maibaum, editors, *Handbook of Logic in Computer Science*, volume 2 (Background: Computational Structures), pages 477–563. Clarendon Press, Oxford, 1992.

[StMe95]   J. Staunstrup and N. Mellergaard. Localized verification of modular designs. *Formal Methods in System Design (Kluwer Academic Publishers)*, 3(6):295–320, 1995.

[StWa74]   L. Staiger and K.W. Wagner. Automatentheoretische Charakterisier- ungen topologischer Klassen regulärer Folgenmengen. *Elektron. Inform- ationsverarb. Kybernet.*, 10:379–392, 1974.

[TaKu93b] S. Tahar and R. Kumar. Implementing a Methodology for Formally Verifying RISC Processors in HOL. In J.J. Joyce and C.-J.H. Seger, ed- itors, *International Workshop on Higher Order Logic Theorem Proving and its Applications*, volume 780 of *Lecture Notes in Computer Science*, pages 281–295, Vancouver, Canada, August 1993. University of British Columbia, Springer-Verlag, published 1994.

[TaKu95]   S. Tahar and R. Kumar. Formal Specification and Verification Tech- niques for RISC Pipeline Conflicts. *The Computer Journal*, 38(2):111– 120, 1995.

[Tayl81]   G.S. Taylor. Compatible Hardware For Division and Square Root. In *Proceedings of the 5th Symposium on Computer Arithmetic*, pages 127– 134. IEEE Computer Society Press, 1981.

[Thom90a] W. Thomas. Automata on infinite objects. In J. van Leeuwen, editor, *Handbook of Theoretical Computer Science*, volume B, pages 133–191, Amsterdam, 1990. Elsevier Science Publishers.

[Toch58]   K.D. Tochter. Techniques of Multiplication and Division for Automatic Binary Computers. In *Quart. J. Mech. Appl. Match*, volume Part 3, pages 364–384, 1958.

[TSLB90]   H.J. Touati, H. Savoj, B. Lin, R.S. Brayton, and A. Sangiovanni- Vincentelli. Implicit State Enumeration of Finite State Machines us- ing BDD's. In *IEEE /ACM International Conference on CAD*, pages 130–132. ACM/IEEE, 1990.

[TZSC96]   S. Tahar, Z. Zhou, X. Song, E. Cerny, and M. Langevin. Formal veri- fication of an ATM switch fabric using multiway decision graphs. In *Proceedings of the Great Lakes Symposium on VLSI (GLS-VLSI'96)*, Iowa, USA, March 1996. IEEE Computer Society Press.

[ViBu92]   B.L. Di Vito and R.W. Butler. Formal Techniques for Synchronized Fault-Tolerant Systems. In C.E. Landwehr, B. Randell, and L. Si- moncini, editors, *Dependable Computing for Critical Applications—3*, volume 8 of *Dependable Computing and Fault-Tolerant Systems*, pages 163–188. Springer-Verlag, Vienna, Austria, September 1992.

[Viss84]   A. Visser. Four valued semantics and the liar. *Journal of Philosophical Logic*, 13(2):181–212, may 1984.

[VLAD92]   F. Van Aelten, S.Y. Liao, J. Allen, and S. Devadas. Automatic Genera- tion and Verification of Sufficient Correctness Properties for Synchron- ous Processors. In *IEEE /ACM International Conference on CAD*, pages 183–187, Santa Clara, California, November 1992. ACM/IEEE, IEEE Computer Society Press.

[WiCo94]   P.J. Windley and M.L. Coe. A correctness model for pipelined micro- processors. In T. Kropf and R. Kumar, editors, *Proc. 2nd International Conference on Theorem Provers in Circuit Design (TPCD94)*, volume 901 of *Lecture Notes in Computer Science*, pages 33–51, Bad Herrenalb, Germany, September 1994. Springer-Verlag. published 1995.

[Wind90]   P.J. Windley. *The Formal Verification of Generic Interpreters*. PhD thesis, University of California, Division of Computer Science, Davis, June 1990.

[WiPr80]   D. Winkel and F. Prosser. *The Art of Digital Design*. Prentice-Hall Inc., 1980.

[Wolp86]   P. Wolper. Expressing interesting properties of programs in proposi- tional temporal logic. In *Proceedings of the Thirteenth Annual ACM*

Symposium on Principles of Programming Languages, pages 184–192, New York, January 1986. ACM.

[ZhBo95]   Z. Zhou and N. Bouleric. Mdg tools user's manual. Technical report, University of Montreal, 1995.

[ZhSe94]   Z. Zhu and C. -J. H. Seger. The completness of a hardware inference system. In David L. Dill, editor, *Proceedings of the sixth International Conference on Computer-Aided Verification CAV*, volume 818 of *Lecture Notes in Computer Science*, pages 286–298, Standford, California, USA, June 1994. Springer-Verlag.

[ZSTC96]   Z. Zhou, X. Song, S. Tahar, E. Cerny, F. Corella, and M. Langevin. Formal verification of the island tunnel controller using multiway decision graphs. In *Proceedings of International Conference on Formal Methods in Computer Aided Design*, pages 233–247, USA, 1996.

Springer
and the
environment

Lecture Notes in Computer Science

For information about Vols. 1–1229

please contact your bookseller or Springer-Verlag